ICONIC
· TREES ·
OF INDIA

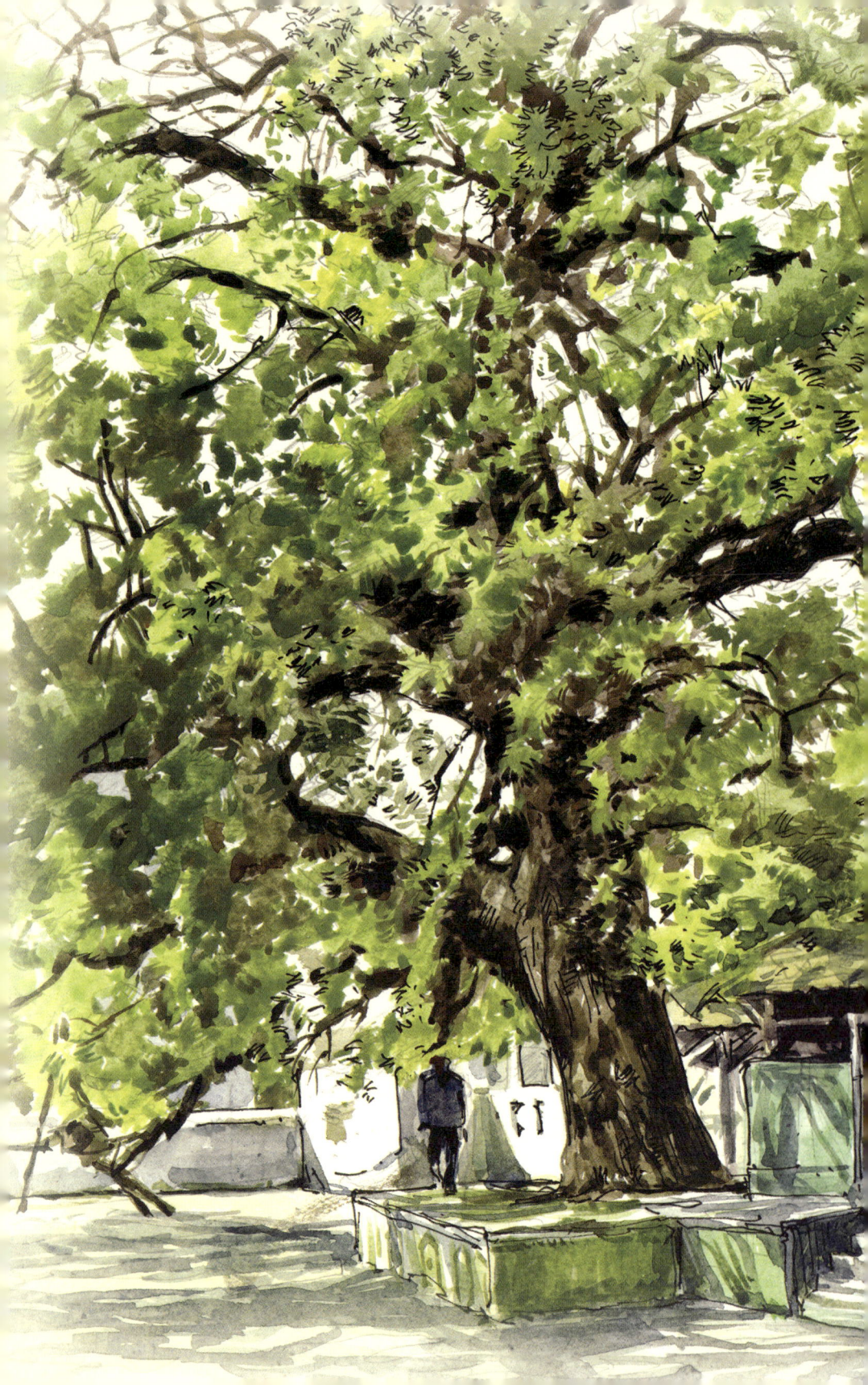

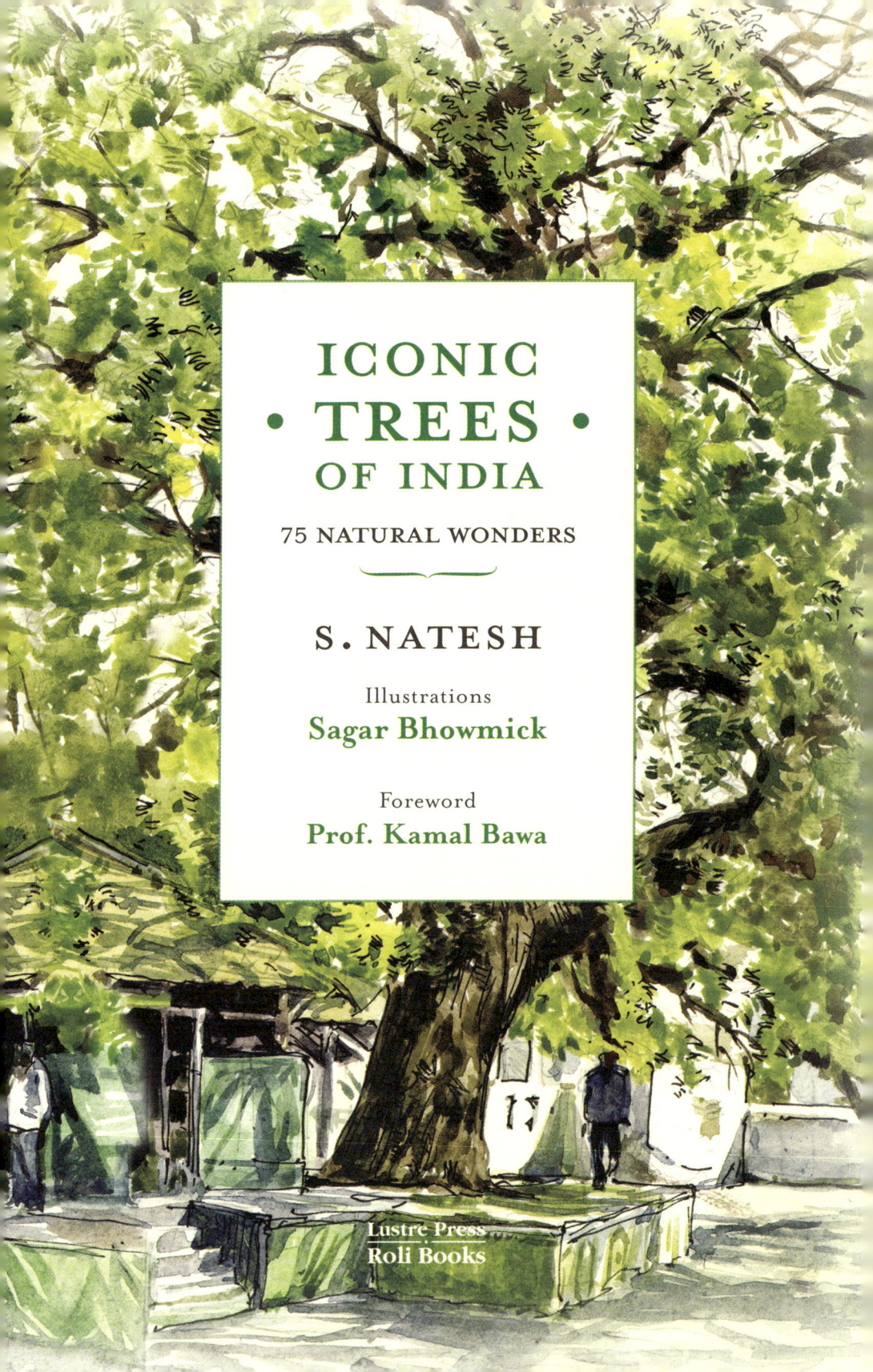

ICONIC
• TREES •
OF INDIA

75 NATURAL WONDERS

S. NATESH

Illustrations
Sagar Bhowmick

Foreword
Prof. Kamal Bawa

Lustre Press
Roli Books

ISBN: 9789392130342

Published in India by Roli Books in 2024
M-75, Greater Kailash II Market
New Delhi-110 048
Ph: +91-11-40682000
info@rolibooks.com
www.rolibooks.com

IMAGE CREDITS
Dr K.G. Misra: page 23
Kerala Forest Research Institute, Peechi: page 24
Dr Balakrishnan Nair: page 25
Dr S.K. Shah: pages 26–27

EDITOR: USHNAV SHROFF
DESIGN: SNEHA PAMNEJA
PRE-PRESS: JYOTI DEY
PRODUCTION: LAVINIA RAO

Printed and bound at
Naveen Printers, New Delhi

 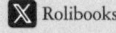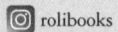

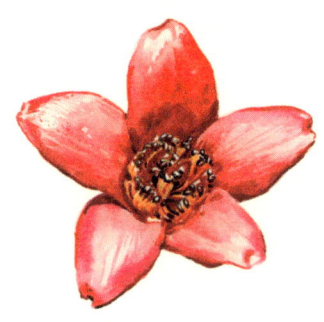

To my Parents
Smt SEETHAMMA
and
Shri B.M. SUBBA RAO
who were also my first teachers
about plants

Table of Contents

Foreword   9

Preface   10

Introduction   12

North

- ### Jammu & Kashmir
1. Qasim Shah's Chinar – India's Oldest and Largest   46
2. The Loneliest Tree in India   49
3. Shah Jahan's Bequest – The Chinars of Naseem Bagh   52
4. Dara Shukoh's Chinar of Bijbehara   55
5. The Toon of Raang Village   58
6. Guru Nanak's Ber Tree   60
7. The Tree that Swallowed a Pillar   62

- ### Himachal Pradesh
8. The Deodar of Chaurasi Temple   64

- ### Punjab
9. Ber Baba Budha of the Golden Temple   67

- ### Uttarakhand
10. The Sacred Mulberry Tree of Joshimath   69
11. The Mahatma's Peepal Tree   72
12. Dehradun's Identity – The Peepal Tree of Ghanta Ghar   74
13. The Giant Toon of Nagarjun   77
14. The Magnificent Deodars of Jageshwar   79

- ### Uttar Pradesh
15. Bareilly's Banyan of Martyrdom   82
16. Kāliya Mardan Kadamb of Vrindavan   84
17. Kunti's Parijata – The Baobab of Kintoor   86
18. Mango Tree with the Most Grafts   89
19. 'Mother' Tree of Dusseheri Mango   93
20. Baba Sheikh Taqi's Toothbrush? The Ancient Baobab of Jhunsi   95

- ### Delhi
21. Khirni Tree of Chirag Dilli   98

- ### Bihar
22. Tree of Enlightenment – Sri Mahabodhi   100

- ### Madhya Pradesh
23. Sleeman's Tree for Hanging Thugs   104
24. The Tree at the Heart of India   107

East

- ### Sikkim
25. Weeping Cypresses of Dubdi Monastery   110
26. The Coronation Cypress of Norbugang   113

- ### West Bengal
27. The Perils of Pregnancy – The Double Coconut Palm of Kolkata   116
28. Kolkata's Oldest Resident – The Great Banyan   119

- ### Assam
29. The Wishing Tree of Assam – Bakhor Bēngana   122

- ### Nagaland
30. The World's Tallest Rhododendron   124

- ### Meghalaya
31. Bridges that Breathe – Living Roots that Connect Meghalaya   127

- ### Manipur
32. The Sacred Fig Tree of the Nagas   131
33. The Sacred Wild Pear Tree of Shajouba   134

South

- **Telangana**

34. Hātiyan ka Jhad of Golconda Fort   136

- **Andhra Pradesh**

35. Thimmamma's Gigantic Banyan   138
36. Horsley's Eucalyptus   142

- **Karnataka**

37. The Sentinel Baobabs of Yogapur   144
38. Sangolli Rayanna's Banyan Trees   146
39. Maulsari Mahavriksha of Dudhgali   149
40. Three Baobabs of Savanūr   152
41. A Sultan's Legacy – The Gums of Nandi Hills   155
42. The Tree that Hosts Millions of Bees   158
43. A King's Scattered Treasure – Nallur Tamarind Grove   162
44. Sir C.V. Raman's Primavera   164
45. The Gentle Giant of Lalbagh   167
46. Gurudev's Weeping Fig Tree   170
47. The Rain Tree of Taj West End   172
48. The Golden Champaka Tree of Pushpagiri   175
49. Dodda Sampigé of Biligiri Rangan Hills   179

- **Andaman & Nicobar Islands**

50. The Speaking Fig Tree of the Cellular Jail   182

- **Tamil Nadu**

51. The Jamun Tree of Kadugamarathur   186
52. The Lofty Blue Gum of Aramby Shola   189
53. The Ugamaram Grove of Omāndur   192
54. The Manjakadambu Tree of Velliangiri   194
55. The Rosewood of Topslip   196
56. The Ancient Malai Naval Tree of Kodaikanal   199
57. The Sacred Vannimaram of Kodumudi   202

- **Kerala**

58. The Chain Tree of Lakkidi   205
59. Conolly's Teak Plantation   208
60. The Green Colossus of Conolly's Plot   212
61. Kannimara, the Pride of Parambikulam   215
62. Dismissed as Extinct and then Rediscovered – After 184 Years!   217

West

- **Rajasthan**

63. The Sacred Rayan Tree of Ranakpur   220

- **Gujarat**

64. The Ancient Limdo Tree of Jaska   222
65. The Towering Simado of Juna Vādiya   224
66. Vallabh Vad of Ras Village   226
67. Kabir's Gigantic Banyan   228
68. The Banyan Tree of Dandi   231
69. *Chalto Ambo* – The 'Walking' Mango Tree of Sanjan   234

- **Maharashtra**

70. The Twin Sen'trees' of Matunga   236
71. Livingstone's Legacy   238
72. Ajanvruksh of Alandi   240
73. Peshwa Bajirao's Mango Tree   242
74. Umaji Naik's Peepal Tree   245
75. The Plumeria of Parvati Hill   248

Acknowledgements   250

Index   252

A note on how the book is organized

There are two main sections in this book. The first section titled *Introduction* provides the context and background in which the rest of the book is to be read. It also outlines the next steps and the path ahead listing policies and actions that are needed to recognize, appreciate, and conserve India's iconic trees.

The second section titled *Living Iconic Trees of India* constitutes the heart of the book with seventy-five sections listing as many iconic trees. These are organized under four regions – north, east, south and west – for convenience as also to provide a sense of their geographical spread across India.

To further alleviate the reading experience, the QR code given below will take readers to the 'Notes and References' section of the book, which help you get to know the material better. I hope these notes provide useful comments and copious references for readers, whether you are a lay person or a professional. Also appended is an index for easy cross-referencing. I hope this note will help you not to just read but enjoy this book.

Foreword

Trees are true sentinels of landscapes, symbolizing protection, stability, persistence and, above all, a secure place for shelter. Unsurprisingly, various cultures have revered these magnificent creations of nature. In India, perhaps no other plant is considered more sacred than the peepal tree (*Ficus religosa*) under which Buddha is supposed to have meditated and attained enlightenment. The symbol of Bharat Ratna, India's highest civilian award, is inscribed on the leaf of the peepal tree.

For a vast majority of these trees, our knowledge about their biology, especially the extent of their distribution and the degree of threat to their existence is extremely limited. Considering that trees constitute almost 25 per cent of our country's rich flora, are the key elements of forest ecosystems, and are important components of our cultural heritage, scarce information about trees is surprising.

Iconic Trees of India is therefore a welcome addition to the scarce but growing literature on trees of India. The book presents a fascinating history of very old and ancient 75 trees belonging to 47 species from all over India. Some of these trees have been introduced from other parts of the world but most are native. The descriptions are highly original, based on author's knowledge and personal observations. Of special interest are trees from the northeast, the region with almost two-thirds of India's biodiversity and trees, about which we know very little. The book thus is the beginning of quest, albeit a significant one, to know more about our oldest trees and to conserve these icons as well as their near and distant relatives in the wild.

Books like these are especially important at a time when we need to reconnect with nature. There is no other better way to know and learn about nature's elegant manifestations, the plants and trees which we can always find around us, which are integral part of our culture. Hopefully this book will inspire others to write more stories about not only iconic trees but also other trees so that our growing curiosity of our bio-cultural heritage reinforces our desire to protect all biodiversity.

KAMAL BAWA

*President Emeritus, Ashoka Trust for Research in Ecology
and the Environment (ATREE), Bengaluru, and
Distinguished Professor Emeritus, University of Massachusetts, Boston.*

Preface

As far back as I can remember, I have been fascinated by plants, although it was not until college that I decided to specialize in botany. After teaching the subject at the University of Delhi for close to a decade, followed by a long career at the Department of Biotechnology, Government of India, I ended up doing a three-year stint at the National Institute of Immunology (NII) as a consultant, helping the institution expand its scientific outreach. Stunningly beautiful, the NII campus seamlessly and creatively blends institutional and residential spaces with its surrounding green and rocky backdrop. There is also an eclectic mix of native and exotic trees and the botanist in me could not help being excited and enthralled by the seasonal changes they heralded. I started writing on one campus tree every month and emailing the write-ups to the institute's scientists. Buoyed by their warm and enthusiastic response to these unsolicited offerings, I also organized campus tree walks. It was all going well, but then I began to be assailed by self-doubt. How much did I really know about trees in India? Which is India's certified oldest tree? Which is our largest, tallest or widest tree? Which ones have had a special place in our historical, cultural and religious life? And what binds the stories of our iconic trees with the people in their ecosystem?

Answers to these and similar questions are not to be found easily. As a rule, neither plant scientists nor historians pay particular attention to exceptional trees. Botanists generally focus on the species as a whole and seldom note individual remarkable trees. Foresters do look out for outstanding 'plus' trees, but their interest is usually limited to stem girth and height; details such as tree age are mostly a matter of conjecture. Moreover, a significant number of heritage trees occur outside forest settings – religious places, village commons, parks and gardens, institutional spaces, private holdings and even crowded sidewalks. Historians generally have an anthropocentric bias and invest little time or interest in trees. Thus, there is no authentic, substantive and easily accessible record of heritage trees in a country endowed with plenteous tree diversity, variegated culture and rich history. Not surprisingly, the number of iconic trees that an average Indian is familiar with can be counted on one hand and knowledge about most remains endemic and unknown beyond the village, district or state boundaries. Compounding this problem further, archaeological and architectural heritage is far more valued in India than green living heritage. This has grave consequences for biodiversity and trees in particular. In any development project, trees are the first casualties, although there is an increasing pushback by civil society in recent times. The challenges were daunting. For one thing, it would involve extensive travel all over India and the Covid pandemic added unanticipated and unwanted complications. I was already on the wrong side of 65, without a supporting team or a grant. How could I even contemplate a work of this magnitude? Even so, once the bug bit me, there was no turning back. By this time I had moved to IIT Delhi. The next several months went by in researching via the internet, newspaper reports, travel blogs, books, scientific journals, official gazetteers, personal information through friends and whatever else I could lay my hands on. Finally, a working list of over 300 potential iconic trees was ready (this list keeps constantly growing and changing).

From the very beginning, I had made up my mind not to write about any tree without personally visiting it in its natural habitat. I have strictly adhered to this self-imposed rule for all but three trees featured here – and those due to reasons beyond my control. Over these past seven years, I have crisscrossed thousands of kilometres across India by air, road, train, boat and on foot, in search of these iconic trees, through cities, remote villages and dense jungles. I have visited and studied hundreds of them in their natural surroundings and interacted with the communities in which they are embedded. I can honestly claim that this book is the result of my toil, sweat and blood: I have suffered from breathlessness, muscle cramps and have fallen and injured myself several times; sundry leeches and mosquitoes have sucked my blood in locations as far apart as Biligiri Rangan Hills in Karnataka and the Japfü mountain in Nagaland. My camera – along with hundreds of as-yet-unsaved images of trees in it – was stolen in a temple in Jageshwar, Uttarakhand. Despite all this, I would not dream of trading any part of the sheer thrill and excitement of finding an age-wizened arborescent centurion or a gentle ligneous giant.

This book is the result of my conviction that India's iconic trees are an inseparable part of our national heritage and deserve to be known and highly regarded by the public. Though the seventy-five living trees featured here are by no means the only ones worth celebrating – there are many others, far more than a volume like this can cover – the list was largely dictated by my desire to chronicle the most interesting stories and provide representative pan-India coverage.

S. NATESH

11

Introduction

To me, there is no greater miracle than the germination of a tiny seed into a tree that rises majestically into the skies as a redwood or spreads over acres of land as a banyan. Trees speak to our souls. Their roots, firmly grounded in the earth, symbolize strength. Their trunk and branches shooting into the air – tough and rigid yet swaying gently in the breeze – signify resilience and endurance. Unlike animals that can run from danger, trees are tenacious in surviving harsh and hostile environments, literally immovable against all odds. Capable of living for hundreds or even thousands of years, they provide a bridge to our past. Many cultures have long believed that trees are sentient beings capable of feeling pleasure and pain. They are majestic in life and dignified in death. As such, trees have inspired dance, music, poetry, prose, painting and sculpture among people of almost every culture across recorded history. They have been held up as models, symbols and analogues. Trees have moved, stimulated and anchored generations of artisans. In many cultures, trees are central to the concept of paradise or, at the least, a stairway to heaven. While the Bible describes paradise as a garden filled with the cedars of Lebanon, Norse mythology identifies the Yggdrasil, an immense ash tree, as the abode of gods, with Valhalla – the paradise of warriors – below one of its roots and hell below another. Religious writings of Hindus, Buddhists, Jains and Sikhs are full of references to trees and their links to gods and gurus.

There is a deep-rooted affinity between human beings and trees. According to a widely held view, human beings have a predisposition towards natural environments. Referred to as biophilia, this instinctive drive impels us to favour certain aspects of natural environments. Though the term has been attributed to the German psychologist and philosopher Erich Fromm who, in 1964, described it as 'the passionate love of life and all that is alive', it was the Pulitzer Prize-winning entomologist E.O. Wilson who popularized it after slight modification in 1984 as 'the urge to affiliate with other forms of life'.

A slew of research publications since then has shown that humans have a deeply ingrained love of nature. For example, people with a view of, or access to, natural environments recover faster from illness, stress and surgery, can better resist mental illness, are more affable, socially better adjusted, can manage their life affairs better and concentrate longer on difficult tasks than those who lack access to nature.[1, 2] There is also recent evidence to suggest that people's desire to be in nature and how they experience it are partially written into their genes.[3] The results of the study also explain why some people have a stronger desire than others to be in nature. However, there are limitations to the biophilia hypothesis. For instance, human beings have a greater capacity for environmental harm than environmental protection, as evidenced by numerous global environmental problems; people do not value *all* life – for example, some animals (like snakes, spiders) and environments (dense forests) evoke fear and aversion. In other words, we value the environment as a resource to be exploited rather than as an asset to be conserved. There is a long history of exploitation of plants and animals, including widespread acts of cruelty.

Nonetheless, the affiliation between human beings and nature seems to extend not only to trees but also to certain individual trees (or groups of such trees) that stand apart owing to some remarkable feature or reason. In this book, I intend to focus your attention on such iconic trees.

WHAT IS AN ICONIC TREE?

Every country has its outstanding trees. Trees may be considered important community resources because of their unique, noteworthy characteristics and values. Such trees have been referred to as heritage-, historic-, iconic-, landmark-, legacy-, significant-, or special-interest trees, indicating the diversity of traits that can be included for recognition. More formal definitions are available.[4] Regardless of the term applied, the meaning is the same: individual trees that stand out in some remarkable, exceptional and uncommon manner. These may inspire wonder and awe, command our respect, or fill us with amazement – but rarely will they leave us unmoved. They matter because people decide they do. They are representative icons and enduring symbols of our shared identity. They deserve our attention, even veneration and must be passed on to our successors for appreciation and safekeeping. In this book, I have preferred the term 'iconic' trees.

Once noticed, an iconic tree remains in our minds. Whether it is a lone arboreal Gulliver in a forest of Lilliputians, a furrowed forest guardian, a wish-fulfilling giant, or a botanical oddball, an iconic tree has an appeal beyond time and space. A living bridge to the past, iconic trees are also part of our collective legacy.

Thanks to India's vibrant history, its rich culture and varied growing conditions, several special trees have flourished here. The Bodhi tree of Bodhgaya, the big banyan of Kolkata, the ancient mulberry tree of Joshimath in Uttarakhand, the Kannimara teak in the Parambikulam forests of Kerala and the old ber trees of the Golden Temple at Amritsar immediately come to mind. Some trees are well known and widely visited, but most are local or regional celebrities.

In this book, I celebrate the joys of locating some well-known, less-known and unknown iconic trees of our country and trace their links to science, people, culture and history. I do hope you enjoy the stories as much as I have. These are by no means the only stories worth knowing or telling. There are many more – far more than can be accommodated in one volume.

WHY ARE ICONIC TREES IMPORTANT?

Trees are of vital importance from ecological, economic and cultural perspectives. They combat climate change, clean up the air, shield us from UV light, cool our cities and towns, conserve energy and water, mitigate pollution, help prevent soil erosion, provide canopy and habitat for wildlife and are sources of food, timber, drugs and other useful products. Trees also beautify spaces; mark the seasons; increase property values; and perform several other useful functions. Their value as tools for understanding basic biological phenomena such as longevity and past climatic events, as components of biodiversity, as features of the landscape, as habitats for terrestrial plant and animal species, as sinks for atmospheric carbon and as critical organisms connecting ecosystems and human health is immense and immeasurable.[5]

Trees, particularly older ones, also help in better biocontrol of invertebrate pests by acting as a larger reservoir of their natural enemies.[6] Older, larger-diametre trees have been shown to sequester disproportionally massive amounts of carbon compared to smaller trees[7] – a feature that underlines their importance in mitigating climate change. Their heritage importance apart, they are not only influencers of earth's changing pasts but also living records of environmental history. While all trees matter as plant taxa, older trees matter in a special sense. The 4,850-year-old Californian bristlecone pine known as Methuselah and the 3,600-year-old alerce tree from Chile are outstanding examples (see Table 1) of life several centuries ago. Such trees make a powerful argument for protecting existing trees and continuing to plant suitable trees wherever possible.

Beyond their scenic and aesthetic value, iconic trees are historical and cultural assets and act as focal points of local traditions. Their presence is evidence of a long relationship between people and places, of memories, myths, emotions and events.[8] The seasons come and go, the rivers rise in floods, epidemics wipe out entire settlements and time passes but trees generate more life and maintain immortality across multiple human generations. Their long life allows humans to imagine them as witnesses to ancestral events. Coming of age, communal trysts, epic romances, agreements of great worth, black moments and gory events have occurred beneath trees.

The famous and the infamous have held forth at specific trees or in special woodlands. Think of Robin Hood whose identity is lost in the mists of time but his relationship with the woods is well remembered and repeated in narratives. King Marthanda Varma (1706–58) of Travancore is said to have been helped by a young goatherd to hide in the hollow trunk of an ancient jackfruit tree to evade powerful enemies in deadly pursuit. The king survived and believing that Lord Krishna himself had saved him in the guise of a goatherd, built a temple to the God near the tree and worshipped both God and the tree. 'Ammāchi plāvu' (grandmother jackfruit tree in Malayalam) as the tree came to be known, bore many fruits until the 1970s. Now its trunk is preserved in a temple near Thiruvananthapuram.

It is no surprise that special trees are regarded as sentient beings, as abodes of gods or spirits and worthy of reverence and devotion. Transcending geographical, temporal and cultural divides, tree worship is common in many polytheistic belief systems.[9] Yet, often their social and bio-cultural value is neglected while designing conservation policies and management guidelines.[10]

A biodiversity crisis is looming, with around one million animal and plant species estimated to be threatened with extinction. A report on the state of the world's trees by the Botanic Gardens Conservation International, an independent UK charity to link botanical gardens of the world in a global network for plant conservation, warns that 17,510 tree species out of a total of 60,000 – corresponding to roughly 30 per cent – are at risk of being wiped out.[11] This is more than twice the number of threatened mammals, birds, amphibians and reptiles put together.

The loss of large old trees is a recognized and growing concern in many ecosystems worldwide. Populations of such trees are plummeting in intensively grazed landscapes in California, Costa Rica, Spain and elsewhere where such trees are predicted to disappear within the next 100 years. A recent study published in *Nature* found that the rate of trees dying in the old-growth tropical forests of northern Australia each year has doubled since the 1980s.[12] Researchers say climate change – particularly the rise in atmospheric water stress, or the drying power of the air, itself a consequence of temperature increase – is probably to blame.

The findings come from an extraordinary record of tree deaths catalogued at twenty-four sites in the tropical forests of northern Queensland over the past 49 years. Large old trees are exceptionally vulnerable to intentional removal, elevated mortality, reduced recruitment (a process by which seedlings create a population or add to an existing population), forest fires, depletion of forests, urbanization, expansion of agriculture, or combinations of these drivers. This is more so in tropical environments. Just as charismatic animals such as elephants, tigers, whales and dolphins have declined drastically in many parts of the world, a growing body of evidence suggests that large old trees could be equally imperilled. Targeted research is needed to understand the key threats and devise strategies to counter them.

DEVELOPED ECONOMIES VALUE THEIR ICONIC TREES

In developed economies, there is considerable interest in exceptional trees. This interest has been greatly fuelled through attractive books (for example, Thomas Pakenham's *Meetings with Remarkable Trees*, 1996 and *Remarkable Trees of the World*, 2003; Robert Van Pelt's *Forest Giants of the Pacific Coast*, 2002; Jenny Landreth's *The Great Trees of London*, 2010; Jared Farmer's *Elderflora*, 2022) and sustained by citizen science projects such as 'champion tree listings' and ecotourism whose chief attractions include tree top walks – for example in Singapore, Australia and continental Europe. Countries such as the UK, USA, Australia, New Zealand and Canada among others attach great value to preserving and maintaining their exceptional trees. Among the criteria used are age, rarity and size as well as the aesthetic, botanical, ecological and historical value of individual trees. Members of the general public are encouraged to nominate candidates for the Champion Tree Register, which is updated annually or six monthly. Public claims are verified by experts before inclusion in the register.

The American Forests' National Champion Trees Program was launched as a competition nearly a century ago to locate the largest living specimens of America's trees. Its goals have widened today into a conservation movement to locate, appreciate and protect the biggest tree species in the United

States and to educate people about the role that these trees and forests play in sustaining a healthy environment.[13] Many states, for example, Arizona, Nebraska, Oregon and Washington, have official registries of champion trees. Similarly, several Canadian provinces including Ontario and Manitoba have their champion tree programmes and registers. Australia and New Zealand also have comparable programmes. The UK has a long tradition of tree appreciation and protection. UK's Tree Register is a founder member of the European Champion Tree Forum, with a large and unique database of notable trees.[14] Additionally, England, Scotland and Wales all maintain and regularly update their tree registers, as does the Republic of Ireland. Singapore and South Africa also have programmes for the recognition of heritage and champion trees. The net result of all this is better awareness and enthusiastic public involvement in maintaining and conserving the earth's biodiversity. Of late, big tree climbing as a sport is receiving an eager response in some quarters, although its implications are still being debated. Meanwhile, it is already bringing awareness to forest conservation.[15]

SOME NOTEWORTHY TREES OF THE WORLD

The era of giant animals ended with the fall of large reptiles millions of years ago, but trees remain among nature's experiments in large size. People are always fascinated by superlative trees such as the oldest, the largest, the heaviest and so on. Some of the oldest and the largest organisms known in the world are clonal colonies of trees. Clonal colonies are different from individual trees in that they are made of thousands of genetically identical plants connected through a massive underground root system, together constituting one organism. The persistent root system of the whole organism will be much older than any individual tree in the colony. All the individuals originate vegetatively through root suckers, without the production of seeds. The oldest clonal colony in the world is that of quaking aspen trees (*Populus tremuloides* Michx.) nicknamed Pando ('I spread' in Latin)[16] in Utah, USA. Occupying ~43 hectares, it is a colony of more than 40,000 stems. Its age has been estimated to be an incredible 80,000 years. However, individual stems in Pando rarely reach 130 years. The second oldest clonal tree colony is a stand of Palmer's oak [*Quercus palmeri* (Engelm.) Engelm.] in the Jurupa Mountains, Riverside County, California, which has survived an estimated 13,000 years through clonal reproduction. A clonal stand of Huon pine [*Lagarostrobos franklinii* (Hook. f) Quinn] on the slopes of Mount Read,

17

Western Tasmania is believed to be over 10,500 years old, although no individual stem is of that age.

Among the single-stemmed, non-clonal, sexually reproducing trees (i.e., through seed), living specimens exceeding 2,000 years are extremely rare worldwide. The oldest known living tree (as determined by counting its annual growth rings) is a bristlecone pine (*Pinus longaeva* D.K. Bailey) now aged 4,854 years. Discovered in 1957 by Edmund Schulman, a scientist at the Laboratory of Tree Ring Research (LTRR), University of Arizona, the tree still graces the White Mountains of Inyo County in eastern California. Schulman named the tree after Methuselah, a biblical figure synonymous with longevity. Seven years later, in 1964, an even older bristlecone pine tree known as Prometheus (a Titan demigod in Greek mythology) was discovered in Wheeler Park, eastern Nevada (now Great Basin National Park). At 4,862 years, it was 73 years older than Methuselah, but tragically its age was revealed only after a graduate student cut it down to count its rings for his research project. It took nearly 50 more years to discover another tree older than Methuselah when, in 2012, LTRR researcher Thomas Harlan found a 5,062-year-old bristlecone pine, dating its innermost ring to 3050 BCE. The tree's identity and location are a secret. The oldest bristlecone pines were already well established by the time the ancient Egyptians built the pyramids at Giza.

Now a challenger has emerged to vie for the record held by the bristlecones. The world's oldest tree may be growing in a cool ravine in Chile's Alerce Costero National Park. It is an alerce tree [*Fitzroya cupressoides* (Molina), I.M. Jonst.] popularly known as Alerce Milenario or Gran Abuelo (great-grandfather) and could be 5,400 years old. Using a combination of computer models and traditional dendrochronological methods for calculating tree age (see Box 1), Jonathan Barichivich, a Chilean environmental scientist who works at the Climate and Environmental Sciences Laboratory in Paris, has estimated that the Alerce Milenario is probably more than 5,000 years old. That would make it senior to Methuselah by at least one century. So far, this finding has been presented only in meetings and conferences and a peer-reviewed publication is awaited.[17] For us, the idea of life spanning millennia may seem alien or even impossible to grasp; but for the earth's oldest trees it appears to be just another day at the office – to stand on the sands of time and observe hundreds of generations flow past them like water.

TABLE 1

THE WORLD'S OLDEST LIVING SINGLE-STEMMED TREES[18]

	Common name	Botanical name	Age in years	Remarks
1	Methuselah, Great Basin bristlecone pine	*Pinus longaeva* D. K. Bailey	4,854	White Mountains, California, USA
2	Alerce	*Fitzroya cupressoides* (Molina) I.M. Johnston	3,622	Western slopes of the Andes, Chile
3	Giant sequoia/ giant redwood	*Sequoiadendron giganteum* (Lindl.) J. Buchholz.	3,266	Sierra Nevada Mountains, California, USA
4	Qilian juniper	*Juniperus przewalski* Komarov	3,053	Zongwulong Mountain, China
5	Bald cypress	*Taxodium distichum* (L.) Rich.	2,624	Black River Preserve, North Carolina, USA. This is the oldest-known wetland tree species and the oldest living trees in eastern North America
6	Rocky Mountain bristlecone pine	*Pinus aristata* Engelm.	2,463	Central Colorado, USA
7	Qilian juniper	*Juniperus przewalski* Komarov	2,230	Delingha, Qinghai Province, China
8	Bodhi tree#	*Ficus religiosa* (L.)	2,217	Mahamewna Gardens, Anuradhapura, Sri Lanka
9	Western juniper or Bennett Juniper	*Juniperus occidentalis* Hook.	2,200	Sierra Nevada, California, USA
10	Foxtail pine	*Pinus balfouriana* A. Murray	2,110	Sierra Nevada, California, USA

*This is an illustrative and not an exhaustive list. #Age based on historic records; age of others determined by dendrochronological and/or radiocarbon dating techniques.

Like age, exceptional dimensions attract attention among the public. Table 2 provides living examples of trees remarkable for their height, girth, volume, or canopy area from across the world.[19] There has been limited work on the global distribution and conservation of the old and large trees of the world. Jiajia Liu at Fudan University, Shanghai along with his colleagues recently mined the International Tree Ring Data Bank database, on nearly 200,000 trees (*Conservation Biology*, 2022, DOI: 10.1111/cobi.13907). The team found 30 trees that exceeded 2,000 years, of which 27 occurred in high barren mountains in cold and arid regions, probably because human activity is limited there. It does not come as a surprise that the oldest trees seem to thrive where human activity is least. The study focused on 95 trees that were at least 500-years-old and showed that about 70 per cent of these are endangered by human overexploitation. Global warming also aggravates the problem, leading the authors to call for new conservation strategies.

TABLE 2
THE TALLEST AND THE LARGEST LIVING TREES OF THE WORLD

	Common name	Botanical name	Trait	Location	Remarks
THE TALLEST TREES					
1	'Hyperion', coastal redwood	*Sequoia sempervirens* Endl.	115.82 m (380 ft)	Redwood National Park, California, USA	Discovered in 2006, its location is kept secret to prevent damage to the tree
2	Cypress	Species unclear; could be either *Cupressus torulosa* D. Don ex Lamb. or *C. majestica* Duch.	102.3 m (335 ft)	Yarlung Zangbo Grand Canyon Nature Reserve, Bome County, Tibet Autonomous Region of China	Discovered in May 2023 by a Peking University research team, the tree towers over the Statue of Liberty (92.96 m)
3	'Menara', Yellow meranti	*Shorea faguetiana* F. Heim.	100.89 m (331 ft)	Sabah Borneo, Malaysia	'Menara' is the Malay name for 'tower'
4	'Centurion', mountain ash	*Eucalyptus regnans* F. Muell.	100.59 m (330 ft)	Tasmania, Australia	Discovered in 2008, it is the third-tallest tree and the tallest Eucalyptus
THE TREE WITH THE LARGEST VOLUME					
5	'General Sherman', the giant sequoia	*Sequoiadendron giganteum* (Lindl.) J. Buchholz	1,487 m^3 (52,500 ft^3)	Giant Forest of the Sequoia National Park, USA	By volume, it is the largest known single-stem tree.

THE TREE WITH THE LARGEST TRUNK GIRTH					
6	El Árbol del Tule ('The Tree of Tule' in Spanish) or ahuehuete ('old man of the water' in Nahuatl) Montezuma cypress	*Taxodium mucronatum* Ten.	Record trunk circumference of 42.0 m (137.8 ft) in 2005, equating to a diameter of 14.05 m (46.1 ft)	Church grounds, Oaxaca state, Mexico	It was placed on a UNESCO tentative list of World Heritage Sites but was removed from the list in 2013.
THE TREE WITH THE LARGEST CANOPY					
7	Thimmamma marrimanu (Thimmamma's banyan in Telugu), the giant banyan	*Ficus benghalensis* L.	World's largest canopy (including gaps). Covering 19,107 m² (4.721 acres)	Gootibylu, near Kadiri, Anantapur district andhra Pradesh	The size of 2.7 football fields, this tree is the largest in the world according to the Guinness Book of World Records in 1989.

TYPES OF ICONIC TREES

A tree could be considered iconic for a variety of reasons. For the sake of convenience, iconic trees in this book are grouped under the following categories:

- Ancient specimens that have survived from a bygone era (longevity/age)
- Champions and giants having record dimensions such as the tallest, the largest, or the widest, etc. (exceptional physical traits)
- Botanical curiosities – rarities, interesting introductions, oddities, or with (features of botanical interest)
- Special groves, plantations, or aesthetic or unique arrangement of trees (special or unique groups or arrangements of trees)
- Trees of religious or cultural significance, or having importance in mythology or folklore (religious/cultural significance)
- Trees of historical importance, being linked to important events, personalities, places or periods in India's history (historical significance)

It is important to bear in mind that while some trees may belong to just one of the above categories, others are likely to fit into more than one.

WHERE ARE INDIA'S ICONIC TREES?

Iconic trees can be found throughout the length and breadth of India. You can see them in the most unlikely places – the land revenue office in the middle of a bustling city, the premises of a hospital, a crowded footpath of a business precinct, a temple, a dargah, a gurdwara, or a monastery. Several iconic trees are found in protected area networks and are usually cared for by the state forest departments. Many important trees survive and even thrive on private landholdings, sometimes even in public parks, village commons and other areas maintained by the central and state governments. Almost every district in every state can boast of heritage trees within its boundaries.

SOME EXCEPTIONAL TREES FROM INDIA

Longevity/age: The age to which a tree survives is typical of the species it belongs to, the environmental conditions of the habitat it occupies and the degree of human interference it experienced. In India, we do not have reliable records of old trees. Overexploitation for timber and other resources in the past has contributed to the loss of old trees. More importantly, the number of scientifically dated trees in India is relatively small. As more and more trees are investigated, our pool of ancient trees will surely expand. As of now, India's oldest-living tree is a 2023-year-old Himalayan pencil cedar (*Juniperus polycarpos* K. Koch), discovered by Ram Yadav and his colleagues at the Birbal Sahni Institute of Palaeosciences (BSIP), Lucknow in Lahaul & Spiti district of Himachal Pradesh in 2016.

Quantifying the age of trees is challenging in India. My experience during field visits has shown that local claims are highly exaggerated. Many trees – especially if they are sacred – are usually claimed to be thousands of years old. For example, an African baobab tree in Kintoor, Barabanki district near Lucknow is revered as 'parijata' and its age is claimed to be 5,000 years. However, radiocarbon dating has shown it to be ~800 years old.[20] Foresters use tree size to estimate age. While this is a good starting point, a large diametre does not always signify old age, as growth rates differ significantly between species. For example, redwoods that routinely live up to 500-800 years (some go on to survive for several thousand years) are the tallest trees around, with massive trunk girth, but are not the longest-lived. On the other hand, the bristlecone pine can live to 5,000 years but barely manages to add one mm to trunk diametre in 10 years. Even within species, such estimates may be inaccurate. Certain individual trees may grow faster than others and younger trees grow faster than older ones. For example, an African baobab near Golconda Fort in Hyderabad, locally known as 'hātiyan ka jhad' has a massive girth of ~28.50 metre, but its age is only 475±50 years; the stem circumference of the 800-year-old baobab tree of Kintoor measures just about 13 metres.

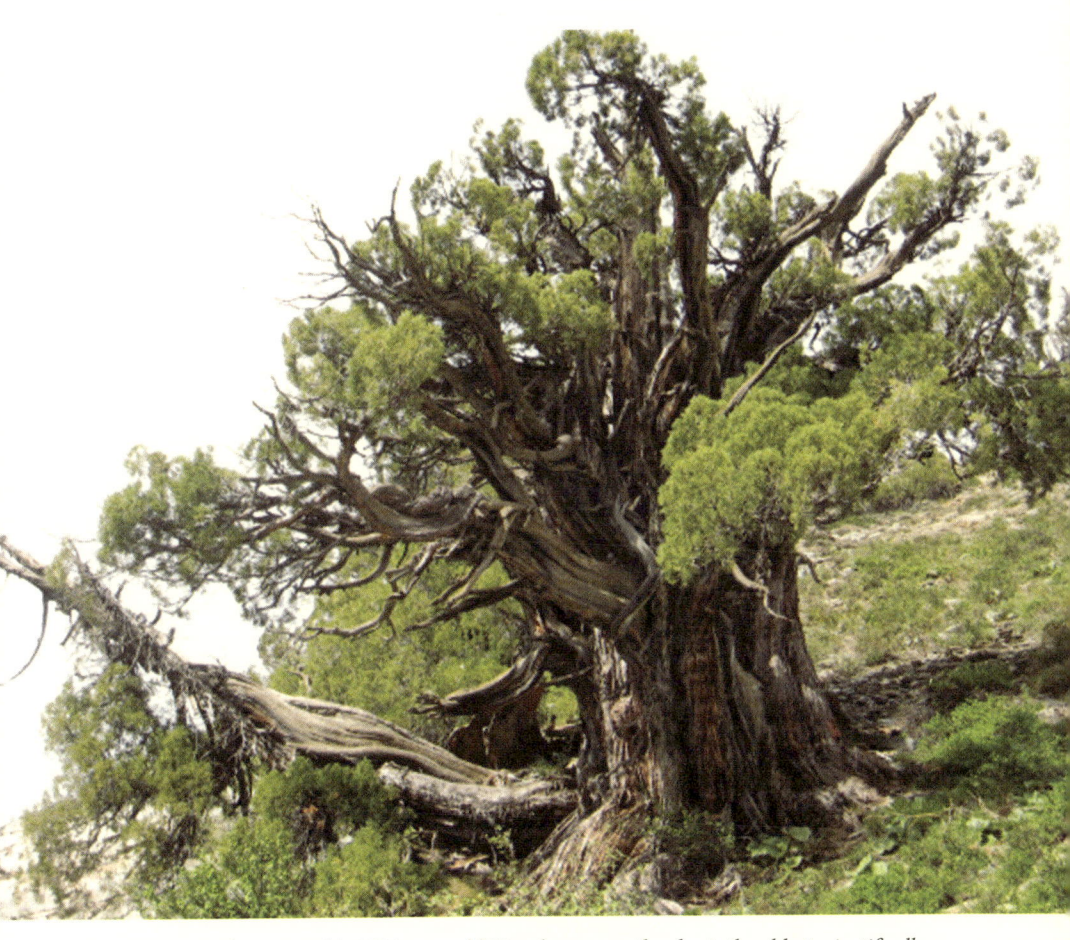

Discovered in 2016, this 2023-year-old Himalayan pencil cedar is the oldest scientifically dated living tree in India from Keylong, Lahaul & Spiti district of Himachal Pradesh.

Factors such as genetic makeup, water and nutrient supply and availability of sufficient energy from the sun, not to mention the degree of human activities around them, matter greatly in this regard.

Scientific methods of accurate tree age determination include dendrochronology and radiocarbon dating. Dendrochronology is the method of choice. It is the scientific study and analysis of dating tree rings (also known as growth rings or annual rings) to the exact year they were formed. Growth of the tree starts in spring, with 'early wood' comprising large light-coloured cells with thin cell walls. As the growing season winds down in autumn 'late wood' is produced, with small dark cells having thick walls. Growth halts in winter, resuming only in the ensuing spring. This pattern is repeated year after year. The line of demarcation between one year's autumn wood and the succeeding year's spring wood is very distinct, sharp and abrupt. When viewed in cross-section, the tree's growth appears to be in concentric 'rings,' with the oldest ring close to the centre of the trunk and the youngest one at its margin. Thus, if a tree was cut after the end of the growing season of 2010, the outermost ring corresponds to that year's growth. Counting the rings back to the centre, the second ring corresponds to 2009, the third to 2008 and so on. It would thus be possible to assign the exact date to each growth ring in a tree.

In good years trees will grow vigorously, producing wider growth rings; in dry and harsh years (years of drought, adverse temperature, or pest attacks, etc.), growth is slow, reflecting in narrow rings. Since no two years are similar, a tree will have a unique pattern of wide and narrow rings over a period. However, in each region, the growth conditions are similar therefore, trees of that region will have distinct ring signatures that are close enough to be synchronized.

Fortunately, to study the growth rings, the tree does not have to be sacrificed or damaged. Dendrochronologists insert an instrument known as an increment borer into the trunk of a tree and withdraw what is known as a 'core sample' to study the ring patterns (See Box 1).

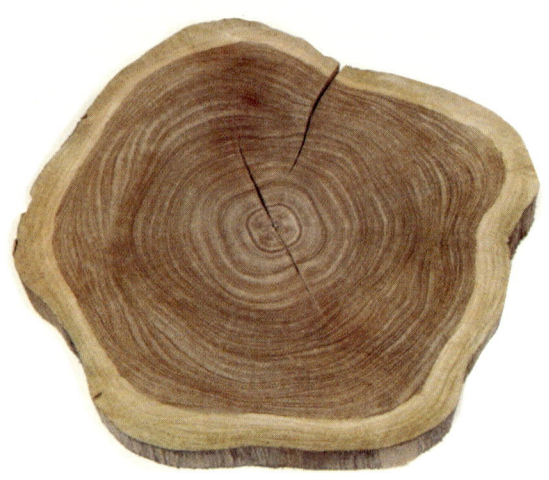

Cross-section of a trunk of teak. Note the distinct growth rings.

BOX 1: TALKATIVE TREE RINGS: MEASURING THE AGE OF A TREE ACCURATELY

Dendrochronologists insert the increment borer into the trunk of a tree usually 130-150 cm from the ground to extract a 'core sample' – a thin cylinder of wood less than the diametre of a pencil. Usually, several trees are sampled in an area. The core samples are finely sanded in the laboratory to bring out the ring pattern visible under the microscope and then cross-dated. Cross-dating results in each ring from a tree being assigned the exact year of formation by matching patterns of wide and narrow rings from the same tree and between trees from different locations to build a living tree-ring chronology. This can be extended farther back for many years by obtaining core samples from dead-standing trees nearby and even from timber from ruins (for example, beams, pillars, etc.).

By cross-matching all these, very long chronologies can be built. These days advanced computer software provides additional help. Master sequences have been only constructed for a few regions and species. For example, the Irish oak (*Quercus petraea*) chronology extends to 5,300 BCE; the California bristlecone chronology in the Southwestern US dates to 6,700 BCE and through cross dating the record has been pushed back to 13,920 years ago and the oak chronology in Germany stretches back to 8,500 BCE without skipping a single year.[21] In India, the longest chronology is of the Himalayan pencil cedar reaching back 2,023 years to 16 BCE.

Not all kinds of trees are useful for dendrochronological purposes. In temperate regions, the growth period is usually one year, in which case, the growth ring may be also an 'annual ring.' In tropical regions, growth rings may not always be discernible or annual. The ring patterns must be clear and distinct. Under certain climatic conditions (for example, moist), ring widths do not vary much through the year. Also, trees growing under stress (for example, water shortage, disease, pest attack) make better candidates than those

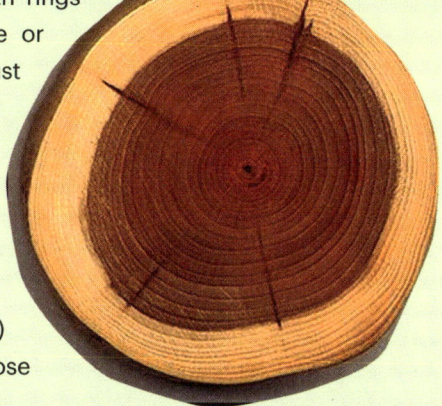

Cross-section of a 50-year-old teak tree showing growth rings.

that stand close to the source of moisture. Some species (for example, willow) have erratic and confusing ring patterns, but others (for example, oaks) are very consistent and reliable. Thus, tree rings are sensitive to both species and climate.

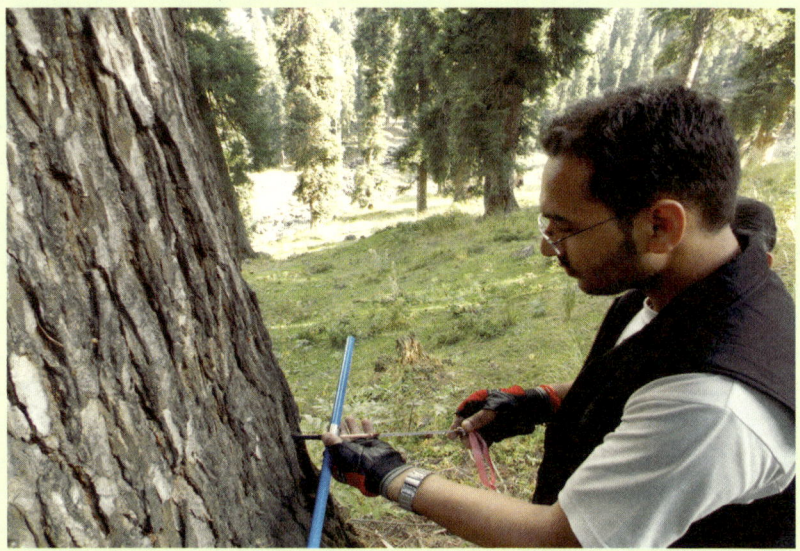

Above: Uttam Pandey of the Birbal Sahni Institute of Palaeosciences (BSIP), Lucknow, collecting a core sample with an increment borer for dendroclimatological work from a fir tree (*Abies pindrow*) in Lidder Valley, Kashmir, during September 2015.

Below: Close view of the tree rings in the core sample of fir (*Abies pindrow*) collected using an increment borer at Lidder Valley, Kashmir Himalaya.

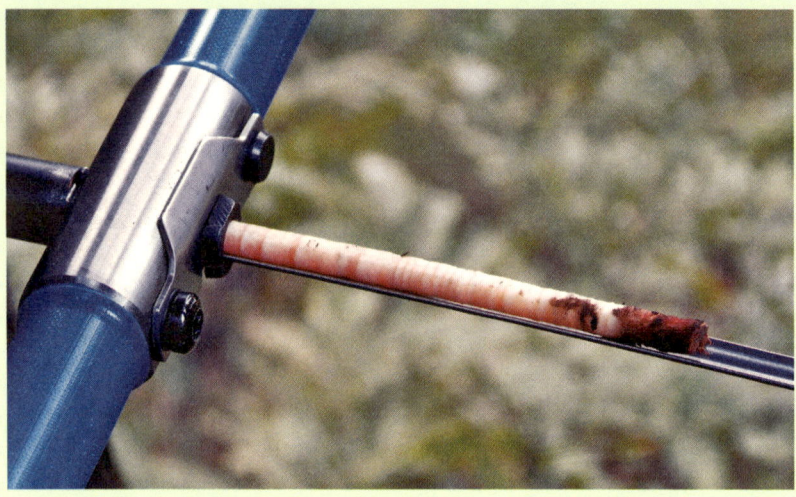

BOX 2: RADIOCARBON DATING: THE TRUSTED WORKHORSE IN ARCHAEOLOGY

Radiocarbon (or ^{14}C) dating[22, 23] is a method of determining the age of a sample containing organic material. It is based on the premise that every living organism assimilates carbon: plants absorb carbon through photosynthesis and animals consume it by eating plants or other animals. Atmospheric carbon exists mainly in the form of stable ^{12}C, but also contains a heavier but less stable form of radioactive isotope ^{14}C, which every living being ends up assimilating in some small quantity. Carbon assimilation stops when the organism dies, but the ^{14}C it has accumulated during its lifespan keeps decaying. Fortunately for scientists, the rate at which radioactive substances decay is specific and measurable. It takes 5,730 years for half of ^{14}C in each sample to decay which is known as 'half-life'. By measuring how much ^{14}C content is left in a sample and checking it against the background ^{14}C level in the atmosphere, it is possible to discern how long ago that organism died. The less radioactivity it contains, the older the sample is. This radiocarbon age is converted to the calendar age of carbon-containing materials such as wood, pollen, food and faeces as well as dead animals and human beings.

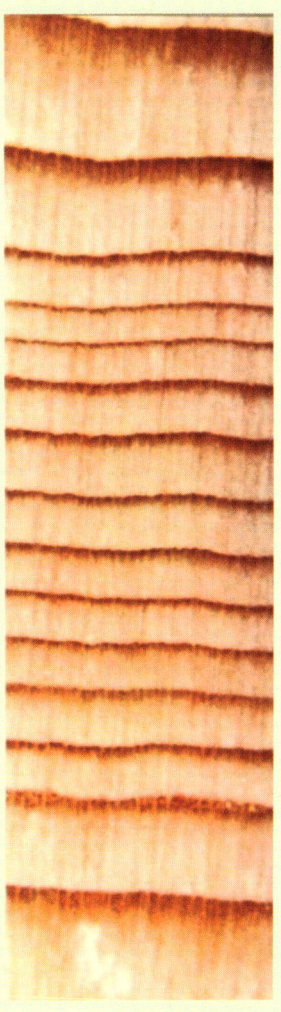

Tree rings of Sikkim larch (*Larix griffithii* var. griffithii), a Himalayan conifer. Note the variation in ring width.

Earlier, the measurement of radiocarbon was done by beta counting devices, but nowadays Accelerator Mass Spectrometre (AMS) is the instrument of choice. It directly counts the ^{14}C atoms in the sample and not just those few that decay during the measurement of decay. It is also faster, more sensitive and can be used with much smaller objects such as individual seeds. Even though a much sought-after archaeological workhorse, radiocarbon dating has so far found very limited use as a tool for determining tree age in India.

Interestingly, all the living trees older than 1,000 years discovered so far in this country are gymnosperm species – such as pines, junipers and cedars that produce cones and seeds instead of flowers and fruits – from the Himalayan region[24] (see also Table 3).

These old trees are usually found in arid environments, where the low moisture content combined with the resinous nature of coniferous wood provides resistance against wood decay.

Because dendrochronology is so precise – ring by ring and year by year – and sits right at the intersection of ecology, climatology and human history, it can tell us a great deal about human and environmental history. Thus, tree ring patterns could be a pointer to past climate; applied to archaeology to date artefacts made of wood; and have applications in anthropology, criminology and many other areas.

The application of dendrochronology and dendroclimatology is growing in India and has made considerable advances over the past decade. There is now a small but dedicated cohort of scientists specializing in the field.

Systematic work commenced only towards the end of the 1970s at the Indian Institute of Tropical Meteorology, Pune. Birbal Sahni Institute of Palaeosciences, Lucknow and Wadia Institute of Himalayan Geology, Dehradun soon followed. The expertise is now diffusing to other institutions across the country, as is the spectrum of species being engaged for study.

Radiocarbon dating, although an indispensable tool in archaeological investigation, has yet to make an impact on studying tree age in India. In the few examples cited in Table 3, radiocarbon processing was carried out abroad.

TABLE 3
EXAMPLES OF THE OLDEST KNOWN LIVING TREES FROM INDIA*

	Common name	Botanical name	Location	Age in years
TREES OLDER THAN 2,000 YEARS				
1	Himalayan pencil cedar	*Juniperus polycarpos* K. Koch	Keylong, Lahaul & Spiti district, Himachal Pradesh	2023 (16 BCE–2006 CE)[25]
TREES BETWEEN 1,000 AND 2,000 YEARS				
2	Himalayan pencil cedar	*Juniperus seravschanica* Kom. (= *Juniperus macropoda* Boiss.)	Keylong, Lahaul & Spiti District, Himachal Pradesh	1584 (420–2003 CE)[26]

28

3	Chilgoza pine	*Pinus gerardiana* Wall. ex D. Don	Barang, Kinnaur, Himachal Pradesh	1555 (457–2011 CE)[27]
4	Himalayan pencil cedar	*Juniperus polycarpos* K. Koch	Keylong, Lahaul & Spiti district, Himachal Pradesh, at an altitude of 4,122 m above mean sea level	1226 (789–2014 CE)[28]
5	Deodar, Himalayan cedar	*Cedrus deodara* (Roxb. ex D. Don) G. Don	Bhaironghati, Garhwal, Uttarakhand	1198 (805–2002 CE)[29]
6	Chilgoza pine	*Pinus gerardiana* Wall. ex D. Don	Purbani, Kinnaur, Himachal Pradesh	1087 (919–2005 CE)[30]
TREES BETWEEN 500 AND 1,000 YEARS				
7	Chilgoza pine	*Pinus gerardiana* Wall. ex D. Don	Sazar, Kishtwar, Jammu and Kashmir	878 (1140–2017 CE)[31]
8	Deodar, Himalayan cedar	*Cedrus deodara* (Roxb. ex D. Don) G. Don	Ratoli, Lahaul Spiti, Himachal Pradesh	787 (1217–2003 CE)[32]
9	*African baobab	*Adansonia digitata* L.	Kintoor, Barabanki district, Uttar Pradesh	775±25 [33]
10	*African baobab	*Adansonia digitata* L.	Jhunsi, Prayagraj (formerly Allahabad) district, Uttar Pradesh	770±25[34]
11	Birch	*Betula utilis* D. Don	Darma Valley, Kumaon, Uttarakhand	687 (1331–2017 CE)[35]
12	Deodar, Himalayan cedar	*Cedrus deodara* (Roxb. ex D. Don) G. Don	Jangi, Kinnaur, Himachal Pradesh	653 (1353–2005 CE)[36]
13	Deodar, Himalayan cedar	*Cedrus deodara* (Roxb. ex D. Don) G. Don	Sazar, Kishtwar, Jammu and Kashmir	625 (1393–2017 CE)[37]

14	Himalayan hemlock	*Tsuga dumosa* (D. Don) Eichler	Talley Valley, Arunachal Pradesh	580 (1421–2000 CE)[38]
15	Deodar, Himalayan cedar	*Cedrus deodara* (Roxb. ex D. Don) G. Don	Jangla, Gangotri, Uttarakhand	553 (1452–2004 CE)[39]
16	*Ceylon ironwood	*Manilkara hexandra* (Roxb. ex Dubard	Vankhandeshwar, Narora, Uttar Pradesh	550±25[40]
17	Deodar, Himalayan cedar	*Cedrus deodara* (Roxb. ex D. Don) G. Don	Malari, Uttarakhand	525 (1456–1990 CE)[41]
18	Teak	*Tectona grandis* L.	Narangathara, Kerala	523 (1481–2003 CE)[42]

Species with an asterisk (*) in column 2 were investigated through radiocarbon dating; all others were through dendrochronological techniques. As of date, the age of trees is more than that indicated here. This is not an exhaustive but an illustrative list.

Several trees are considered ancient because of their height, size, appearance, or historical or other associations (Table 4). However, their exact age is not determined by scientific methods and should only be regarded as guestimates.

TABLE 4
EXAMPLES OF LIVING TREES IN INDIA BELIEVED TO BE ANCIENT*

	Common name	Botanical name	Location	Age in years	Remarks
1	Sal	*Shorea robusta* C.F. Gaertn.	Satrenga village, Korba Forest Range, Chhattisgarh	1,400	This tree is considered locally as a godly being and protector of the village. The trunk circumference is 6.70 m and height 8.53 m

2	Sacred mulberry	*Morus serrata* Roxb.	Joshimath, Chamoli district, Uttarakhand	1,200	With an enormous 'pedestal-like' trunk and large canopy, this is a most beautiful male tree. The Indian Vedic scholar Adi Shankaracharya is said to have meditated under the tree
3	Qasim Shah's Chinar	*Platanus orientalis* L.	Chattergam, Badgam district, Jammu & Kashmir	650	Syed Qasim Shah, a Sufi preacher, is believed to have planted the tree in 1374
4	Dodda Sampigé	*Magnolia champaca* (L.) Baill. ex Pierre	BR Hills, Chamara-janagara district, Karnataka	600	The Soligas, who are among the earliest settlers in India, consider the tree with its gnarled and callused trunk as an embodiment of Shiva with his matted locks
5	Rayan, Khirni	*Manilkara hexandra* (Roxb.) Dubard	The Jain temple, Ranakpur, Pali district, Rajasthan	>580	Associated with Adinatha (or Vrishabhadeva), the first teerthankara, the species is especially revered by Jains
6	Kabir vad, banyan	*Ficus benghalensis* L.	Kabirvad sland, Bharuch district, Gujarat	500–600	Considered the oldest banyan in India, it is associated with the saint, poet and reformer Kabir Das (1440–1518). Local people believe it is 3,000 years old

7	Thimmamma marrimanu, the giant banyan	*Ficus benghalensis* L.	Gootibylu, near Kadiri, Anantapur district andhra Pradesh	500–600	Considered the second-oldest banyan in India, this tree is associated with a 16th-century woman Thimmamma who is said to have committed sati
8	Kannimara teak	*Tectona grandis* L.f.	Parambi-kulam Tiger Reserve, Palakkad district, Kerala	>500	This is considered to be the oldest teak tree existing in the wild
9	Guru Nanak's ber tree	*Ziziphus mauritiana* Lam.	Panjvaktra Mahadev Temple, Jammu, Jammu & Kashmir	500	According to legend, Guru Nanak stayed at the temple around 1514 and gave a discourse under the tree
10	Malai naval	*Syzygium densiflorum* Wall. ex wight & Arn.	Upper Shola Road, Bombay Shola, Kodaikanal, Tamil Nadu	500	This tree's description as the 'largest tree on Palani Hills' may well be true. Its hollow trunk can easily accommodate half a dozen people within.
11	Baba Budha ber tree	*Ziziphus mauritiana* Lam.	Golden Temple Amritsar, Punjab	>450	This tree is believed to be the oldest of the three sacred ber trees within the temple complex. It is actually older than the temple itself

12	Grove of tamarind trees	*Tamarindus indica* L.	Nallur Village, Bengaluru rural district, Karnataka	>400	The grove is believed to have been established in the thirteenth century during the reign of the Chola dynasty

*This is an illustrative list.

Trees with exceptional physical traits: Some trees have been recognized for their exceptional dimensions such as canopy area, height, girth and so on (Table 5). Some of these features have entered the *Guinness Book of Records*. A more systematic study across the country is likely to reveal several more examples.

TABLE 5
TREES WITH RECORD DIMENSIONS FROM INDIA*

	Common name	Botanical name	Trait	Location
1	Thimmamma marrimanu (Thimmamma's banyan), the giant banyan	*Ficus benghalensis* L.	Record for the largest net canopy (not including gaps), of 19,107 m^2 in the Guinness Book of World Records, 1989	Gootibylu, near Kadiri, Anantapur district andhra Pradesh
2	Kabir vad, the giant banyan	*Ficus benghalensis* L.	Second-largest net canopy of 17,520 m^2	Kabirvad Island, Bharuch district, Gujarat
3	Hātiyan ka Jhad, African Baobab	*Adansonia digitata* L.	Largest baobab outside of Africa, with a trunk circumference of 25.48 m	Naya Qila, Golconda Fort, Hyderabad
4	Būruga, white kapok tree	*Ceiba pentandra* (L.) Gaertn.	It is the stoutest kapok at 23 m among eight contenders from Cape Verde, Costa Rica, Cuba, Guinea-Bissau, Mexico, Senegal, Singapore and USA	Lalbagh, Bengaluru

5	Himalayan cedar/ deodar	*Cedrus deodara* (Roxb.) G. Don	Forest officials claim that this tree with a girth of 14.50 m is the world's largest (and perhaps, the oldest) deodar tree	Chanti Bala in the Bhalessa Forest Range, Doda district, Jammu & Kashmir
6	Qasim Shah's chinar	*Platanus orientalis* L.	The largest girth recorded in India at ~15 m. It is also believed to be the oldest chinar tree in India	Chattergam village, Badgam district, Jammu & Kashmir
7	Rhododendron	*Rhododendron delavayi* Franch.	World's tallest rhododendron at 32.9 m; unconfirmed reports claim that in 2002 the height was 38.1 m. Finds honourable mention in the Guinness Book of World Records, 1993	Japfū Mountains, near Kohima, Nagaland

*This is an illustrative list.

Trees of botanical interest: Botanically interesting trees are waiting to be discovered in various places across India and a few are featured in this book. A single tree of attilippa [*Madhuca diplostemon* (C.B. Clarke) P. Royen] from Kerala considered extinct and forgotten, was rediscovered after 180 years. A banyan tree near Bengaluru hosts more than 600 beehives on its branches – a record. Many exotic trees have been introduced into India in the past and several have enriched our environment. However, there are a few interesting species of which only single representatives have managed to survive (for example, a specimen of giant sequoia in Kashmir). Some trees are remarkable because of where they stand (a baheda tree that stands in Karaundi village in Madhya Pradesh, at the exact geographical centre of India); or what they 'did' (a peepal tree near Jammu that 'swallowed' a border post at the India-Pakistan border). Then some trees are interesting because of the unique ways in which they are used: Displaying an amazing feat of native engineering skills, the local community of southern Meghalaya fashions durable living bridges across rivers and streams from the roots of the Indian rubber tree. Doubtless, others deserve to be brought to public attention.

Special/unique groups or arrangements: Much has been discussed about sacred groves, but there are other types of special groves or plantations. These are usually well-known locally. Invariably, they have some historical significance due to association with some personality or event and are important to their locals. Examples include Naseem Bagh, established as a garden of chinars around 1635 by the Mughal Emperor Shah Jahan and now part of the campus of the University of Kashmir on the western shore of Dal Lake. An exceptionally well-preserved grove of tamarind trees on the outskirts of Bengaluru is reputed to have originated during the early thirteenth century when the imperial Chola Dynasty ruled over this region. Another notable example is the earliest surviving teak plantation in the world, part of which is now preserved near Nilambur in Kerala. Locally known as Conolly's Plot, it is named after Henry Valentine Conolly, the then district collector and magistrate of Malabar, who was instrumental in establishing it in 1842 with the help of his Indian protégé Chathu Menon.

Religious and cultural association: In a country where tree worship is common, it is not surprising to find individual trees that have been singled out for reverence and adoration. Everyone knows Bodh Gaya's hallowed Bodhi tree that occupies a unique position among Hindus, Buddhists and people of other faiths. Fewer would have heard of the ancient khirni tree (*Manilkara hexandra* Dubard) associated with Adinatha, the first Jain *teerthankara*, that stands in the exquisitely beautiful marble temple in Ranakpur, Rajasthan. I have already referred to the 800-year-old African baobab tree in Kintoor, near Lucknow which is worshipped as the wish-granting pārijāta of mythology. Pilgrims from all over Maharashtra visit the temple in Alandi near Pune and pay obeisance to an Ajānvruksh (*Ehretia aspera* Willd.) which is said to emerge from the samadhi of Sant Dnyaneshwar who, at a very young age, wrote a philosophical masterpiece known as *Bhāvārthdeepika* in Marathi. And the list goes on.

There are also specific trees that are culturally linked to institutions and communities. Two sapthaparni [*Alstonia scholaris* (L.) R. Br.] trees, which are sadly no more, are inseparably associated with Shantiniketan, established as a spiritual retreat by Maharshi Debendranath Tagore. An ancient pilkhan tree (*Ficus virens* Aiton) is associated with the primaeval mother of Naga tribes and stands in Makhel, Senapati district, Manipur as a symbol of Naga culture. The *chalto ambo* or 'the walking mango tree' at Sanjan in southern Gujarat, which was the first port of call in India for the Parsi community, is said to have been planted by the early settlers. Several others exemplify unique cultural bonds between people and nature.

Historical significance: With India's colourful past, it is but natural that the country is dotted with trees that have played a role in shaping our history, whether happy and delightful (for example, the coronation of a king) or gruesome and macabre (for example, the beheading of criminals and political opponents, or hanging of freedom fighters during colonial rule). Yet others are linked to India's historical personalities (for example, Shah Jahan, Dara Shukoh, Mahatma Gandhi, Rabindranath Tagore, Sardar Vallabhbhai Patel). Some were planted to commemorate a landmark occasion (for example, the visit of Queen Elizabeth of England or Sir C.V. Raman's death). Unfortunately, several of them, including those associated with India's freedom struggle, have fallen prey to ignorance or apathy.

INDIA IS STILL TAKING BABY STEPS

Although India's cultural origins date back thousands of years, we do not have meticulous and systematic records of our remarkable trees. Doubtless, several studies are available on ecologically important regions such as the Western Ghats or the Himalayas, or on the association of trees with gods and saints and the wish-granting abilities of trees. However, the approach – with a few notable exceptions – is usually generic and the intention is not to identify individual exceptional trees. Our regional floras barely mention exceptional trees. There is no national programme on inventorying – let alone conserving – our trees. For this reason, a study carried out during 2008–09 by Y.D. Bar-Ness, a Fulbright US scholar to document and geo-tag 962 'landmark trees' of India is noteworthy. Bar-Ness developed an online database (*Landmark trees of India. https://outreachecology.com/landmark/*) of such trees (barring the states of Arunachal Pradesh, Mizoram, Tripura, Manipur and Nagaland). While not all the trees included in this work are of heritage value, his work offers a very good platform for future research.

Exceptions apart, most Indian states do not even have inventories of their iconic trees. Following the establishment of State Biodiversity Boards under the Biological Diversity Act 2002, some of them initiated preliminary work in that direction. The Government of Sikkim, for example, officially notified in 2016 two lists – trees having a girth between 20 and 25 feet (6.096 and 7.62 metre) and those whose girth exceeds 25 feet (7.62 metre). The forest departments of Gujarat (2010) and Tamil Nadu (2017) have publications on heritage trees. Haryana has reportedly compiled a list of special trees. These state-sponsored initiatives apart, there are also books published by non-government organizations for example, *Heritage Trees in and Around Bengaluru* by Bengaluru Environment Trust (2011) and *Living Landmarks of Chennai* by Nizhal (2015).

Many states and some private organizations offer activities centred on nature appreciation, biodiversity conservation, or reconnecting trees and people in our cities. Attractive ecotourism packages are regularly on offer by forestry departments in protected areas across the country. Several dedicated academic groups and non-governmental organizations promote tree outreach programmes. The *Hindu* (18 April 2022) reported that Indian National Trust for Art and Cultural Heritage (INTACH) is tracing interesting stories and memories associated with the grand old trees in the villages of Palakkad district of Kerala. The aim is to educate the younger generation about the close affinity between people, trees and the natural world. Tree walks are becoming increasingly popular, especially in urban centres. Citizen science projects in ecology, however, are still taking baby steps in India, with 17 projects across the country, with just one devoted to trees during 2017–18.[43] Print and electronic media play a very significant role in spreading knowledge and information about India's remarkable trees, although the information may not always be verified or accurate. Increased incomes and the rise of social media have enabled anyone of means with a smartphone to upload text and images easily on the go on nature trails in remote places, making them instantly available. A lot of useful information on unusual trees can be gleaned from blogs and travelogues penned by itinerant nature lovers. Unfortunately, most of the information is unorganized and widely scattered across domains. Obviously, there is much work to be done in India.

HOW SAFE ARE OUR ICONIC TREES?

Trees, being living things, must eventually die. However, disconcertingly large numbers of them disappear from our forests and urban landscapes both due to natural calamities and human-related causes. Storms, wildfires, pests and diseases are the major natural drivers of tree morbidity and mortality. Work carried out by Professor C.Y. Jim of the University of Hong Kong provides some pointers.[44] In Guangzhou, China between 1986 and 1995 mortality rates among the 209 heritage trees were 21.5 per cent. The same increased to 29 per cent over the period 1995 to 2007. A similar phenomenon was seen in Hong Kong between 1993 and 2005 when 14 per cent of the heritage trees were lost. Exact figures of heritage tree mortality over time are not available in India, but sporadic and scattered information is available on some individual trees. The banyan tree under which Mahatma Gandhi regularly held prayer meetings before famously breaking the British salt law at Dandi in 1930 was laid low in a severe storm in 1982. A branch of the fallen tree was replanted at the original site on 6 November of that year by Morarji Desai, the former prime minister of India. Another giant banyan tree known as Pillalamarri ('children's banyan') in Mahabubnagar district of Telangana was on the brink of collapse from a severe infestation of termites in 2018. It was treated by forestry officials first with a chemical pesticide chlorpyrifos through drips into its trunk and branches, much like saline drips to patients and then with biopesticides. After a long battle with death, it seems to have revived now. More recently, cyclone Amphan struck Kolkata on 20 May 2020 and wrought havoc on the city's iconic Great Banyan, damaging 44 of its prop roots.

Human-related reasons (logging, agriculture expansion, deforestation, extension of infrastructure) are far worse culprits. In India, built heritage is valued over and above living heritage. Perhaps, the following example will illustrate this. Eucalyptus was first introduced to India by Tipu Sultan, the ruler of the Mysore kingdom, around 1790 on Nandi Hills near Bengaluru. In the winter of 2018, officials in charge of the garden cut down three of the ancient trees as they were leaning toward a dilapidated building used occasionally by the Sultan as a summer guesthouse. It is not clear whether they were unaware or unmindful of the heritage value of these trees. Then there is the iconic big banyan tree in Kethohalli, near Bengaluru – reputed to be 400 years old and much-favoured spot for photoshoots. Environmental experts worry that its days are numbered because of the concretization of the ground around it by government agencies that are tasked to protect it.

Far too many trees fall prey to the development of road, rail and metro services all over India. Examples of the direct and indirect environmental impacts of roads include deforestation and fragmentation, chemical pollution, noise disturbance, increased wildlife mortality due to automobile collisions, changes in population gene flow and facilitation of biological invasions.[45] The *Economic Times* (13 November 2022) reported that according to an affidavit filed by the Forest Department of Delhi in a contempt case regarding the preservation of trees in the city, 77,420 trees were permitted to be cut down during 2019–21. That amounts to three trees cut down every hour! Similar cases are available in almost all the states. The Indiatimes (30 July 2019) reported that while responding to a question in the Lok Sabha, the Minister of State for Environment stated that between 2015 and 2019, the Ministry of Environment,

Forest and Climate Change permitted the cutting of 10,975,844 trees across the country for various development projects. Amid this dismal scenario, it is very heartening to note that a Supreme Court bench led by the Chief Justice of India in February 2021 has suggested that road alignments ought to be made in such a way as to minimize the number of trees to be cut. The bench had also sought a report on how to rationally assess the value of trees so that they can be monetized and factored into the project costs while making a cost-benefit analysis. Heritage trees in urban and peri-urban settings suffer from soil compaction, degradation, topsoil loss, or pollution. The oft-seen response in our crowded cities (even religious places) is to install impermeable concrete or stone slabs that can cause even more damage to subterranean parts. The fact that India still manages to host many iconic trees is due more to accident or good fortune – and, in several instances, religious sentiment – than good planning or design!

To be sure, there are instances, where spirited resistance by concerned local citizens and non-government organizations has saved our iconic trees. In one curious example, the Brihanmumbai Municipal Corporation suddenly announced its intention in 2007 to transform the then 148-year-old Rani Bagh at Byculla into a 'Singapore-style' international zoo at a cost of some ₹430 crores. There was already a small zoo coexisting with the garden that was famous for several heritage trees, with virtually no treeless space for any kind of construction. The 'Save Rani Bagh' group that started in April 2007 turned into a powerful movement with support from the concerned public. After a protracted battle lasting almost 15 years, the master plan was reworked and thanks to the foundation's monumental efforts, several centurion trees at Rani Bagh have been saved for posterity. A lavishly illustrated book presenting Rani Bagh's rich botanical, historical and cultural heritage was brought out in 2012 to commemorate the garden reaching the significant landmark of 150 years.[46] In the absence of organized state protection, there is another kind of informal community protection afforded to trees that are endowed with religious or spiritual values or considered intrinsically holy for some reason. Many of them are located within places of worship (for example, temples, gurdwaras, dargahs) in urban or rural settings and their management and protection are implicitly the responsibility of the local religious group or administration. At any rate, people avoid lopping or mutilating them and they are usually safe. I have noticed that across India, most trees that survive well in public places are those under which religious shrines are built.

NEED FOR SPECIAL PROTECTION FOR ICONIC TREES

Action to protect trees is mandated by our regional and national laws and international policies and legislation to which India is a signatory. However, there is no specific focus on special trees. Existing tree acts offer some legal protection, but heritage trees do not find adequate coverage under these, simply because there has been no serious nationwide discussion on the subject. Recent months have seen interest among certain state governments to protect their old trees. For instance, the Government of Maharashtra plans to introduce provisions for the protection of heritage trees[47] equal to or older than 50 years. The age of the heritage tree will determine the number of trees to be replanted in case it is felled. For example, the felling of a 64-year-old tree entails 64 trees to be planted in compensation. Similar provisions are being mulled over in Uttar Pradesh and Haryana and lists of heritage trees estimated to be over 100 years are under compilation. Chandigarh has also earmarked some of its remarkable trees for protection. While the intent to save old trees is good, emphasis on age alone would leave out several other categories of iconic trees that deserve appreciation, conservation and protection. Moreover, it is not clear what scientific methodology will be used to determine tree age and which agency will be entrusted with that task. As mentioned earlier, age has implications for compensatory planting.

THE NEXT STEPS AND THE PATH AHEAD

The inexorable march of human development has often been at odds with the natural world of which we are but a part. The pushback is now palpable and we are already experiencing the fallouts in many ways – the latest manifestation being global warming. There can be no argument against development, but nowadays there is grudging acknowledgement that the cost of ignoring its environmental impact would be untenable. The slogan of the United Nations Environment Programme (UNEP) 'Development without Destruction' has never been more relevant. It is, therefore, time that we appreciate, prioritize, secure and protect our natural environment. A very significant part of our terrestrial environment is made of trees, including those of special value to the community in terms of their cultural and heritage value and we need to recognize that it deserves at least as much importance as built heritage.

These are some of the steps necessary in this direction.

BUILDING CONSENSUS THROUGH PUBLIC CONSULTATION AND SCIENTIFIC DOCUMENTATION

Trees have been valued across the world for diverse sets of attributes and programmes are in place in several countries to acknowledge and protect unique examples. However, a consensus on the key attributes to be used in designating them has eluded a solution: some 60 terms have been employed in global publications, with the term 'heritage tree' being the most used.[48] In this book, I have used six broad criteria to describe iconic trees as a starting point for discussion Perhaps other criteria/sub-criteria could be added to these (for example, aesthetic value, important landmark, role in the life of rural or urban communities). Subjectivity can be avoided through a transparent process of wide and systematic consultation with heritage tree experts and stakeholders across India at different geographical regions and scales (for example, national, state and district). A recent global attempt made to evolve a consensus identified 16 common criteria shared across tree programmes and could serve as an initial template.[49]

The authors rightly noted that: (a) the study should be replicated to enlarge the number of experts consulted; (b) some of the criteria should be evaluated for their importance in various geographical regions and at different scales; and (c) case studies should be carried out in different countries to test the data. The importance of evolving a consensus in a vast country like ours cannot be overemphasized. Such an exercise with public consultation and access to documents that can verify a tree's relationship with a given criteria would add transparency and robustness to both the process and the outcome. The latter is especially important in cases dealing with 'historical', 'rare', or 'endangered' criteria. An indicative but not exclusive list of stakeholders should include heritage tree specialists, environmental experts, forestry officials, historians, social scientists, representatives of the National Green Tribunal, town planners, architects, landscape developers, tourism specialists, farmers, voluntary organizations, legal experts and policymakers.

INVENTORY, MAPPING AND PREPARATION OF ICONIC TREE REGISTERS

It is very important to track, inventory, map and geotag iconic trees to produce iconic tree registers at different geographical scales – district, state and national. Registers must be electronic and updated at predetermined periodicity. The information gathered should include (i) the botanical identity

41

of the tree, (ii) common and local name(s), (iii) geotagging (GPS location including latitude, longitude and altitude), (iv) features such as age, height, girth, crown shape and size, etc., (v) reasons for recognizing it as an iconic tree, with associated information along with documents where available, as well as (vi) adequate digital images of the tree and its ambience. Inventory and scientific data can reinforce community awareness and improve management. Citizens should be encouraged to propose candidate trees under various categories and an expert committee appointed by a designated professional agency should validate the proposal after site visits and verification of the data and records. Information on the person(s) who proposed the tree for inclusion in the register and that of the approving authority should also be recorded. The designated agency should regularly and periodically publish and update the iconic tree register(s).

CONSERVATION AND MANAGEMENT OF ICONIC TREES

The fact that current legal provisions for the protection and management of iconic trees in India are inadequate and that an appropriate legal framework should be put in place urgently has been emphasized earlier in this chapter. The framework must be backed with statutory and administrative measures to ensure the protection and safety of our outstanding trees at the ground level. Systems to periodically monitor tree health and assess and mitigate risk (especially in urban and peri-urban sites) should be put in place vis-à-vis environment-driven events (for example, wind, snow, cyclones, diseases), wood strength and species differences.[50] It is now possible to use more sophisticated 'precision arboriculture' to analyze and monitor individual trees using remote sensing and artificial intelligence techniques for this purpose.[51] Any conflicts with developmental programmes should be settled by in situ preservation and tree transplantation should be the last resort.

AWARENESS GENERATION, PUBLIC PARTICIPATION AND INVOLVEMENT

Public participation and involvement can reinforce community awareness and buttress conservation efforts of our green national treasures. Assessing their economic and ecological value could muster support for public-funding justifications. Engaging citizens, educational institutions, voluntary organizations and the business sector could cultivate ownership and involvement. Members of the community should be coached in tree management techniques so that they can become 'citizen tree wardens'.

Signages should be placed near iconic trees, providing information on their importance and the reasons for designating them as such. Preferably, the signages should be of uniform design and size across the country (for example, as is being done near archaeological sites) so that there is brand recognition. Adequate publicity must be provided through public lectures, attractive books, heritage tree walks, websites, print-, electronic- and social media. Ecotourism should be planned and organized around iconic trees. Outreach programmes should be organized bringing out the stories and legends associated with the trees.

EVOLVING A TIME-BOUND PLAN AND FIXING RESPONSIBILITY

A time-bound action plan at local, state, regional and national levels should be evolved as well as responsibility fixed for the protection, monitoring and regular and periodic review of iconic tree morbidity and mortality. The monitoring group should have adequate civil society representation and skill upgradation for tree protection and monitoring of tree health must be ensured.

MAKE ADEQUATE BUDGET PROVISION

Experience has shown us that the most well-intentioned programmes come to naught for failing to provide adequate budget provisions. The designated agency should, in addition to having its allocation from the government, be eligible to receive grants from philanthropic agencies and individuals, industry and public-spirited members of the civil society.

Finally, there is enough knowledge, experience and skillset in India and worldwide to recognize, appreciate and protect the iconic trees. We must study, adopt and implement the global best practices after suitably modifying them, where necessary, to fit our national needs.

A LAST WORD

As custodians of our environmental security, witnesses to India's history, envoys of our culture and much more, the trees in this book are of concern to each one of us. I will feel well rewarded if it helps improve awareness of India's living and breathing monuments and sparks greater public participation in their inventory, management, appreciation and protection.

In the pages that follow, you will discover gnarled ancient champions, dendroid behemoths that have achieved unmatched dimensions, those that remind us of historic personalities or events, benevolent wish-granters and even a few that bring back memories of gory deaths and dark punishments. Each narrative includes humans but, invariably, there is a tree at the heart.

LIVING

ICONIC

TREES

OF

INDIA

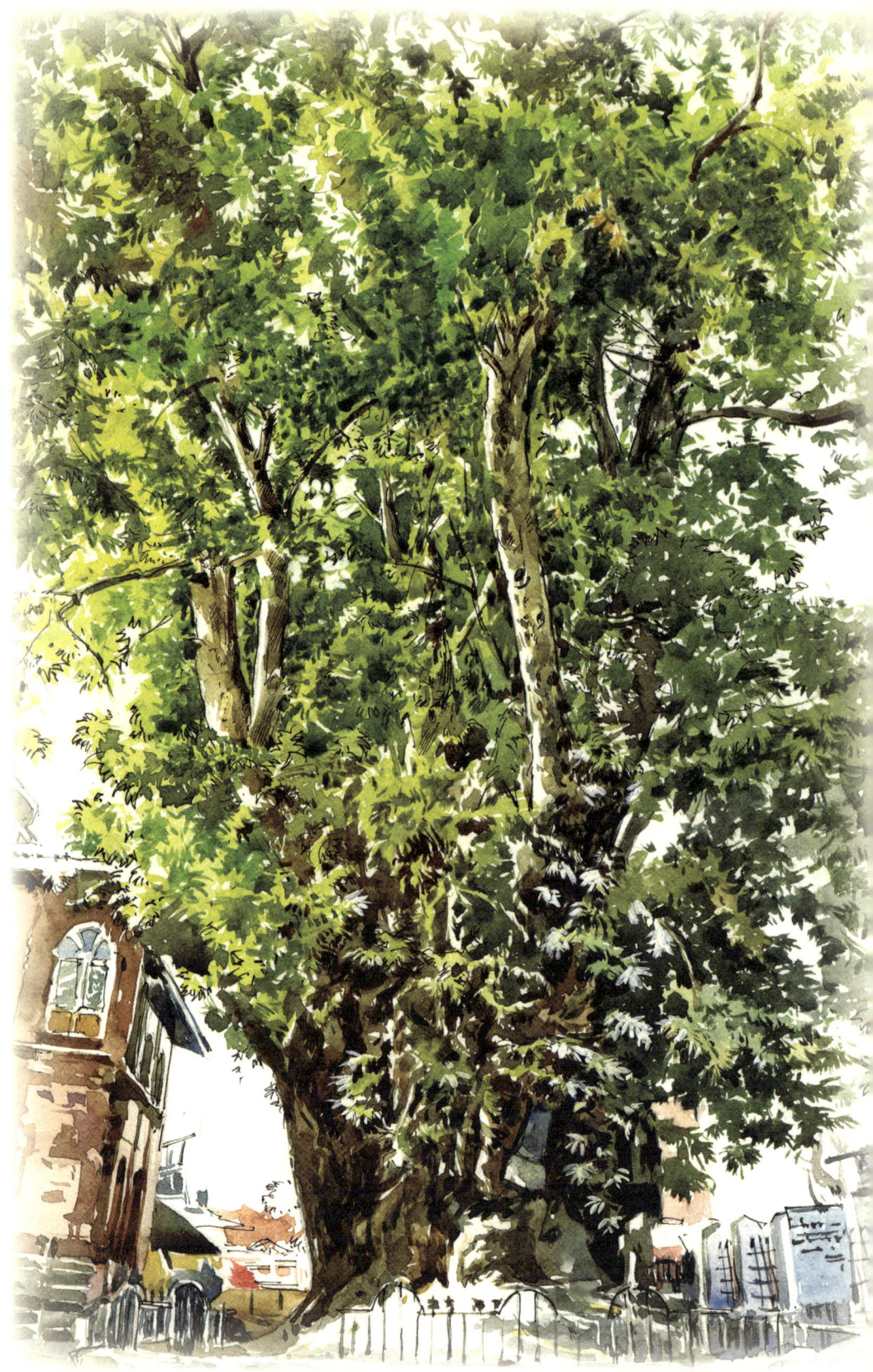

1

Qasim Shah's Chinar –
India's Oldest and Largest

COMMON NAMES: Oriental plane (English); Bouin, Booen,
Booni, Cheenar, Chinar (Kashmiri)

SCIENTIFIC NAME: *Platanus orientalis* L. | FAMILY: Chinar (Platanaceae)

WHERE TO SEE: Mosque of Hazrat Syed Qasim in Chattergam village,
Chadoora tehsil, Badgam district, Jammu & Kashmir

LATITUDE: 33.5901° N; LONGITUDE: 74.50090° E; ALTITUDE: 1,591 metres

In Kashmir, it is impossible to ignore the stately chinar. With its maple-like leaves and lofty height, it is everywhere. Prominently featured in Kashmiri art and craft, literature, religion, politics and romance, it is no surprise that Kashmiris claim it as a symbol of *kashmiriyat*, the essence of Kashmir. Except that the chinar is not native to Kashmir but an immigrant; its natural home is between the Eastern Mediterranean region and Iran.

The popular belief is that the tree came to the valley with the Mughals, but its arrival here predates the Mughal era. The Mughal emperor Akbar annexed Kashmir in 1586. The *Akbarnāma*, the official history of Akbar's reign, written by Abul-I-Fazl ibn Mubarak records the occurrence of large chinar trees in Kashmir even at that time.[1] One particular hollow tree was so large as to accommodate thirty-four soldiers in its commodious trunk. Other historians have noted that two centuries prior to Babar's arrival in India, Lalleshwari, Lalla or Lal Ded (1320–92), the Kashmiri mystic and poet, lovingly revered by Hindus and Muslims alike, compared a virtuous and loving wife to the cool and refreshing shade of the boin (chinar) tree on a hot summer day. Obviously, this imagery must have been based on the tree's common occurrence and extensive cultivation back then. Moreover, archaeological evidence provides conclusive evidence for the early arrival of chinar in Kashmir.[2] The Mughal emperors, especially Jahangir (r. 1605–27) and Shah Jahan (r. 1627–58), used the chinar as an essential element of Mughal garden architecture along with waterways, fountains and fruit trees. Jahangir also declared the chinar a royal tree and banned its felling. All the rights to its produce were vested in the emperor. Later, it was declared the state tree of Jammu & Kashmir (now a union territory).

Kashmir has many grand specimens of the chinar, but what is claimed to be the oldest and largest stands in the village of Chattergam, Chadoora tehsil, in Badgam district.[3] The tree was discovered by M.S. Wadoo, a forest officer of the

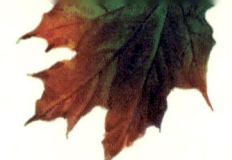

Jammu & Kashmir cadre. In 1372, during the Sultanate era (Shah Mir dynasty), Hazrat Mir Syed Ali Hamdani, a great Sufi reformer, poet and widely travelled scholar arrived in Kashmir to spread the message of Islam and convert the people to its faith. Several of his comrades followed him and travelled to different corners of the valley to accomplish the mission. One of them, Syed Qasim Shah, chose Chattergam as his home. He is believed to have planted the chinar tree in the year 1374.[4] If this is true, the tree could be roughly 650 years old.

The large and majestic chinar stands next to a small mosque. Its huge trunk has a girth of fifteen metres, the circumference of Qutub Minar at the base. Of the six main branches spreading out from the trunk in all directions, one was cut off some years ago to avoid damage to the roof of an adjacent madrasa. The other five branches are still intact and support a massive crown, home to hundreds of birds that maintain a cacophonous refrain. A low circular boundary has been built at the base of the tree. Locals say that Qasim Shah's last resting place is located next to the chinar, making it a sacred spot. Over time, however, the tree has grown so much that no sign of the shrine is visible today.

So far, the local government has done very little to protect the oldest living chinar tree in Kashmir, even probably in the whole of South Asia. The global record for age is shared by two plane trees – one in Gedelme village, Kemer, Turkey and the other in Tnjri in the village Ghirmizi Bazar, Georgia. The latter also holds the record for trunk girth (twenty-seven metres).[5] Both these trees are believed to have been planted in 50 BCE ±500. Repeated petitions by concerned citizens to develop a park around the champion tree have gone unheeded. In 2007, a local politician donated a small amount of money to spruce up the area, but work was left incomplete.

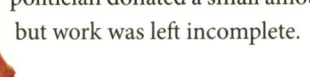

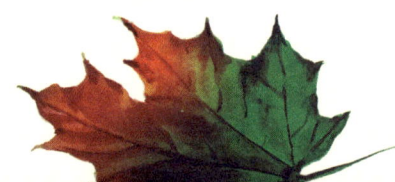

48

2

The Loneliest Tree in India

COMMON NAMES: Giant Sequoia, Big Tree (English)

SCIENTIFIC NAME: *Sequoiadendron giganteum* (Lindl.) J. Buchholz

FAMILY: Cypress (Cupressaceae)

WHERE TO SEE: Experimental Station of Indian Institute of Integrative Medicine, Yarikha, Tangmarg, Kulgam district, Jammu & Kashmir

LATITUDE: 34.447° N; LONGITUDE: 74.2537° E; ALTITUDE: 2,123.51 metres

Tangmarg – literally 'Peach Town' in Kashmiri dialect – is considered the gateway to the popular skiing resort of Gulmarg, just 13 kilometres further east. It is the last hurried stop for tourist buses and other heavy vehicles before Gulmarg as there are no shops in between. Few visitors would prefer to stay in Tangmarg and fewer still would care to venture into Yarikha village close by – unless they are looking for the experimental farm of the Indian Institute of Integrative Medicine (IIIM) headquartered in Jammu. Here, at an elevation of 2,560 metres, the institute, a constituent of the Council of Scientific and Industrial Research (CSIR), conducts experiments on high-altitude medicinal and aromatic plants. As you negotiate the last bend up the winding road of the hillock on which the 10-acre farm is situated, you come across something you never expected here or anywhere else in India, probably even South Asia: You are in the imposing presence of a giant sequoia. Giant sequoias, for the uninitiated, represent the most massive single-stemmed trees on earth. The most famous and the largest known tree, nicknamed 'General Sherman' (in honour of the American Civil War hero) has an estimated trunk volume of 1486.9 m³ (not including the branches or roots).

The tree stands out because of its height (approximately 25 metres). Towering over the surrounding deodars and Himalayan white pines with its tapering crown as well as dense foliage, its trunk is characteristically thick (5.85 metres) with spongy reddish-brown bark having deep furrows. While the trunk is straight and clear of branches for up to six metres from the ground, the branches are horizontal and get progressively shorter towards the apex, giving the crown a conical shape. The scale-like leaves remind you of a juniper.

Although the tree bears both male and female cones, no one has ever seen a seedling here, possibly because the tree cannot produce enough pollen to successfully pollinate the female cones.[6] Natural reproduction in sequoia is known to be weak: Average seed viability is very low. Moreover, in its natural habitat, giant sequoia regeneration benefits from fire – both wildfire and prescribed, a factor that is absent here. Sadly, although the giant sequoia is acclimatized to Kashmir's climate, it is unable to regenerate naturally.

How did this lone tree – restricted to high elevation (1,220–2,440 metres) on the western slopes of the Sierra Nevada mountains of California – come to be in this neck of the woods? It seems no one took the slightest notice of the tree for the first three or four decades of its life. Discovered in 1975 by Dr D.L. Dhar,[7] a botanist then at the University of Kashmir, Srinagar while on a survey expedition, the tree was nearly 22 metres tall with a trunk girth of 2.8 metres at that time. Dhar estimated it to be 40-45 years old. Thus, its current age could be roughly 85–90 years.

While there is no record of who introduced it here, residents state that whenever the British Resident visited Gulmarg for skiing, members of his staff usually camped at Yarikha. It is possible that one of them planted the seedling here. My guess is that it was probably Sir Peter Clutterbuck. Having just retired from service as Inspector General of Forests in India, he had joined service in the princely state as Chief Conservator of Forests at the invitation of Maharaja Hari Singh of Kashmir in February 1933. He could have easily sourced the seeds either from Europe or the US and got them planted in the high-altitude Yarikha forest zone, which was under his direct jurisdiction.

The International Union for Conservation of Nature (IUCN) has placed the giant sequoia under the 'endangered' category. One cannot but feel a sense of sadness for what must be India's loneliest tree – located so far away from its natural home, in an alien habitat in the middle of nowhere, forgotten by those who were responsible for its journey here, devoid of kin, practically unknown to the outside world – underappreciated and unable to reproduce. There is, however, one small consolation. Aware perhaps of its unique treasure, in 2014, IIIM commissioned two paintings of the tree by the well-known Jammu-based artist K.K. Gandhi. One of these is proudly displayed at the institute's headquarters in Jammu and the other at CSIR's Central Drug Research Institute, Lucknow.

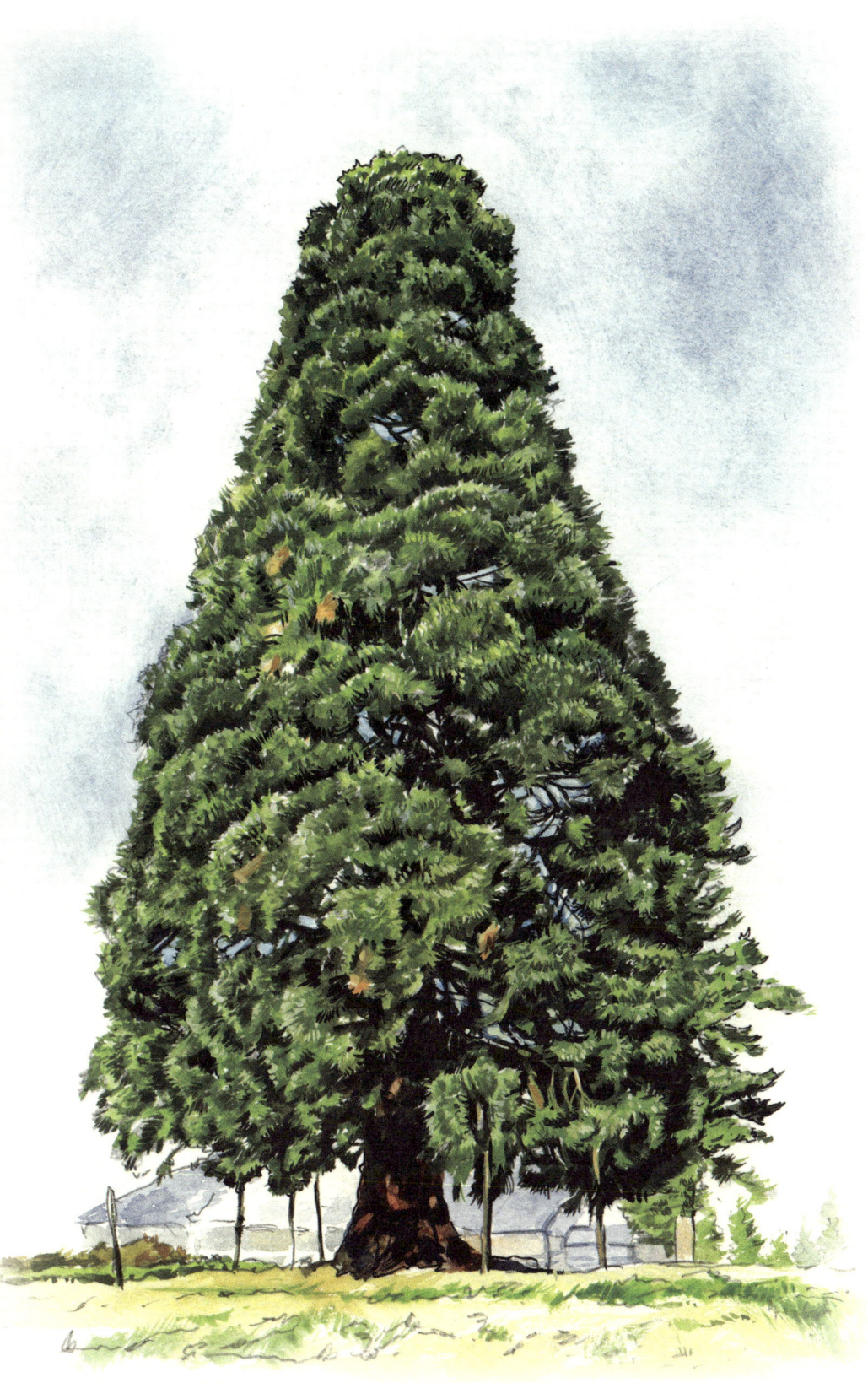

3

Shah Jahan's Bequest –
The Chinars of Naseem Bagh

COMMON NAMES: Oriental plane (English); Boiun, Booen,
Booni, Cheenar, Chinar (Kashmiri)

SCIENTIFIC NAME: *Platanus orientalis* L.

FAMILY: Chinar (Platanaceae)

WHERE TO SEE: University of Kashmir,
near Hazaratbal Mosque, Srinagar

LATITUDE: 34.1280° N; LONGITUDE: 74.8365° E;
ALTITUDE: 1,583 metres

Although the Mughals did not introduce the chinar to Kashmir, they were smitten by its beauty and majesty. It must have reminded them of their Persian home. Even the name 'chinar' is derived from Persian, meaning 'what a fire!' and alludes to the autumn colours of the tree.

Emperor Akbar took a particular fancy to the chinars and ordered a residential garden to be laid out in 1586 on a terraced terrain just north of Srinagar on the western shore of Dal Lake.[8] Almost five decades later, in 1635, emperor Shah Jahan had a new garden made at the same site. Twelve hundred saplings of chinar were planted on a rectangular patch, creating a unique chinar garden. The saplings were reportedly irrigated with pure milk for six months. A special channel was dug to bring water to the site from the Zakura stream to embellish the garden with watercourses and fountains, finally winding its way into the Dal Lake. Once the trees were established, a blissful gentle breeze began to blow through the garden. This is referred to as *baad-e-naseem* ('zephyr of early dawn' in Urdu). Hence the garden came to be named *Bagh-e-Naseem* (or 'garden of the gentle wind'). Kalim Kashani (1581/85–1651), one of the leading poets of the seventeenth century and poet laureate in Shah Jahan's court, penned odes on Naseem Bagh.

In 1838, Godfrey Thomas Vigne, an independent traveller and talented artist with an interest in botany and geology[9] found the average girth of the chinars in the garden to be 3.96 metres and guessed their age to be 248 years. The largest trees, close to the water, averaged a girth of 6.1 metres.[10] Today, Naseem Bagh is among the oldest Mughal gardens in India and still has several chinar trees planted by Shah Jahan. However, the waterways, fountains and the original boundary wall have long disappeared through centuries of neglect following the decline of the Mughal Empire. Sir Walter Lawrence[11] noted in *The Valley of Kashmir* in 1895 that

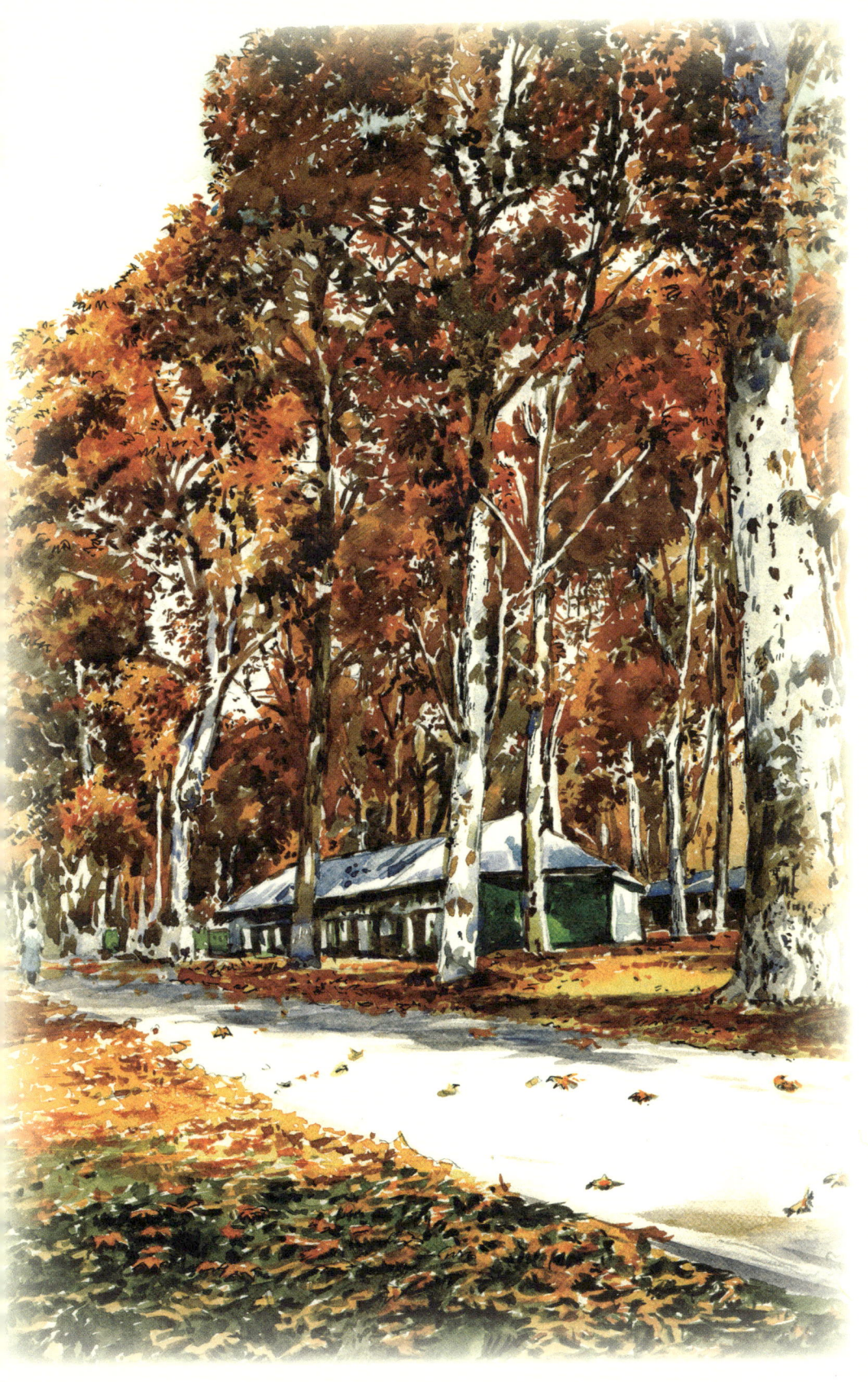

several fine chinar trees in Naseem Bagh were hollowed out. During the 1950s, the Dogra royal family who owned the land handed over the garden to the civil administration which, in turn, decided to set up the University of Kashmir on the site. Over the years, the chinars started disappearing, their places replaced by ugly concrete constructions.[12] A study carried out in 2004 by the university's Post-graduate Department of Environmental Sciences found 733 surviving chinars, mostly damaged due to lack of proper care and maintenance and their tree girths to be in the range of 2.44 to 4.88 metres. Since then, the number of chinars has further declined to less than 700, occupying an area of a little over 22 hectares.

Naseem Bagh continues to inspire generations of scholars, artists, poets and people from all walks of life. Maulana Abul Kalam Azad, a great stalwart of the Indian freedom movement and the first education minister of free India as well as an outstanding Urdu writer has made Naseem Bagh memorable in his scintillating literary work *Ghubar-e-Khatir* (Sallies of Mind). Bollywood directors are now returning to Kashmir and Naseem Bagh has also featured prominently in Hindi films (for example, the Tabu-starrer *Haider*, 2014). Fallen chinar trunks in the garden have also become a medium to vent anguish through art for the students of Kashmir University. They gather around these tree trunks to sing, draw and paint and get to know each other.[13]

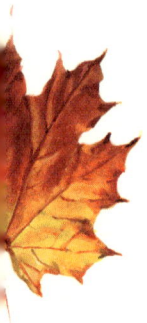

A society grows great when old men plant trees whose shade they know they shall never sit in. An ounce of practice is worth more than tons of preaching.

Mahatma Gandhi

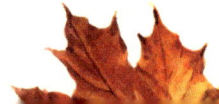

4

Dara Shukoh's Chinar of Bijbehara

COMMON NAMES: **Oriental plane (English); Boiun, Booen,
Booni, Cheenar, Chinar (Kashmiri)**

SCIENTIFIC NAME: *Platanus orientalis* L.

FAMILY: **Chinar (Platanaceae)**

WHERE TO SEE: **Badshahi Bagh of Bijbehara town in Anantnag district**

LATITUDE: **33.8° N**; LONGITUDE: **75.1° E**; ALTITUDE: **1,591 metres**

Bijbehara, the ancient and historic town in Jammu & Kashmir close to the capital Srinagar, has two chinar gardens – Padshahi (Badshahi) Bagh and Dara Shukoh Garden – that can be traced back to the Mughal era. Work on the gardens commenced around 1634 on the orders of Prince Dara Shukoh[14] (1615–1659), the ill-fated favourite son of Emperor Shah Jahan, his Queen Mumtaz and heir-apparent to the Mughal throne, Dara was an unorthodox Mughal who has been both vilified and venerated. Though he has been (sometimes) portrayed as an effete prince, utterly incompetent in military and administrative affairs, he was undoubtedly a visionary thinker; a talented poet and a prolific writer. He was defeated by his pugnacious younger brother Prince Muhiuddin (later the Emperor Aurangzeb) in a bitter struggle for the throne and executed in 1659.

The two gardens were planned so as to be situated on the opposite banks of the Jhelum – here popularly known as Veth – with a connecting Padshahi Bagh bridge across the river and took two years for completion. Both these gardens have several magnificent chinar trees from this period, justifying Bijbehara's sobriquet of 'the town of chinars'.

The Padshahi Bagh is spread over approximately 1.85 hectares and originally had 25 chinars. At present, 19 of these can still be seen and several of them are stately. Until recently, one of these was held to be the oldest chinar of the subcontinent.[15, 16] However, with the discovery of a larger and older chinar tree in Chattergam village of Badgam district, it now becomes the second oldest.[17] Possibly planted by Dara Shukoh himself in 1636, this majestic tree is more than 380 years old. In 1987, its trunk had a circumference of 13.30 metres.[18] Currently, the girth has increased slightly to 13.75 metres. It is interesting to note that the main trunk is completely hollowed out since several years[19] and about five metres above the ground it is also broken or cut off. Near the broken end, there are seven large and healthy branches spreading out in all directions. It is a wonder that the hollow bole bears the weight of the entire aerial portion of the tree. A pretty circular platform with a wooden fence constructed around the tree further

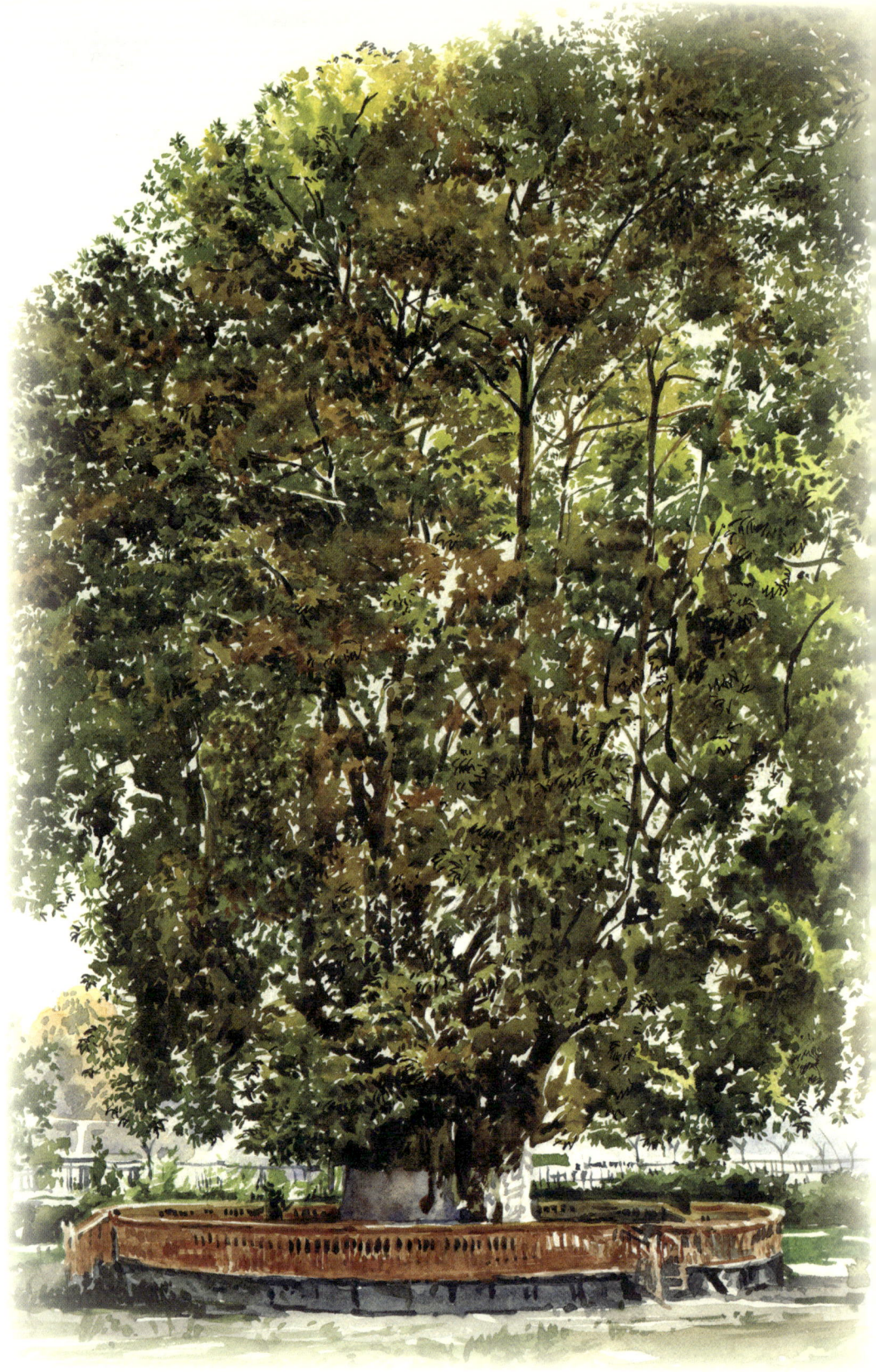

enhances its beauty and majesty. On the trunk of the tree, a painted board informs visitors of its girth and age. Assuredly, the tree is the showpiece of the garden.

On 11 November 1972, a Delhi daily announced that the historic Bijbehara chinar was dead as it had been wantonly felled to provide firewood to the people. The news was promptly reproduced without verification in the March 1973 issue of the forestry journal, *The Indian Forester*. Both were, of course, wrong – the damage having been caused probably by the first snow of the winter.[20]

Dara Shukoh's life might have been prematurely and tragically cut short, but this tree hulk of Bijbehara has survived the vagaries of time for close to four centuries.

5

The Toon of Raang Village

COMMON NAMES: Toon Tree (English); Kacchapah, Tuna,
Nandi, Nandikah (Sanskrit); Tun (Hindi)

SCIENTIFIC NAME: *Toona ciliata* M. Roem.

FAMILY: Neem or Mahogany (Meliaceae)

WHERE TO SEE: Open field in Raang village,
Ramnagar block, Udhampur district, Jammu & Kashmir

LATITUDE: 32.91° N; LONGITUDE: 75.14⁰ E; ALTITUDE: 575 metres

In a remote but scenic village known as Raang in the Udhampur district of Jammu and Kashmir is a large toon tree, noticeable from a distance in an open meadow even before you catch sight of the village. Rising to a height of around 18 metres, it has a short trunk, just about a metre from the ground. Nine branches take over at that point and spread out to approximately 14 metres in all directions. At breast height, the circumference of the trunk is 6.5 metres. A square platform bordered with neatly arranged stones has been constructed by the residents around the base of the tree. The crown is large and the bark on the trunk and branches are festooned with moss and lichens. Mistletoe (*Dendrophthoe sp.*), a hemi-parasite and pilkhan (*Ficus virens*) can be seen hanging from the upper branches. The locals say the tree is 200–250 years old.

Although the species occurs naturally in this region, no other toon tree is seen in the neighbourhood, as most have probably been cut down for timber. The villagers say that not even dried wood from the tree is used by anyone here. How then was this particular tree spared from such a fate? The answer is simple: it is considered sacred by both Hindus and Muslims.

Long ago, a revered Sufi *pir,* whose name no one in the village can recall today, apparently made his home under the tree. Green flags can be seen near its trunk. The local Hindu goddess Chownthara Mata is also worshipped annually under the tree. Her idol is ceremoniously brought down in procession to Raang village from her temple on the hill on an auspicious day in the month of *Jeth* (*Jeshtha*) as per the Hindu almanac, usually during July. This signals the inauguration of the annual fair. Hindus and Muslims from nearby villages gather at Raang for this fair. The chief attractions, along with the *pir* and the goddess are the stalls of food, colourful trinkets and local produce set up in the vicinity of the tree. Sweetmeats and *sherbet* are enthusiastically offered to one and all. Embraced and accommodated into their lives with great enthusiasm, the toon tree has remained protected over the years by Raang and its people.

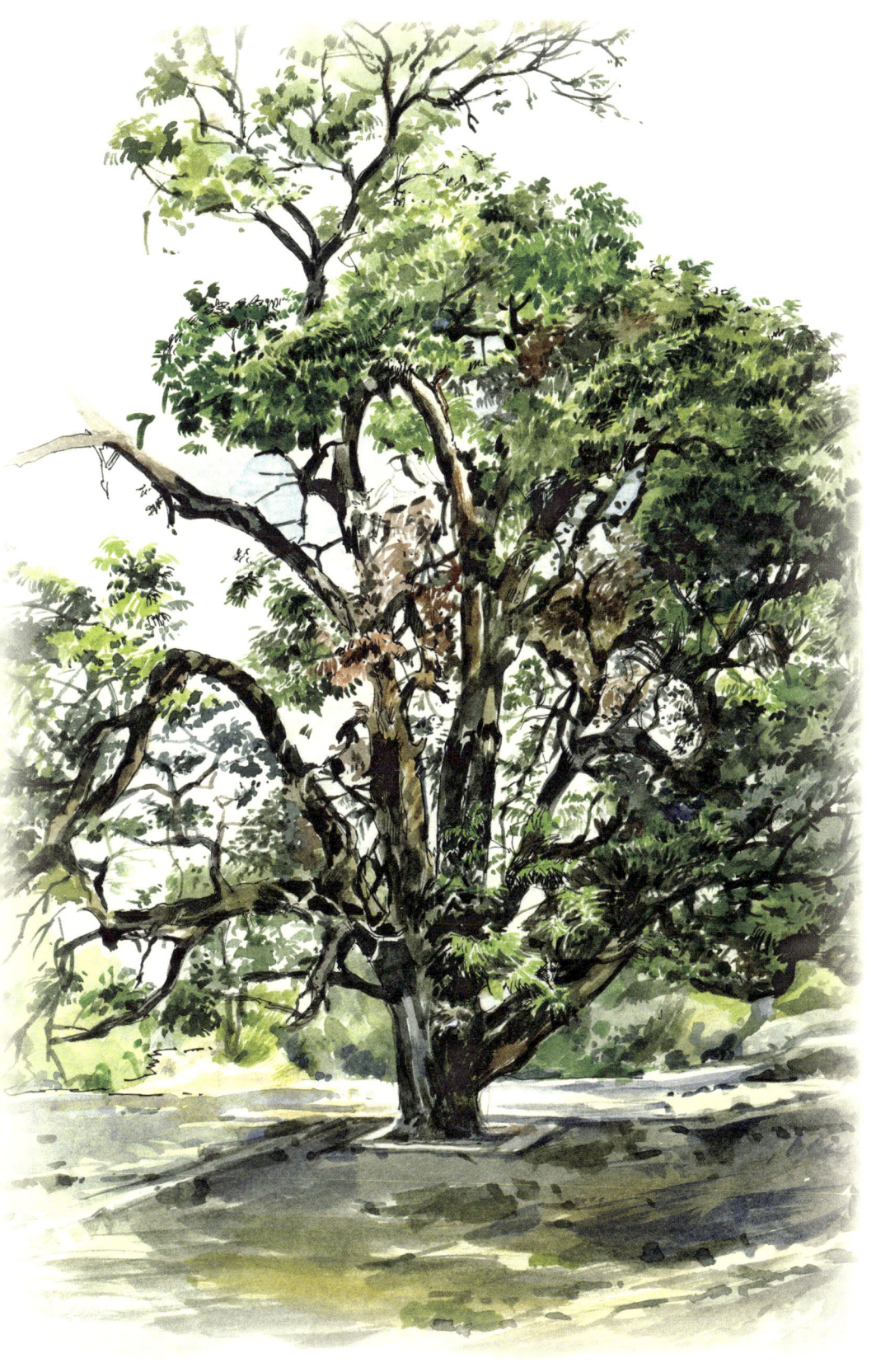

6

Guru Nanak's Ber Tree

COMMON NAMES: Jujube (English); Ber (Hindi/Dogri)

SCIENTIFIC NAME: *Ziziphus mauritiana* Lam.

FAMILY: Ber (Rhamnaceae)

WHERE TO SEE: Panjvaktra Mahadev Temple, Raghunath Bazar, Jammu Tawi, Jammu & Kashmir

LATITUDE: 32.731° N; LONGITUDE: 74.867° E; ALTITUDE: 350 metres

In the heart of Jammu city, in one of the narrow bylanes of the crowded Raghunath Bazar, is a very ancient temple known as Panjvaktra Mahadev Mandir dedicated to God Shiva.[21] The term 'Panjvaktra' denotes the five aspects of Lord Shiva: Sadyojāta, Vāmadeva, Aghora, Tatpurusha and Ishāna. They also symbolize the five elements: earth, water, light, air and space. Thus, Shiva is represented as Panchamukhi, a five-faced being. In the Jammu region, Panchamukhi Shiva has been worshipped since ancient times.

According to local legends and oral history, Adi Shankara stayed here when he visited Jammu. Others believe that the temple was discovered in the fourteenth century during the reign of Raja Mal Dev. Whatever the truth may be, there is a very old ber tree associated with Guru Nanak (1469–1539), the founder of Sikhism and the first of its 10 Gurus, on the premises.

A cursory examination confirms that the tree is indeed very old. The trunk emerges from a small gap in the marble-tiled flooring at an angle of 450 degrees from the ground and is without branches for nearly 4.6 metres under the cemented awning of the temple roof. Upwardly-bent branches emerge where the awning ends and make a lush green umbrella-like canopy. The basal portion of the trunk is hollow; its inner tissues decayed and degenerated long ago. Its dark brown bark has deep longitudinal furrows. Temple authorities have placed iron rings at different points along the trunk to keep it together, along with metal props to hold the bent trunk aloft. Brass bells have been suspended from iron chains along the trunk. Kumkum-dipped thread and red cloth adorn the trunk as a mark of worship. The tree bears small yellow flowers in bunches that give rise to shiny globose or oval ber fruits.

It is believed that Guru Nanak visited these parts around 1514 and stayed in the temple for three days. The then king of Jammu. Raja Khokhar Dev and his queen along with their entourage, called on Guru Nanak at the temple to pay their respects and seek his blessings. Guru Nanak, seated under the ber tree in the temple precincts, gave a discourse expounding on *raj dharma* (the duty of the rulers) and emphasized that the first and foremost duty of the king is to be fair to his subjects, even at the cost of his own comforts; that the king should be devout and rule the state with compassion and justice towards all. A beautiful mural commemorating the guru's discourse sitting under the ber tree can be seen decorating one of the temple walls.

It is reasonable to assume that at the time these events occurred the tree was already large enough for Guru Nanak to be seated under its canopy. If so, the ancient ber tree could now be over 500 years old, making it the oldest recorded ber tree in India. However, scientific confirmation of the age would, of course, be necessary to assess the veracity of this claim.

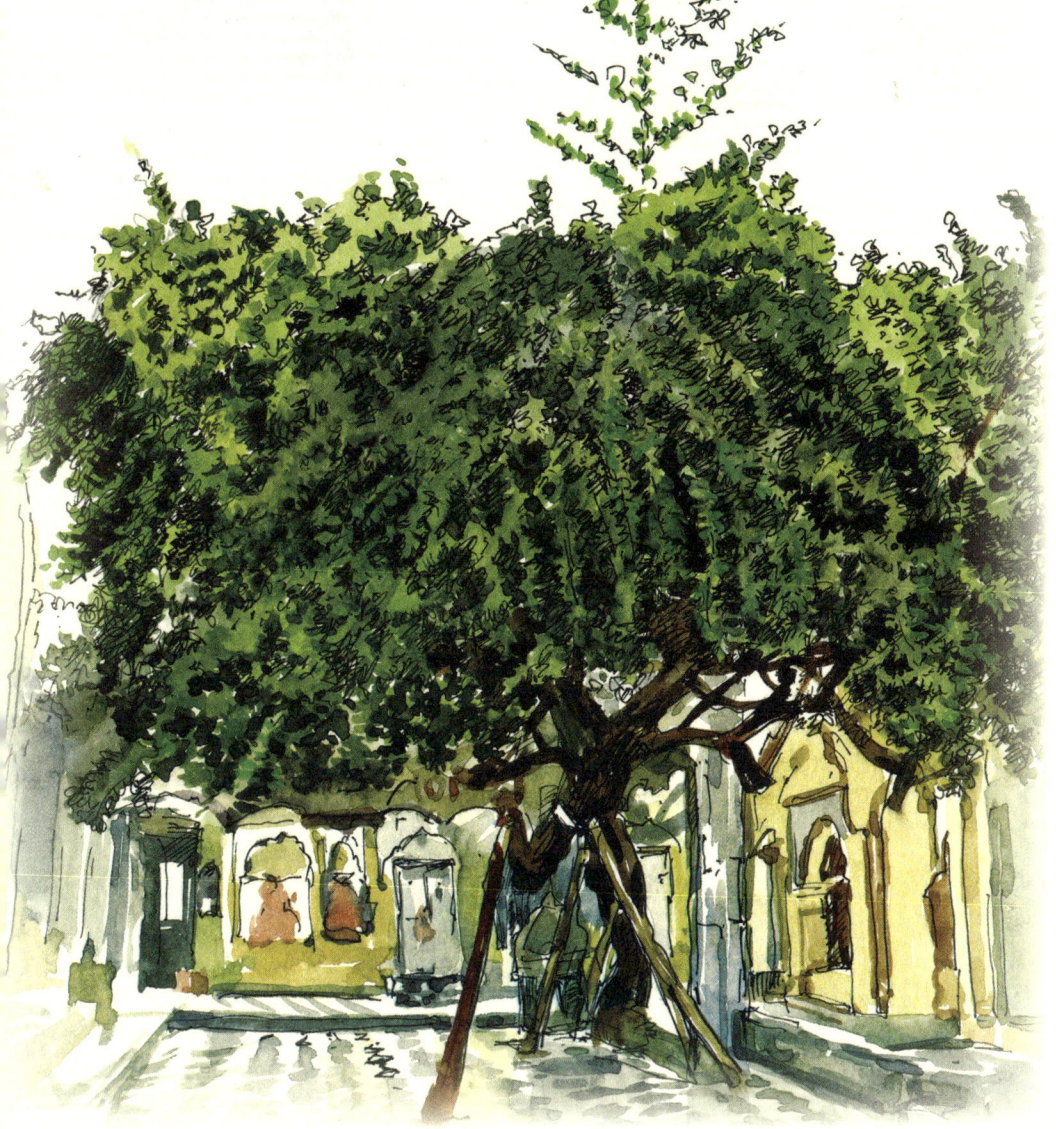

7

The Tree that Swallowed a Pillar

COMMON NAMES: **Sacred Fig Tree (English); Ashwattha (Sanskrit);**
Peepal Tree, Bodhi Tree (Hindi)

SCIENTIFIC NAME: *Ficus religiosa* L.

FAMILY: **Fig (Moraceae)**

WHERE TO SEE: **India and Pakistan border at Suchetgarh,**
R.S. Pura block, Jammu district, Jammu & Kashmir

LATITUDE: **32.5660° N;** LONGITUDE: **74.6740° E;**
ALTITUDE: **Approximately 333 metres**

I f I hadn't seen this for myself, I would have put this headline down as pure clickbait. Let us start at the beginning. The international boundary between India and Pakistan is managed by the Border Security Force (BSF) at a checkpost at Suchetgarh, a small village barely twenty-eight kilometres from Jammu city. Known as BSF's Octroi Border Outpost, it has drawn its name from an octroi collection point in the village. Once inside the high-security entrance, visitors can walk straight ahead towards the international boundary demarcated in 1947 by the so-called Radcliffe Line, named after its architect Sir Cyril Radcliffe. The visible indicator of this line is in the form of short, white, pyramid-shaped pillars painted with their serial numbers in black. One such pillar – bearing the serial number 918 – happens to be on the 'zero line' within the checkpost. Another pillar bearing the serial number 919 can be seen nearby. Close to pillar 918 stands a peepal tree. Due to seasonal growth every year, its trunk kept increasing in girth and started pushing against pillar number 918. Gradually, it surrounded the pillar and, in due course, completely engulfed it. A scrutiny of the tree's trunk still reveals vestiges of the pillar but that too will soon disappear. Fortunately, the Indian BSF and the Pakistan Rangers guarding their respective borders did not cut the tree. Instead, they painted serial number 918 on its trunk, accepting it as the new boundary pillar. This is probably the only living border pillar on the international boundary between the two countries, making the tree even more attractive to visitors who swarm this place.

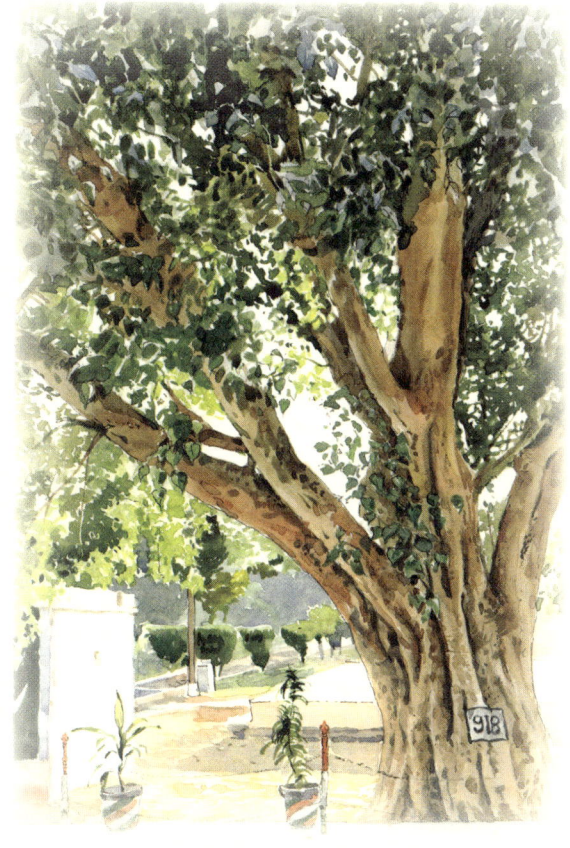

The peepal is erect and handsome but does not yet have the dignified look of an ancient tree, though it could be close to a hundred years. Its trunk is quite fissured. Three main branches bend towards India and two others lean towards Pakistan. Its characteristic crown of dark green leaves is a delight to behold. On either side of the tree, soldiers built concrete rectangular platforms.

The Jammu & Kashmir government intends to develop this border checkpost into a tourist hub, but progress has been slow. Marking the 41st World Tourism Day, the state's tourism department launched an open-roof luxury bus between Jammu city and Suchetgarh border on 27 September 2020.[22]

Life for the tree – as for the villagers of Suchetgarh – is as tough as it is unpredictable. Although an all-out war has not broken out in decades, the tension is always palpable. Residents are used to the sudden burst of gunshots or the exploding thud of mortar shells as they rip through houses, vehicles or livestock in their way. The ancient Raghunath temple built in 1837 near the checkpost was partially damaged due to a similar gunfire incident in 1971. For now, the tree is fine. The guns have fallen silent, especially after the two countries signed a peace treaty in February 2021. Hopefully, tranquillity and amity will prevail so that tourism takes root here and the unique tree that swallowed an entire pillar continues to be admired and celebrated.

8

The Deodar of Chaurasi Temple

COMMON NAMES: **Himalayan Cedar (English);
Devadaru (Sanskrit); Deodar (Hindi)**

SCIENTIFIC NAME: *Cedrus deodara* (Roxb. ex D. Don.) G. Don

FAMILY: **Pine (Pinaceae)**

WHERE TO SEE: **Chaurasi Temple complex, Bharmour,
Chamba district, Himachal Pradesh**

LATITUDE: **32. 2632° N**; LONGITUDE: **76. 3214° E;**
ALTITUDE: **2,133.6 metres**

Bharmour is one of the beautiful sleepy towns dotting the Pir-Panjal and Dhauladhar ranges of the Western Himalaya straddling Himachal Pradesh. Located about 65 kilometres southeast of Chamba, the state's district headquarters, Bharmour – known as Brahmapura in the sixth century – was the seat of power of the Maru dynasty for over 400 years,[23, 24] until the capital was shifted to Chamba in 920 CE.

Today, life in Bharmour revolves around the Chaurasi temple complex that owes its name to the 84 shrines built in varying architectural styles between the seventh and tenth centuries. The centrepiece of the complex is the Manimahesh (a form of Shiva) temple, reputedly built during the early part of the seventh century, likely by King Meru Varman who reigned around 700 CE. Partially shading the temple's *shikhara* (tower) is an ancient and mighty deodar tree, which, if local legend is to be believed, existed even when the temple was constructed. This would make it somewhere between 1,300 and 1,400 years old but a more pragmatic estimate would place its age at 400 years. Only a scientific study could validate its true age.

Deodars are evergreen conifers native to the western Himalayan region, usually occurring between the altitudes of 1,500 and 3,200 metres. In Sanskrit devadaru literally means 'wood of the gods'. Deodars are held to be dear to Shiva.

The Chaurasi deodar has a particularly massive main trunk, with a girth of 7.10 metres. The crown spreads across 13.4 metres and gradually tapers towards the apex, rising majestically like a 12-story tower (approximately 39 metres). The trunk splits into a dozen main branches that rise almost vertically into the sky. The bark is rough and cracked with dark and light brown patches. Birds have made nests in the main branches. As the tree is revered, some of its branches have festoons of yellow, saffron, red and gold cloth as well as floral garlands as votive offerings. The tree bears both male and female cones regularly.

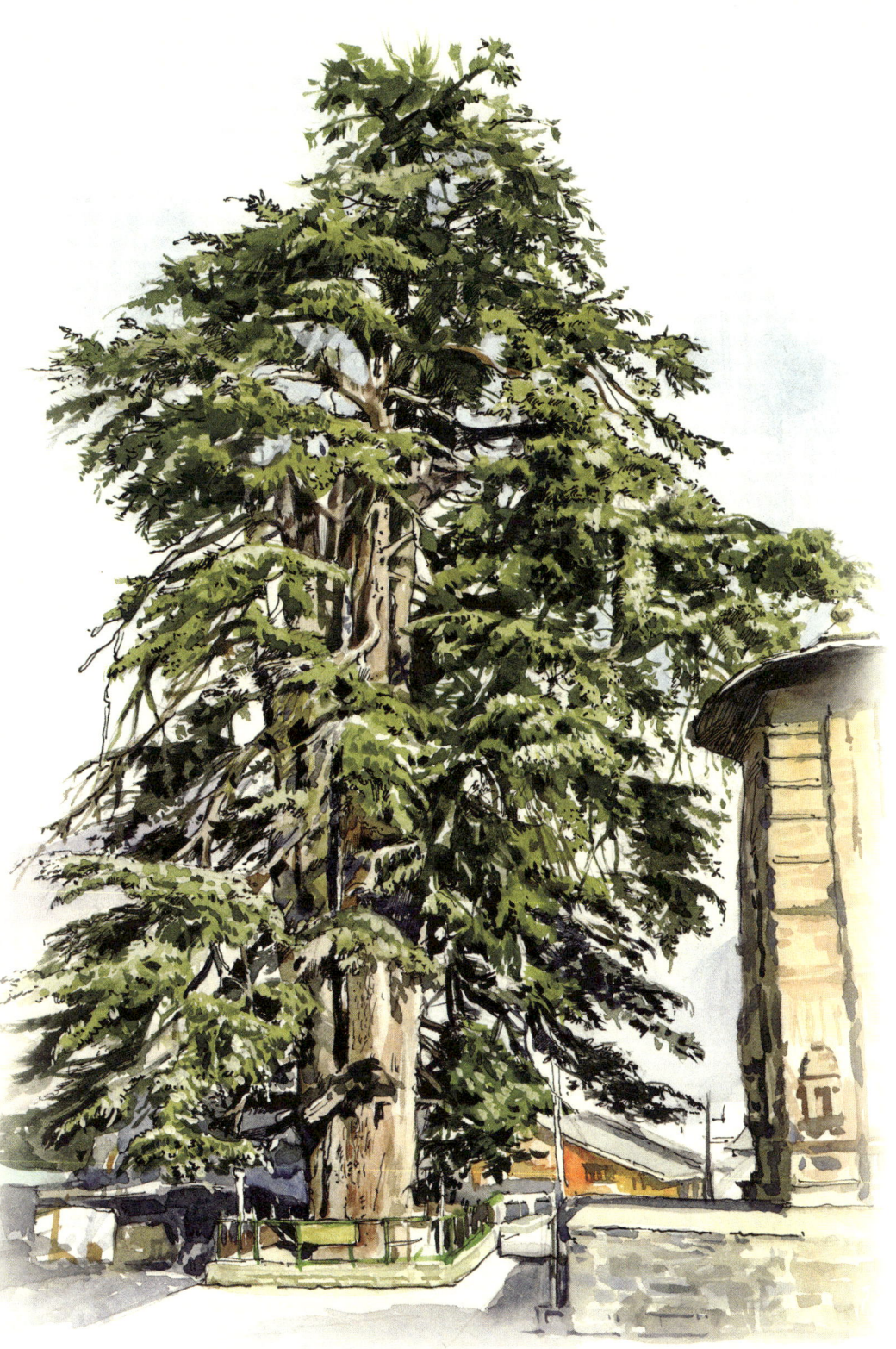

The deodar of the Chaurasi temple complex is closely associated with the annual Manimahesh Kailash yatra that traditionally kicks off on Janmashtami and concludes about two weeks later on Radhashtami, generally in August and September each year. The pilgrimage officially begins when a devotee designated by the temple's head priest climbs atop the sacred deodar tree on Janmashtami to hoist a saffron flag at its summit. This is a signal for devotees thronging Bharmour to commence their pilgrimage to the Manimahesh Lake which is located 4,415 metres in the mountains above. Taking a holy dip in the lake, they hope to catch a glimpse of the hallowed Mount Kailash, considered to be the abode of Lord Shiva.

The centurion deodar of Bharmour is thus an integral component of the annual religious festival. The temple of Manimahesh would never be the same without its majestic presence next to it.

A tree is a wonderful living organism
which gives shelter, food, warmth
and protection to all living things.
It even gives shade to those who
wield an axe to cut it down.

Gautama Buddha

9

Ber Baba Budha of the Golden Temple

COMMON NAMES: Jujube, Indian jujube, Chinese Apple (English); Badari, Karkandhu, Vadari (Sanskrit); Ber (Punjabi/Hindi)

SCIENTIFIC NAME: *Ziziphus mauritiana* Lam. | FAMILY: Ber (Rhamnaceae)

WHERE TO SEE: Golden Temple complex, Golden Temple Road, Atta Mandi, Katra Ahluwalia, Amritsar

LATITUDE: 31.6207° N; LONGITUDE: 74.8766° E; ALTITUDE: 232 metres

There are three ancient *ber* trees in the famous Golden Temple complex in Amritsar – the Dukh Bhanjani Ber, the Laachi Ber and the Ber Baba Budha, all of them on the *parikrama* (path of circumambulation) around Amrit Sarovar, the holy temple tank.[25]

Of these, the Ber Baba Budha is perhaps the oldest, older than the temple itself. It is associated with Baba Budha, a devout companion and go-to-man of the Sikh gurus, who supervised the digging and lining of Amrit Sarovar when

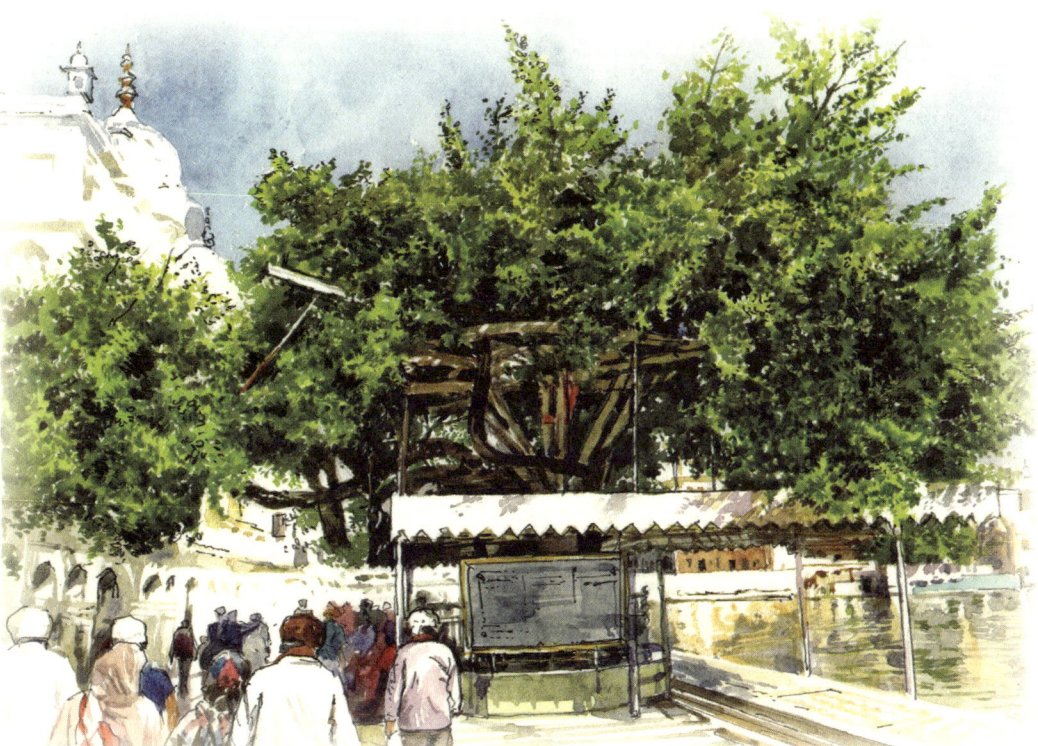

construction began in 1570 during the time of Guru Arjan Dev, the fifth guru. Throughout the seven years it took to complete the work, the baba oversaw the operations sitting in the shade of a ber tree that was at the tank's northern edge. In 1604, Guru Arjan Dev installed the newly compiled and edited *Adi Granth*, meaning 'The First Book', containing Sikh scriptures in the temple and appointed Baba Budha as the first *granthi* or ceremonial reader of the holy book. Baba Budha also later supervised the construction of the Akal Takht, which symbolized the gurus' temporal authority over Sikhs, along with Bhai Gurdas. He is thus one of the most venerated and central figures of Sikhism. Baba Budha lived to be 125 years (d. 1631) and is said to have performed several miracles. The ber tree under which he sat serves as a symbol of his piety and dedication and is regarded as sacred by visitors to the temple. A hardy smallish tree with crooked branches, the ber inhabits the Indo-Malaysian region. Punjabi folktales have numerous references to ber and several gurdwaras and Sikh personages are associated with it. The ber tree under which the Baba used to work must have already been of some size then. The usual lifespan of a ber tree is about a hundred years, but this one has lived four times longer. Although no scientific attempts have been made to date the tree, its association with Baba Budha places its current age at over 450 years. It has also been considered the oldest ber tree in India.[26]

During the 1990s, the ber tree started showing signs of decay and death. Alarmed temple authorities sought help from the Punjab Agricultural University, Ludhiana. An expert team led by Dr J.S. Bal carried out a careful examination of the tree and discovered that the devotees used to touch the main trunk with greasy hands after partaking of the prasad. The ghee in the prasad, said the experts, impeded the flow of sap by clogging the passages, resulting in the drying up of the branches. The tree was also severely infested by lac insects. The team pruned off the dead branches during May – ber is summer deciduous – and sprayed it with a combination of insecticides. The tree was fenced off to prevent devotees from either placing prasad or flowers near the main trunk or touching it with their ghee-smeared hands. Lastly, the team got the thick layer of concrete and marble close to the tree removed to ease pressure on the root zone and facilitate air circulation. The road to recovery was grindingly slow but steady. In the succeeding years, new leaves began to appear and finally, after close to two decades, the tree bore healthy fruits.[27, 28]

In its long life of over four centuries, the ber tree has witnessed momentous upheavals – historical, military, political and social – within the temple complex. It seems to have weathered them all including a battle for its own survival and has come up trumps.

10

The Sacred Mulberry Tree
of Joshimath

COMMON NAMES: **Mulberry (English); Kinpu/Shehtoot (Garhwali/ Hindi)**

SCIENTIFIC NAME: *Morus serrata* **Roxb.**

FAMILY: **Mulberry or Jackfruit (Moraceae)**

WHERE TO SEE: **Joshimath, Chamoli district, Uttarakhand**

LATITUDE: **30.5555° N;** LONGITUDE: **79.5596° E;**
ALTITUDE: **Approximately 1,907 metres**

O n a hillock presiding over the scenic town of Joshimath in the mountain state of Uttarakhand stands an amazing-looking Himalayan mulberry tree. Pilgrims proceeding to Badrinath and other holy places always break their journey at Joshimath, but few stop to admire this giant tree under the shade of which the great Hindu sage Adi Shankaracharya is said to have meditated. Most scholars agree that the great philosopher lived during the eighth century (788–820 CE). That would make this tree over 1,200 years old.

The mulberry tree is located not far from Jyotirmath, the northernmost of the four cardinal institutions established by Shankaracharya. When the Badrinath shrine in the hills above closes for winter every year, an idol of that deity is brought down to the temple within the Jyotirmath premises and worshipped for six months.

The sacred mulberry – a large deciduous tree – was first studied and described in 1967.[29] Undoubtedly, its most outstanding feature is the enormous trunk with a circumference of over 21.6 metres. This huge 'pedestal-like' trunk is assumed to be due to the fusion of several adventitious stems around the main trunk – over time combining into one massive body.[30] This trunk remains unbranched up to a height of roughly three metres, at which point other principal branches fan out and support a wide canopy with a spread of some 15 metres. The leaves are usually trilobed with a heart-shaped base, margins with sharp teeth and a short tail of an apex.

The locals believe that the tree is barren as it has never produced fruit. This is not surprising since male and female trees are separate in this species and this one happens to be a male, bearing only male flowers in bunches. A shrine for Jyotishwar Mahadev occupies a large portion of the platform built around the tree. The tree is still sound and healthy and continues to produce a new flush of leaves and flowers in season.

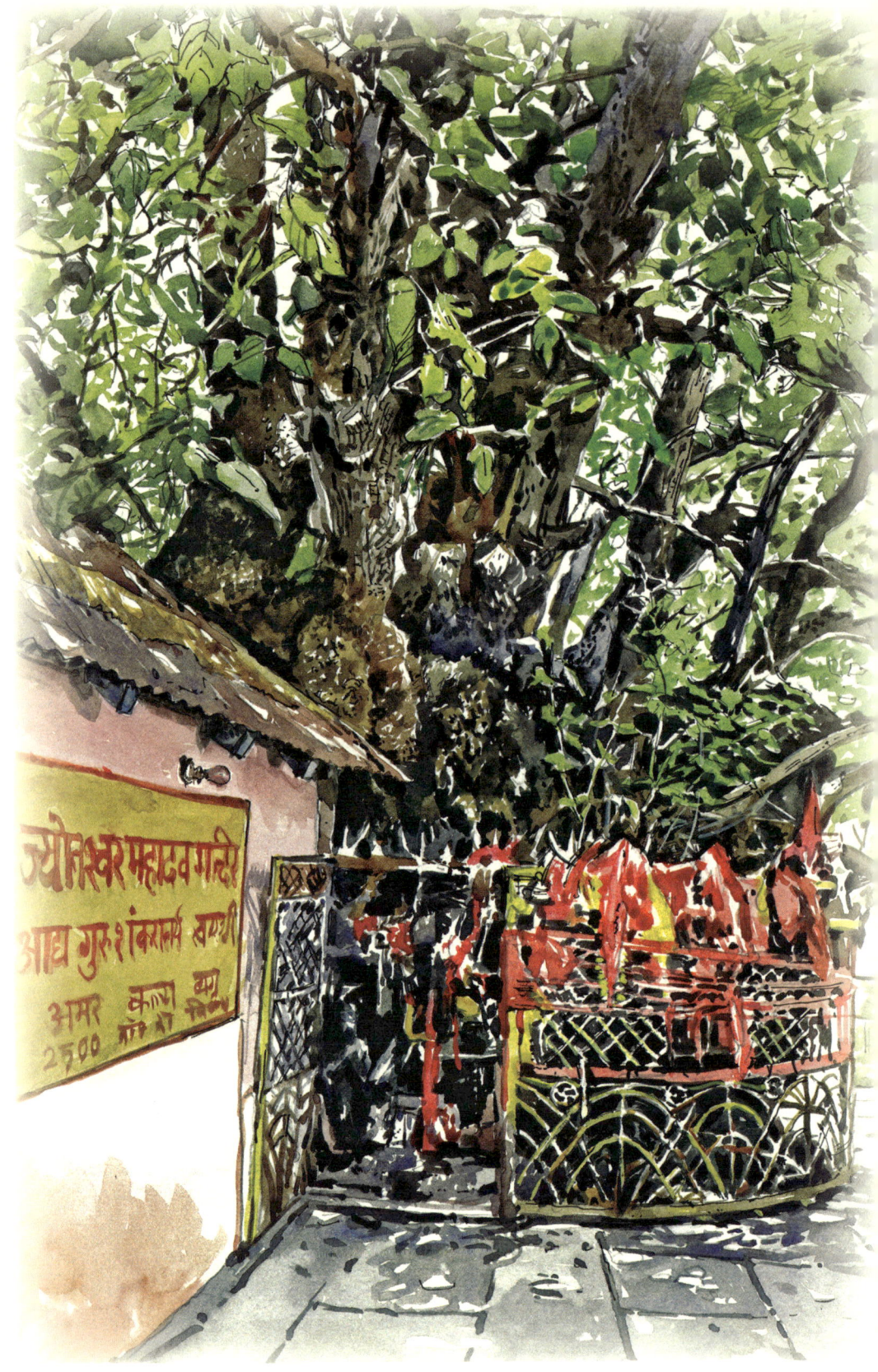

Native to the Western Himalayas[31] and Northeast India, until some time ago, this particular tree was considered the oldest in the country, based on its association with Adi Shankaracharya, but an accurate scientific estimation of its age is still awaited (see Table 4).

Close your eyes for a moment, blow away the unsightly fencing, banish the surrounding macadam-paved road and visualize a densely-wooded hillock – and you can easily imagine the great preceptor meditating serenely under the shade of this majestic tree. Shankaracharya was no ordinary sage. He was the first person to distil and integrate the diverse thoughts of Hinduism into a common philosophy based on the Vedic dictum of 'One Truth, Many Expositions'. Like Shankaracharya's Advaita philosophy, the ancient mulberry tree has stood the test of time well, even surviving the land subsidence that occured in January 2023.

11

The Mahatma's Peepal Tree

COMMON NAMES: Sacred Fig Tree (English);
Ashwattha (Sanskrit); Peepal Tree (Hindi)

SCIENTIFIC NAME: *Ficus religiosa* L.

FAMILY: Fig (Moraceae)

WHERE TO SEE:Entrance of the Christian Retreat and
Study Centre, Old Mussoorie Road, Near Moravian Institute,
Rajpur, Dehradun, Uttarakhand

LATITUDE: 30.402° N; LONGITUDE: 78.094° E;
ALTITUDE: Approximately 640 metres

From 16–24 October 1929, Mahatma Gandhi was on a tour of Dehradun and Mussoorie to attend a series of meetings.[32] On 17 October, he visited the Manav Bharti Vidyalaya (today, the Christian Retreat and Study Centre is located there) in Rajpur to unveil the portrait of Pandit Keshava Deva Shastri, a well-known educationist. He also planted a peepal tree on the premises. After planting the tree, Gandhiji had lunch with the school children and spoke to them about education as well as the freedom struggle.

Gandhiji's tree is now about 12 metres tall. Its fluted trunk is 2.8 metres thick, straight and unbranched up to a height of 3.5 metres, at which point it bifurcates. It looks generally healthy, putting out new leaves and branches. A circular stone platform has been built around the tree, with a trimmed hedge of clerodendron at the base.

In the recent past, Dehradun's nature-loving citizens discovered that the tree has been infested by the honeysuckle mistletoe (*Dendrophthoe falcata*). It is a partial parasite – technically referred to as hemi-parasite – that is dependent on water and minerals from its host plants but can convert these into sugars and other nutrients through photosynthesis on its own. Perched high up in the air on the host tree's branches, such hemi-parasites produce special structures known as 'haustoria' that penetrate and invade the xylem (water-conducting tissue) of the host plant. A heavy infestation could gradually, but surely, lead to the death of the host tree over a decade or so. As an interim measure, the forest department applied copper sulphate to the affected areas to prevent the spread of the parasite, though a long-lasting solution is yet to be found. However, it is not clear who will initiate it – the owners of the premises on which it grows or the government authorities. Meanwhile, a piece of India's freedom struggle legacy is in danger of being lost.

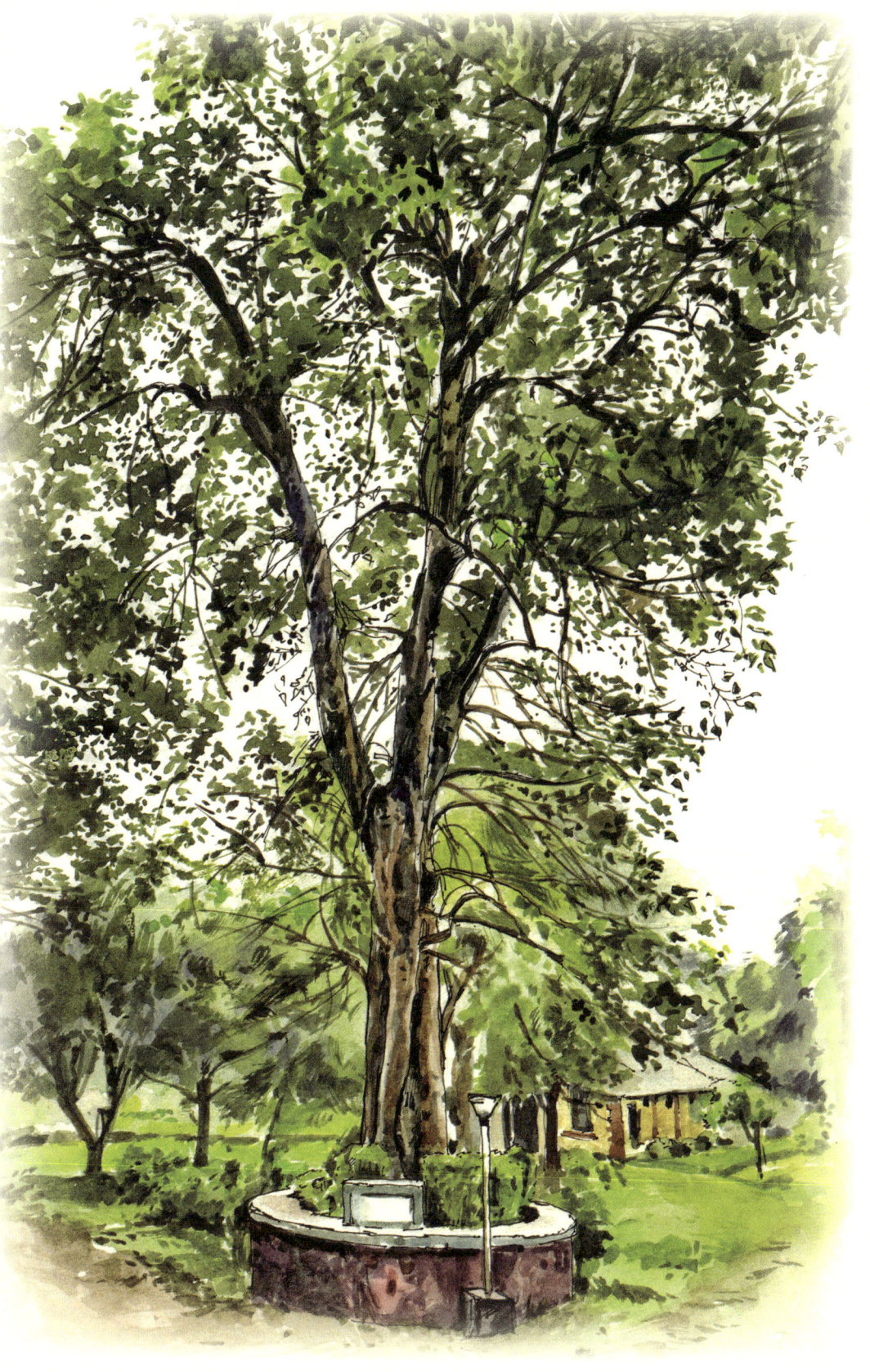

12

Dehradun's Identity –
The Peepal Tree of Ghanta Ghar

COMMON NAMES: **Sacred Fig Tree (English); Ashwattha (Sanskrit); Peepal Tree (Hindi)**

SCIENTIFIC NAME: *Ficus religiosa* **L.**

FAMILY: **Fig (Moraceae)**

WHERE TO SEE: **Clock Tower/Ghanta Ghar roundabout, Rajpur Road, Near Paltan Bazar, Dehradun, Uttarakhand**

LATITUDE: **30.324° N;** LONGITUDE: **78.042° E;**
ALTITUDE: **Approximately 672 metres**

A peepal tree at Dehradun's Ghanta Ghar Chowk, reputed to have been planted by Sarojini Naidu,[33] has been in the eye of the storm for more than a decade. Attempts by the civic authorities to axe it, or shift it elsewhere to 'beautify' the Clock Tower – claimed by some to represent the identity of Dehradun – have been met with stiff resistance by citizen activists who have seen too much of the city's green cover destroyed over the years in the name of development.

For more than a decade, civic authorities have targeted the peepal tree for damaging the foundation of the Clock Tower, as well as for obstructing its view. The tree has also been repeatedly seen as 'green trouble' and indeed, the main reason for delaying the renovation and beautification of the tower.[34] However, attempts to cut it down or shift it elsewhere have been met with considerable resistance over the years.

Inspired by the Raksha Sutra Andolan,[35] Dehradun's local green group – Citizens for Green Doon – spearheaded a sustained protest. *Raksha sutra* means a thread for protection. One day in August 2015, in Dehradun, protesters wearing colourful festive dresses first prayed to the peepal tree and then tied *rakhis* (sacred threads) around its trunk, pledging to protect it. Thanks to their effort, the tree still survives. Although the 'beautification' has incurred much criticism from heritage and environmental experts, the foundation of the Clock Tower has been strengthened and the area has been 'beautified' with the tree still standing in its place.

Judging by its girth and general appearance, the tree appears to be older than 70 odd years (as it should be if it was planted in 1948) and seems to have existed on the site at the time the Clock Tower was being built. Two water tanks and a

74

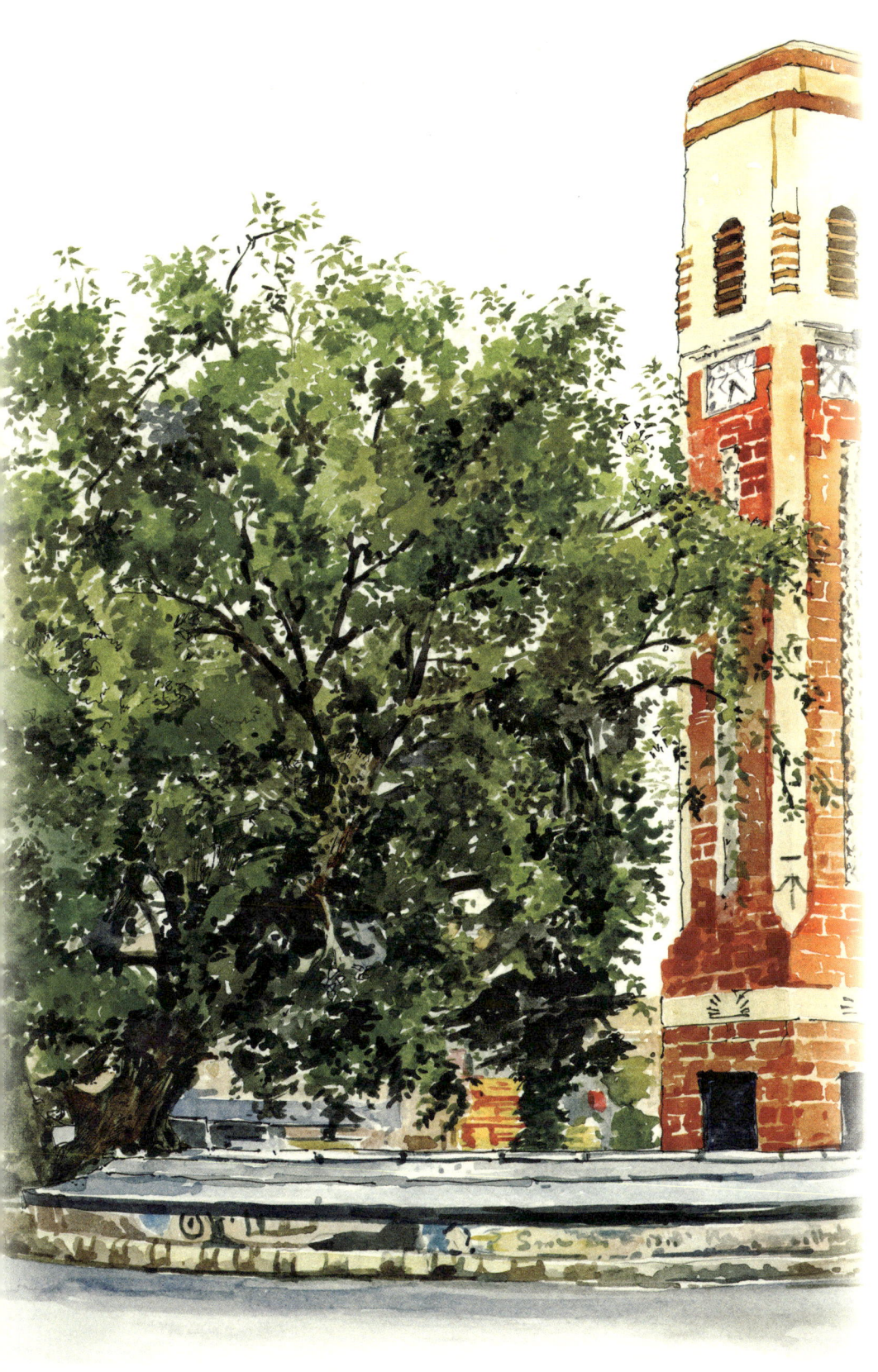

tonga stand also stood at that site. The cool shade of the tree must have been a welcome gift for the horses. A photograph – claimed to have been taken in 1881 – shows the presence of the tree at the place where the Clock Tower was eventually built.[36] Even at that time, it seems to be considerably large and spreading.

Senior residents of Dehradun recall that the tree has been standing on the spot for a long time. One of them is Ruskin Bond, who has spent much time in these parts and recalls that the peepal was indeed there even when he was a child.[37] As of now, it is probably closer to 175 years and thus has even greater heritage value than has been attributed to it. Hence, it symbolizes the identity of Dehradun even more than the Clock Tower. Sadly, its story is yet another illustration of the overriding importance of built heritage in India vis-a-vis living, green cultural heritage.

13

The Giant Toon of Nagarjun

COMMON NAMES: **Toon tree (English); Kacchapah, tuna, Nandi, Nandikah (Sanskrit); Tun (Hindi)**

SCIENTIFIC NAME: *Toona ciliata* **M. Roem.**

FAMILY: **Neem or Mahogany (Meliaceae)**

WHERE TO SEE: **In Vishnu Mandir, Nagarjun village, Dwarahat, Almora district, Uttarakhand**

LATITUDE: **29.8049° N;** LONGITUDE: **79.3592° E;** ALTITUDE: **345 metres**

T he only thing that is noteworthy in Nagarjun in Almora district is a small but ancient temple dedicated to Vishnu and Shiva. What catches the eye though is the giant toon tree looming over the temple tower. Being native to this part of the world, toons are not uncommon in this hilly terrain. However, this particular specimen commands attention as it soars more than 20 metres into the sky. Its massive trunk (7.5 metres girth) is clothed with a grey-to-brown fissured and flaky bark. About 10 metres from the ground, the trunk splits itself into five main branches, bearing a rounded crown and spreading outwards to nearly 19 metres in all directions. It is usual to find people sitting on the small

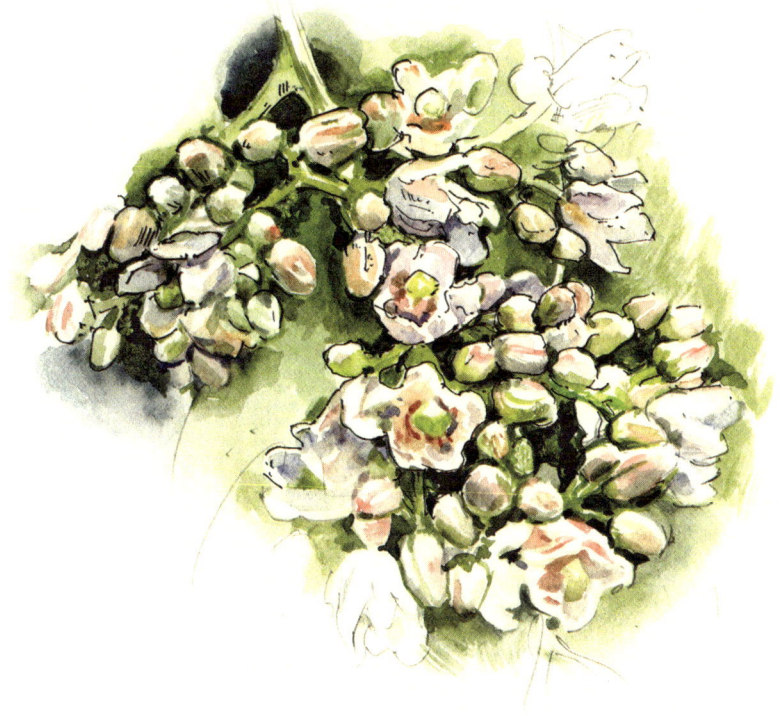

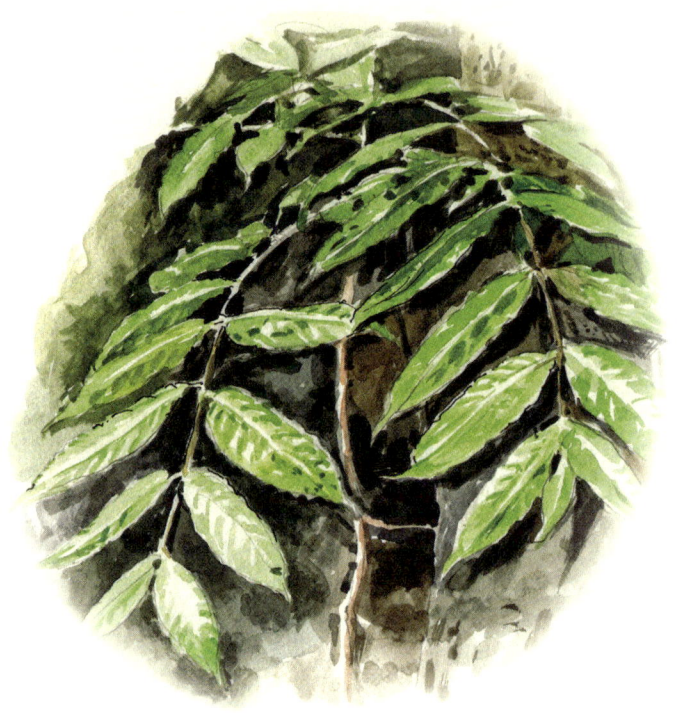

cement bench by the temple under its shade. As is characteristic in the neem family to which the tree belongs, the leaves are typically feather-like and divided into leaflets.

The temple itself is said to be over 400 years old, having been built in the late sixteenth or early seventeenth century by the Chand dynasty. No trace of that original structure is visible today. According to a local legend,[38] it is said that about 75 or 80 years ago one Devi Datt Upreti set out to build a wooden roof for a hall next to the sanctum. Since its wood is greatly valued as timber, he hired a contractor to saw off a large branch of the toon tree conveniently growing next to the temple. Unfortunately, when he reached home in the evening, he suddenly collapsed and died. Since then, the villagers have taken it as a sign against cutting the tree's branches or otherwise harming it in any manner.

If this is to be believed, the tree ought to be close to 400 years. This is unlikely as toons are fast-growing species and attain impressive proportions relatively quickly. Fortunately, they also lend themselves to dendrochronological studies with clear growth-ring patterns and variations in ring width.[39] It will therefore be possible to ascertain the tree's exact age with a targeted study. As of now, what we do know is that ever since it sprouted as a young seedling, it has stood in the same village, much like most of the Upretis themselves – braving good seasons and bad ones with equanimity and acting as a faithful chronicler of the times past in this rather remote spot.

14

The Magnificent Deodars
of Jageshwar

COMMON NAMES: **Himalayan Cedar (English);**
Devadaru (Sanskrit); Deodar (Hindi)

SCIENTIFIC NAME: *Cedrus deodara* **(Roxb. ex Lamb.) G. Don**

FAMILY: **Pine (Pinaceae)**

WHERE TO SEE: **Temple complex, Jageshwar Dham,**
Almora district, Uttarakhand

LATITUDE: **29.638031° N;** LONGITUDE: **79.85559° E;**
ALTITUDE: **1,815.1 metres**

In Jageshwar, you are never too far from a deodar tree. The pilgrimage town is located in the narrow but scenic valley of the Jataganga rivulet surrounded by magnificent deodars. The Nageshwar shrine of Jageshwar finds mention in Hindu texts before the tenth century as a *tirtha* (pilgrimage site) for one of the 12 *jyotirlingas* of Shiva,[40] making it one of the most important pilgrimage sites in the Kumaon region of Uttarakhand. Indeed, the town's raison d'être is the temple complex comprising a cluster of 124 stone temples, most dedicated to Shiva, but a few to Vishnu, Durga and other deities. They were built in various sizes and architectural styles between the seventh and thirteenth centuries – with some buildings continuing even into the twentieth century.[41]

Although the surrounding hills feature oaks, rhododendrons and pines, it is the deodars that dominate the landscape. Some trees around the temple complex have attained huge proportions, with girths touching nine metres and must be centuries old. Walking through the groves of deodars, you are filled with peace – the silence occasionally interrupted by the call of a bird or the peal of the temple bell.

Though the deodars have stood here perhaps for centuries, experts believe that there are no natural strands of deodars in the Kumaon region, as this species is restricted to areas receiving winter snow and summer monsoon rains, with Garhwal representing the outermost limit.[42] Yet close to 324 hectares of deodar forests occur in Kumaon, chiefly around temples.[43] Were they then intentionally planted around temple premises, as deodar is considered sacred and is a favourite tree of Shiva?

One way of testing this postulate is to ascertain if the age of the deodar groves is contemporaneous with the construction of the Jageshwar temples. As noted, construction of the temples commenced around the seventh century and continued up to the thirteenth century and beyond. Is it possible that the deodar trees were also planted around the same period to form a sacred grove around them? In 2012, a team of scientists led by Dr Ram Yadav from BSIP investigated the very old trees in and around the Jageshwar temple complex. Using dendrochronological methods of dating they found the oldest sampled tree to be 477 years, tracing back the chronology to 1536 CE.[44] In other words, the age of the oldest tree recorded by the group at the Jageshwar forests extends the date back to the early sixteenth century (similarly, the oldest deodar tree from the nearby Gangolighat was aged 346 years old, planted around 1668 CE). More in-depth studies might reveal the period of plantation to be even earlier. Indeed, during the study, researchers discovered several dead upright trees left to decompose naturally at the site. These had much larger circumferences than those of the sampled trees, suggesting that they were much older. It appears reasonable to conclude that the relatively younger age of the trees sampled at Gangolighat indicates that the deodar plantations could have commenced in the Jageshwar temple area, gradually spreading to other regions in Kumaon.

It is not known who thought of establishing a grove of sacred deodars around the temples of Jageshwar, but whoever it was should receive our collective gratitude, because it is simply impossible to imagine the temple complex without the surrounding forest. Doubtless, it is the deodars that enhance the sense of serenity and majesty of Jageshwar and invite pilgrims and tourists alike for another visit.

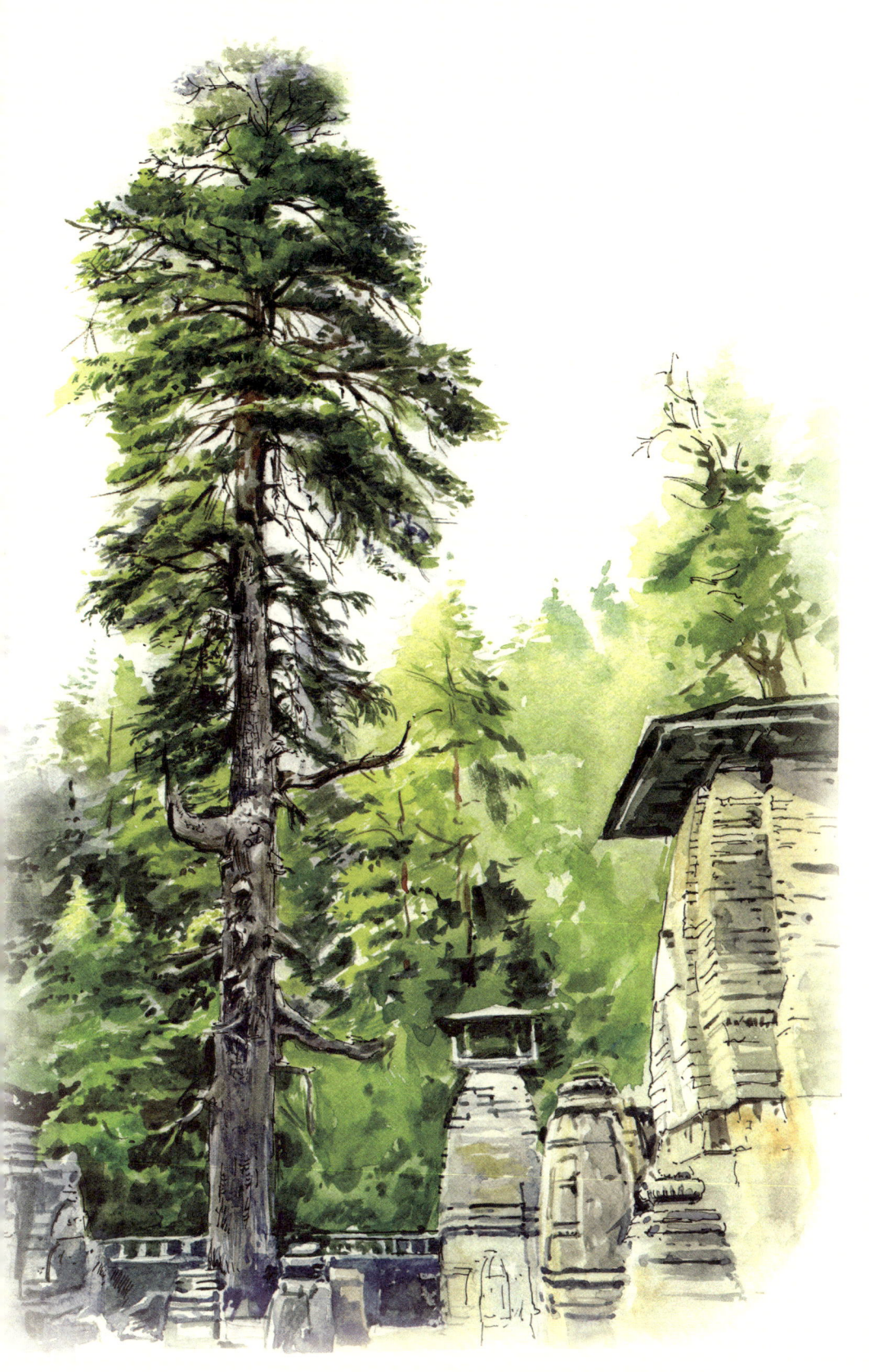

15

Bareilly's Banyan of Martyrdom

COMMON NAMES: **Banyan (English); Vat, Bahupada (Sanskrit);**
Vat, Bat (Hindi)

SCIENTIFIC NAME: *Ficus benghalensis* L.

FAMILY: **Fig (Moraceae)**

WHERE TO SEE: **Divisional Commissioner's Office, Bareilly,**
Bareilly district, Uttar Pradesh

LATITUDE: **28.34694° N;** LONGITUDE: **79.42028° E;**
ALTITUDE: **179 metres**

Squeezed into a small soil bed on a large red sandstone platform, the banyan suffocating in the divisional commissioner's office in Bareilly has nowhere to strike new roots. Years of human neglect and nature's vagaries have destroyed large parts of the tree, leaving it with just two remnant stems and disjointed branches. The banyan should be the star attraction on the platform but the Amar Shaheed Stambh, an obelisk in black, has usurped that privilege. Over 150 years ago, this banyan tree must have been at the heart of the town, majestic and spreading – its pillar-like roots anchoring its crown and branches that were strong enough to serve as gallows for 257 freedom fighters who were hanged to death in the aftermath of the 1857 rebellion, widely regarded as India's first war of independence. Historical records to corroborate this claim, however, are hard to come by.

Bareilly became the headquarters of the 1857 revolt in the Rohilkhand region, now part of north-western Uttar Pradesh. The revolt had started on 29 March in Barrackpore, Calcutta (now Kolkata) and quickly spread to Meerut. On 10 May, thousands of soldiers left Meerut for Delhi to request Bahadur Shah Zafar, the aged Mughal monarch, to assume leadership of the war for freedom.

News of these events reached Bareilly on 14 May 1857. Inspired to join the revolt, the freedom fighters in Bareilly rallied around Khan Bahadur Khan, the grandson of Hafiz Rahmat Khan, the ruler of Rohilkhand who had died in battle fighting the British in 1774.[45] On 31 May, the rebels killed several British officers (including the magistrate, jail superintendent, civil surgeon and the principal of Bareilly College) and captured the treasury of the East India Company. By evening, Bareilly had been liberated from the British. They crowned Khan Bahadur Khan as the nawab of Bareilly the next day. Khan Bahadur even issued silver coins from Bareilly as an independent ruler, the last ones to be minted here. (Incidentally, Bareilly has been known for minting coins for a long time, starting from the

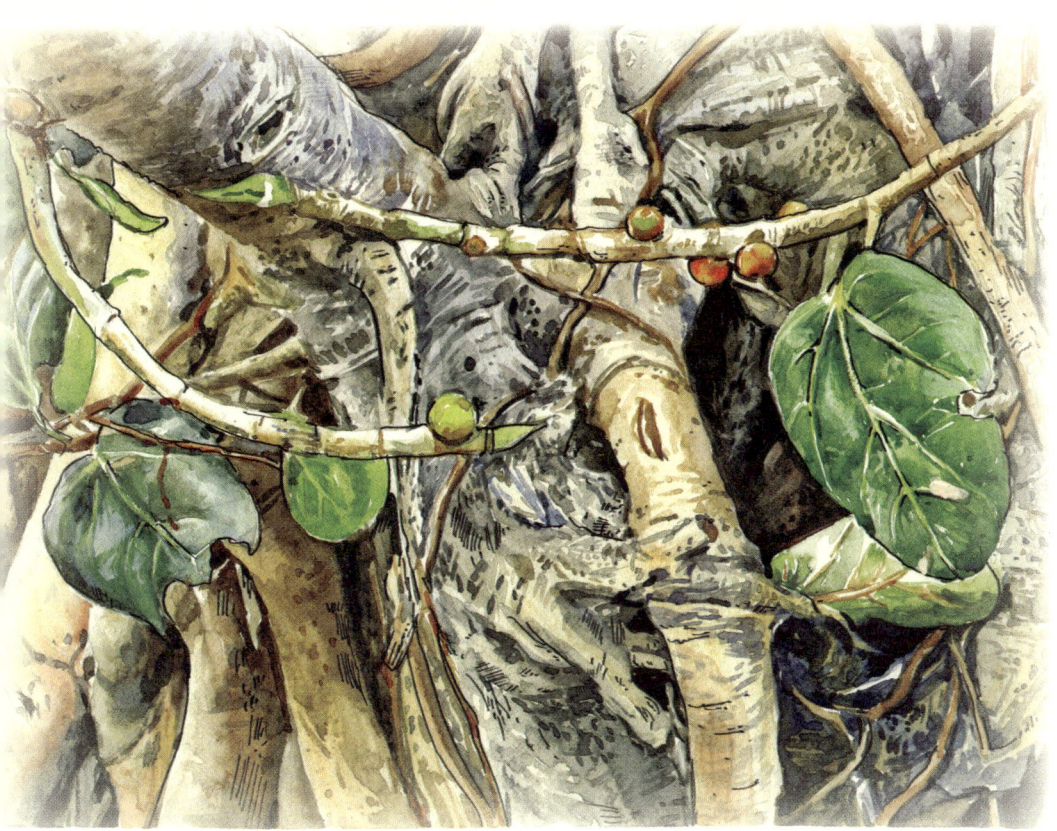

Panchal period (176–166 BCE)[46]. Bareilly's independence lasted 11 months and five days. The British, under the leadership of Commander Colin Campbell – fwho having defeated Tantia Tope, the most vigorous rebel of 1857 – attacked Bareilly. Khan Bahadur's army gave a spirited fight but wilted under the superior might of Campbell's Highland regiments, who marched 'in red coats, kilt and feather bonnet under the blazing sun, showing 112 degrees Fahrenheit (44.44⁰C) under shade'.[47] The British forces occupied Bareilly on 7 May 1858.[48]

It was a common practice of the colonial rulers to force the local people to watch the gory spectacle. The use of religious trees surely exacerbated the horror. There is, of course, no proof to suggest that the British – in a display of racial arrogance and callous brutality – repeatedly chose banyan or peepal trees as gallows because of their religious significance, but they could not have been blind to the significance these trees held for Indians – for British writings from the colonial period abound with descriptions of people praying at and worshipping peepals and banyans.[49] The 1857 uprising in Bareilly and other places failed but such violence ultimately undermined colonial rule by alienating the Indian population and turning its victims into martyrs of the national movement.[50]

This iconic banyan of Bareilly is but one such sacred tree believed to have been used by the British across India as an instrument of death, grimly reminding us that freedom comes at a cost.[51]

16

Kāliya Mardan Kadamb
of Vrindavan

COMMON NAMES: **Kadamba (Sanskrit); Kadam, Kaim (Hindi)**

SCIENTIFIC NAME: *Mitragyna parvifolia* **(Roxb.) Korth**

FAMILY: **Coffee (Rubiaceae)**

WHERE TO SEE: **Kāliyaghat, Vrindavan, Mathura district, Uttar Pradesh**

LATITUDE: **27.5781° N;** LONGITUDE: **77.6824° E;**
ALTITUDE: **174.60 metres**

The Braj heartland, where Lord Krishna grew into adulthood, holds many attractions for Krishna devotees. There are not just temples to see here, although there are hundreds of those – the attractions cover the varied places where the young god worked miracles such as the Govardhan hillock, which he uprooted to shelter from the rain god Indra, or Vrindavan, where he performed the Ras Lila with other cowherds.

For me, the chief attraction of the Braj area is Vrindavan's Kāliyaghat, where Krishna subdued Kaliya, the venomous snake that the wharf along the Yamuna is named after. The episode is described in the *Bhāgavata Purāna* colourfully. Kaliya's presence in the waters polluted the entire river system and its neighbourhood and young Krishna is said to have jumped into the Yamuna to subdue the serpent. The people of Vrindavan and other devotees believe that the ancient kadamb – locally known as kaim – that stands on Kāliyaghat today is the same tree that Krishna used as a diving board to plunge into the poisonous waters of the Yamuna to subjugate Kaliya. They claim that the tree is over 5,000 years old, but there have been no scientific attempts to ascertain its age. The trunk and lower branches are gnarled and fluted, covered with pale grey bark and at places quite hollow. The tree bears flowers and fruits regularly. Pilgrims come in thousands to pray at this tree. The Yamuna is a much-polluted river today and the verdant forests of kadamb in Vrindavan (and indeed much of Braj Bhoomi) are nowhere in evidence.[52] This degradation of a sacred landscape is not merely an ecological problem, it represents a steady erosion of heritage and a weakening of ties that bind the community to its natural habitat, which are the river, trees and vegetation.[53] Notwithstanding all this, the sacred kadamb tree has remained unharmed because it is revered. Fortunately, this has inspired several environmental groups, who are currently working to restore Vrindavan's rivers and trees.[54] That will take some time and effort, without divine intervention.

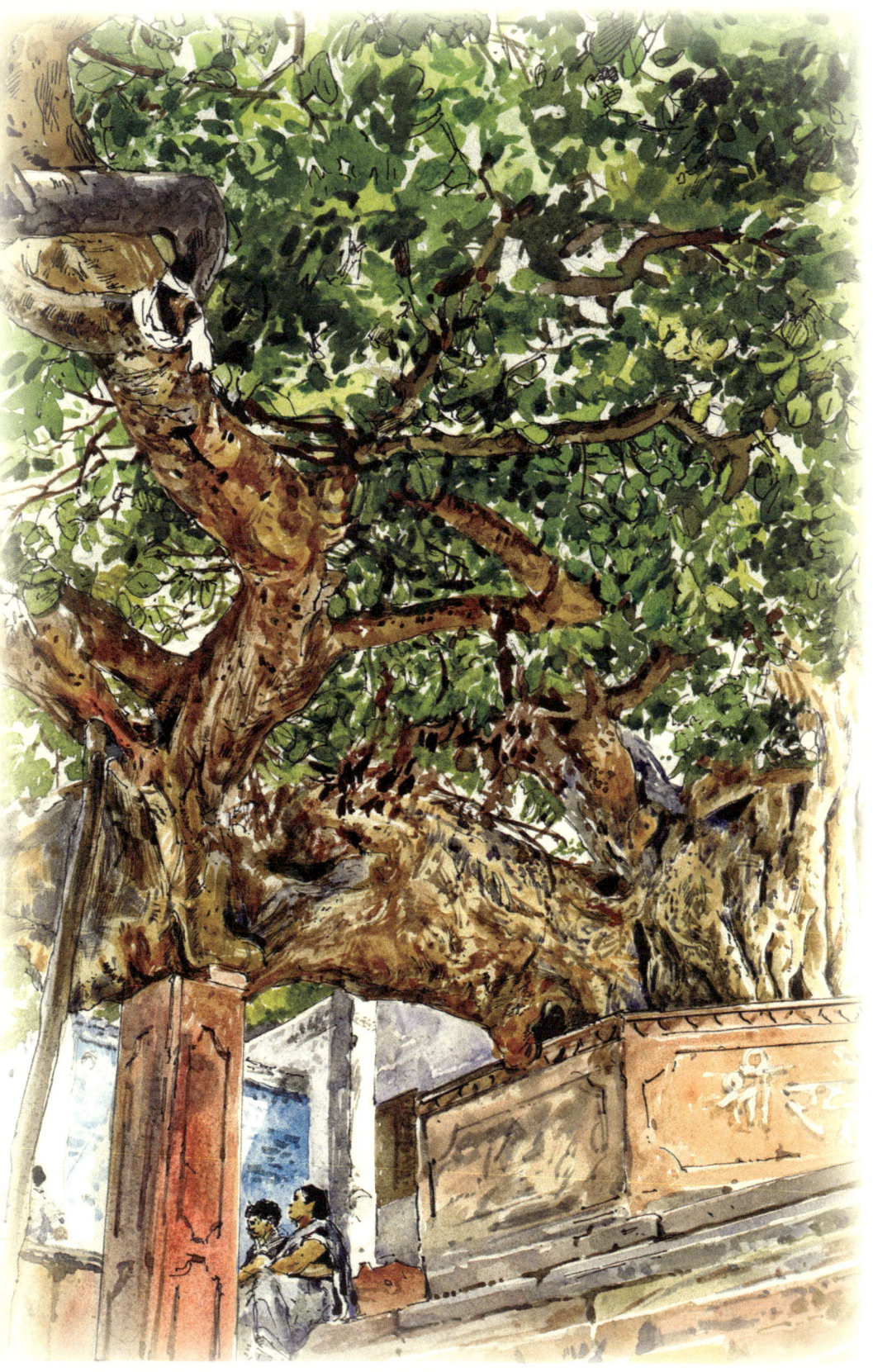

17

Kunti's Parijata – The Baobab of Kintoor

COMMON NAMES: **Baobab (English); Gorakshi (Sanskrit); Gorakh imli, Parijat, Parijata (Hindi)**

SCIENTIFIC NAME: *Adansonia digitata* L.

FAMILY: **Hibiscus (Malvaceae, Bombacoideae)**

WHERE TO SEE: **Kintoor village, Barabanki district, Uttar Pradesh**

LATITUDE: **27.004° N;** LONGITUDE: **81.482° E;**
ALTITUDE: **106 metres**

A large and ancient African baobab tree in Uttar Pradesh's Kintoor village is venerated as parijata by local people. Considered to be a 'kalpavriksha',[55] the mythical parijata is said to grant all objects of desire. The parijata has considerable significance in Hindu mythology and finds mention in several puranic texts, with interesting but contradictory narratives associated with it.[56]Botanists commonly identify it as *Nyctanthes arbor-tristis* L. (family Oleaceae). It is also known as harsringar in Hindi. However, people hereabouts consider it to be the African baobab and believe that it was brought to earth from Indra's celestial garden by Arjuna (one of the heroes of *Mahabharata*) to fulfil his mother Kunti's wish to worship Lord Shiva with its flowers. They also believe that the tree is 5,000 years old and can grant their hearts' desires. It is held to be powerful: In April–May 2020, locals were sure that the pandemic would die down 'now that the tree had flowered'.[57]

Kintoor's baobab has been cordoned off with two concentric metal fences. At the base of the tree, within the inner metal fence, is a small shrine devoted to Lord Krishna. The squat trunk has a conical shape with a massive girth of approximately 13 metres. As is characteristic of this species, the trunk comprises four or five primary stems fused perfectly together. Seven large primary branches arise from the trunk but some of them are either broken or have been cut off. The crown is huge. Although flowers bloom regularly, fruit set has not been observed for a long time.

In 2015, the baobab was found to be suffering from fungal infection as well as opportunistic bacterial infection, no doubt due to the offerings of sweets and milk by the devotees. The state forest department was contacted and they enlisted the services of the CSIR – National Botanical Research Institute, Lucknow. A team of scientists there studied the problem and started treating the trunk and leaves

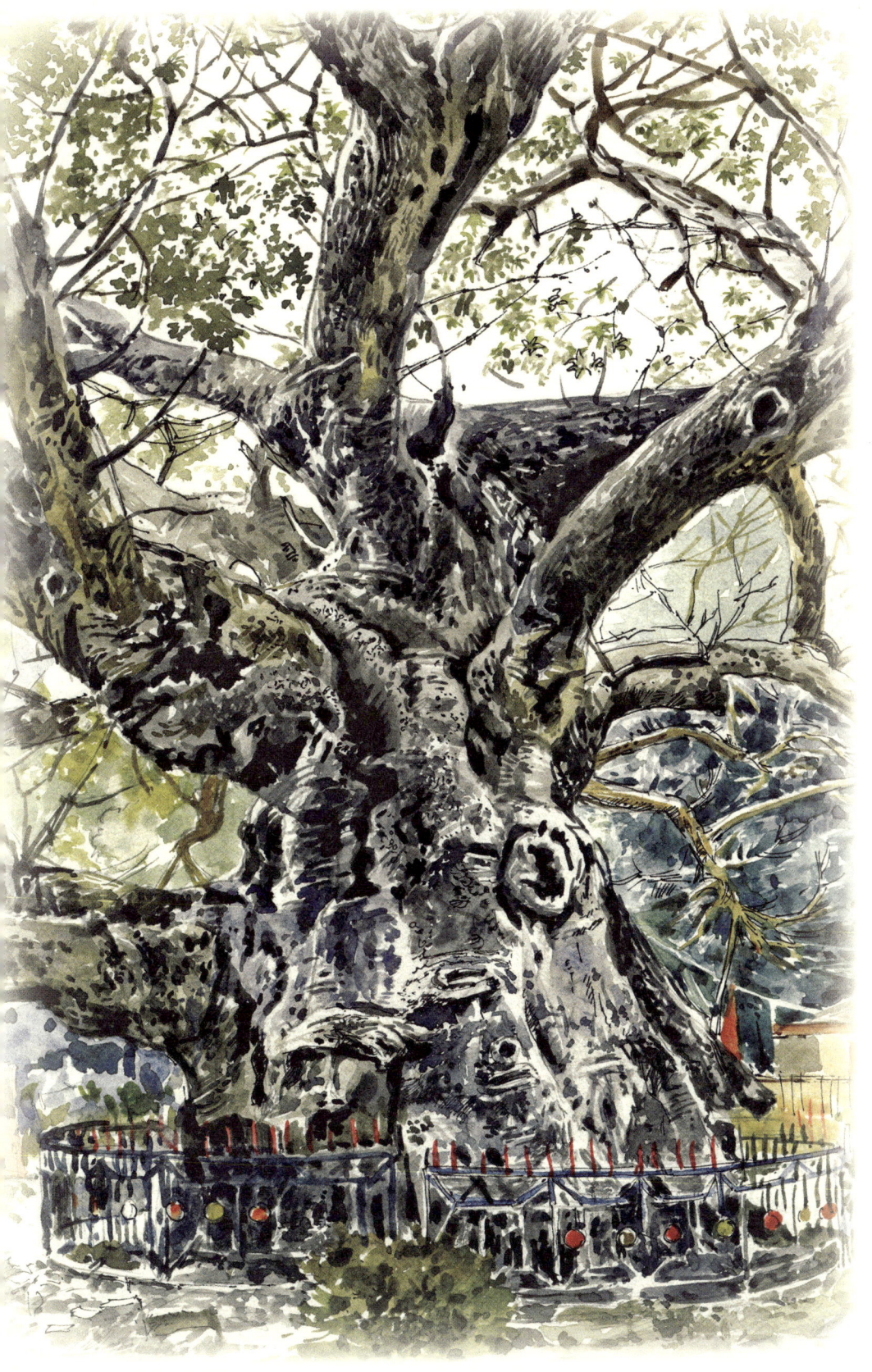

of the plant with a combination of chemicals and microbial biopesticides to keep infections at bay. After months of their ministrations, the tree was on the road to recovery.

Using radiocarbon technique, the age of the tree was found to be about 800 calendar years.[58] Thus it would have started its life around 1200 CE. Indeed, this is one of the two oldest and accurately dated baobab trees outside of Africa.[59] Incidentally, the oldest-dated baobabs in Africa turned out to be over 2,000 years old,[60] becoming the longest-living flowering plants (angiosperms).

An analysis of the Kintoor baobab showed an extremely low level of water content – in fact, as little as 39.7 per cent.[61] Baobab wood usually has a high level of water content, up to 79 per cent. Sadly, this suggests that the tree is close to the end of its life cycle and may topple over in the near future. Hopefully, this prediction will not come true anytime soon.

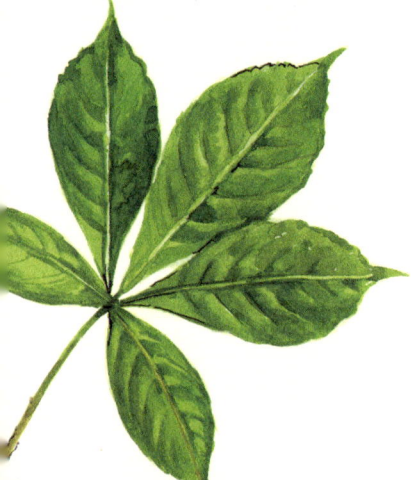

18

Mango Tree with the Most Grafts

COMMON NAMES: **Mango (English); Aamra (Sanskrit); Aam (Hindi)**

SCIENTIFIC NAME: *Mangifera indica* L. | FAMILY: **Mango (Anacardiaceae)**

WHERE TO SEE: **Abdulla Nursery, Malihabad, Lucknow district, Uttar Pradesh**

LATITUDE: **26.9272° N**; LONGITUDE: **80.7140° E;**

ALTITUDE: **128 metres**

If you are in Malihabad, India's mango capital, you wouldn't need to know the address of Abdulla Nursery. Just ask any passer-by and they are sure to tell you, if not lead you straight to it. Why is it so famous? Because the owner is the 'mango man', Haji Kaleemullah Khan, whose nursery maintains several mango (and other fruit) trees. What is remarkable about them are the numerous varieties.[62] Khan is best known for grafting more than 300 different varieties of mangoes onto a single tree and for commercially propagating several new ones through grafts. He was decorated with the Padma Shri in 2008 for his contribution to horticulture.

A graft comprises two parts: a lower part known as the rootstock, usually selected for its rooting ability and an upper part known as the scion, selected for some desirable quality of its stem, leaves, flowers or fruits. For grafting to be successful, the vascular tissues of the stock and the scion have to join together. In this case, Khan used one single tree as rootstock and all the selected varieties as scions – a good technique to maintain many variations at a relatively low cost in a minimum space.[63]

The tree on which he has carried out this extraordinary feat is named *Al Muqarrar* (the decision) and occupies a pride of place in Abdulla Nursery. Although it belongs to the Dusseheri variety, it is a tree like no other. Now probably a 100 years old, the tree can be approached on all sides by a cemented circular walkway to facilitate easy grafting. Almost every emerging branch seems to be adorned with a skilfully carried out graft, each graft site covered with a thin polythene sheet. They look like so many medals for achieving some extraordinary feat. Khan has lost count of the exact number of grafts the tree bears. He finds that question unimportant as, 'theoretically, one could make even 3,000 grafts'.[64]

Making use of the variations, Khan has been producing commercial grafts of his selections. Over the years, he began to be recognized as a mango-grafting magician. Awards, recognition and fame followed soon. Some of his varieties are quite novel: 'Anarkali', for example, has two skins, an outer orange one that tastes like Chausa and an inner yellow one that resembles Dusseheri.

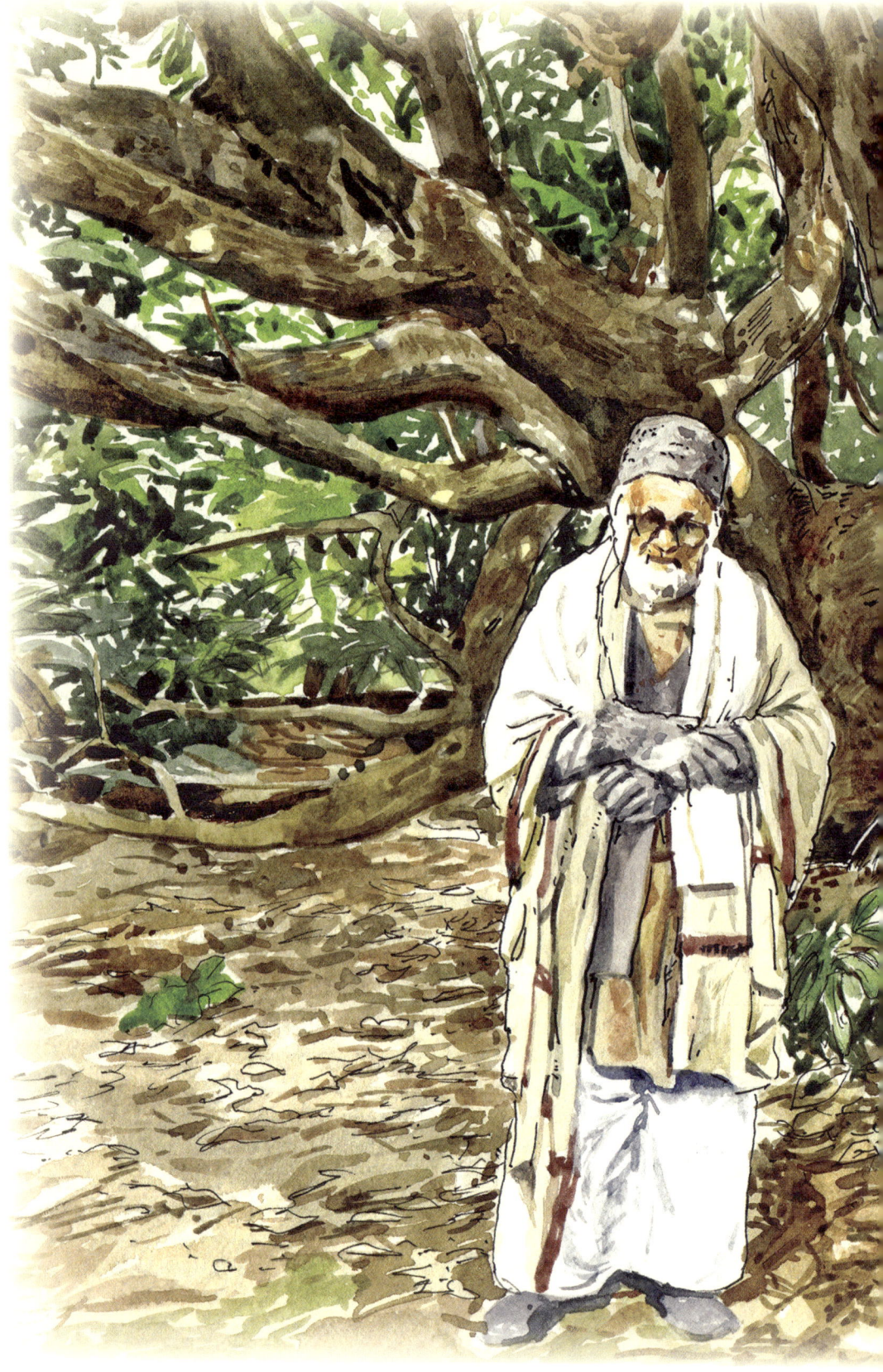

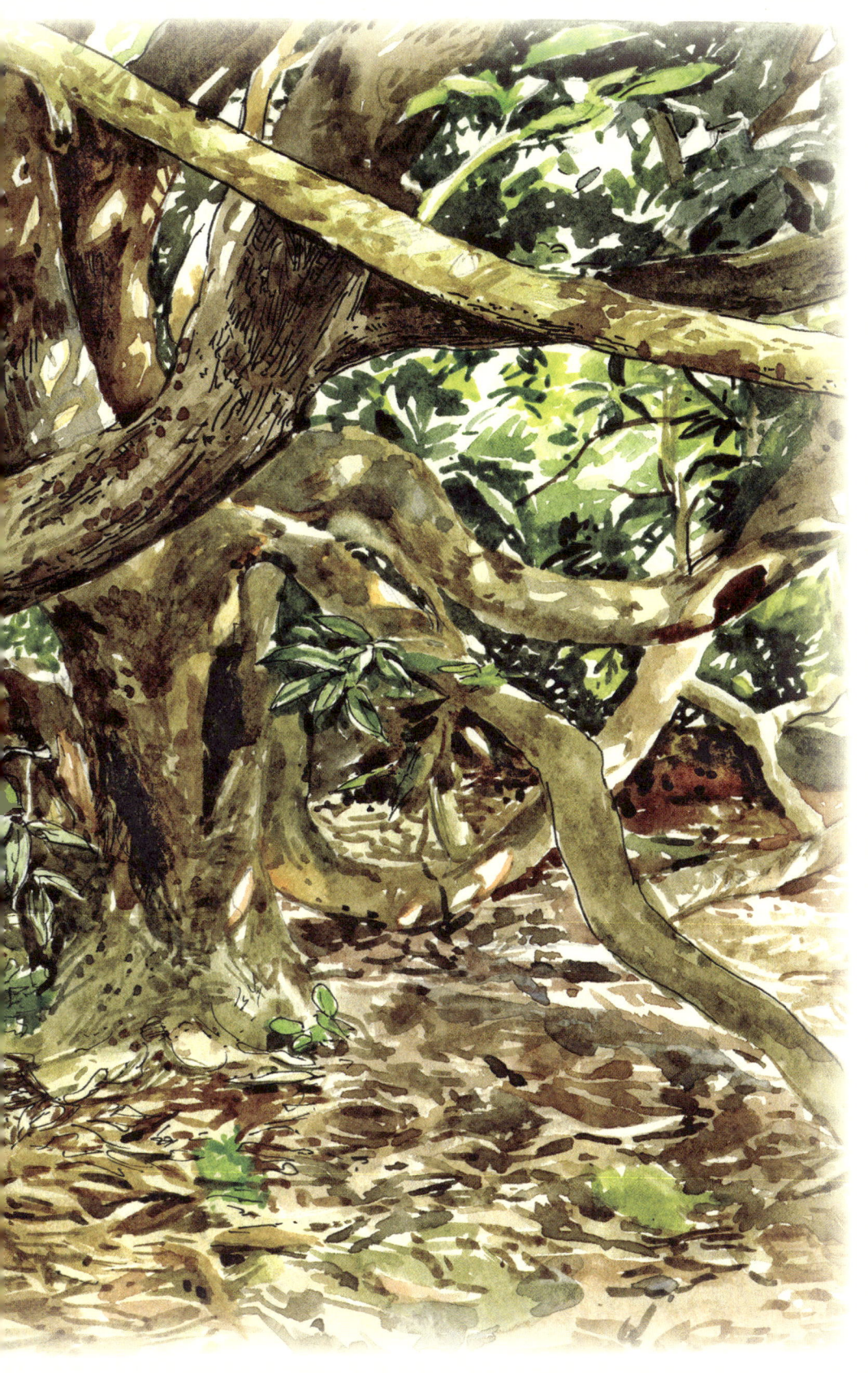

Many of his varieties are named after film stars (for example, 'Aishwarya', 'Amitabh'), sports personalities (for example, 'Sachin') and politicians (for example, 'Akhilesh', after former chief minister Akhilesh Yadav and 'NaMo' after prime minister Modi). In recent years, Khan has been looking into possible medicinal uses of mango sap. Currently, he is working on the beneficial application of the sap in cataracts.

Khan's magical mango tree – *Al Muqarrar* – carries so many varieties that differ widely in terms of colour, taste, texture, flavour and many other traits. Yet, they do so in complete harmony as a living example of the unity in diversity that is India's USP. It also stands as a powerful beacon demonstrating that people of differing habits, views and persuasions too can co-exist in peace and harmony as a nation and as responsible citizens of the world community, if only we set our minds to it.

Teach our children to love nature and the
rest will happen on its own.

Jadav Payeng
Environmentalist activist

19

'Mother' Tree of Dusseheri Mango

COMMON NAMES: **Mango (English); Aamra (Sanskrit); Aam (Hindi)**

SCIENTIFIC NAME: *Mangifera indica* L.

FAMILY: **Mango (Anacardiaceae)**

WHERE TO SEE: **Dusseheri Village, Kakori block, Lucknow district, Uttar Pradesh**

LATITUDE: **26.8602° N;** LONGITUDE: **80.8252° E;**
ALTITUDE: **124 metres**

If I were to produce a catalogue of the best Indian varieties of the mango, it would be safe to bet that the Dusseheri is most likely to feature in it. Undoubtedly among the preeminent varieties of north India, it is known for its exquisitely sweet taste and pleasant aroma and firm, yellow, non-fibrous pulp. The variety is protected under the Geographical Indications tag[65] under the label of 'Mango Malihabadi Dusseheri' for its uniqueness and importance to the regional economy.

The Malihabad belt in Uttar Pradesh – not far from the state capital Lucknow – is the acknowledged mango capital of northern India, producing and exporting its precious harvest to many national and overseas destinations. Of the varieties cultivated here, the Dusseheri occupies nearly 80 per cent of the area. It was also born right here and the original 'mother' tree that gave birth to the variety is still alive and standing in the small village that is reputed to have more mango trees than people – and lent the fruit its name. No one knows the exact age of the tree, but the consensus is that it is probably close to 200 years.

The tree is cordoned off with barbed wire to prevent theft of its fruit. Its trunk, some three metres in girth at chest height, rises to 1.5 metres before branching commences. A dozen branches then fan out, most of them growing parallel to the ground. The crown is dense and spreading, providing ample shade even in scorching summers. The tree's yield – typically between 80 kg and 190 kg[66] – is said to have fallen slightly below par over time. However, its taste, colour and aroma have remained intact. As is characteristic of mango, the 'mother' tree is also a typical alternate bearer – bearing a heavy crop one year and little to no crop the next. Sturdy for its age, it is remarkably free from termite and stem-borer attacks, problems common in mango.

Stories on the origin of the variety vary slightly. In essence, it boils down to the following: In the nineteenth century CE, a mango grower – probably a Pathan from one of the Malihabad villages – was transporting his best mangoes

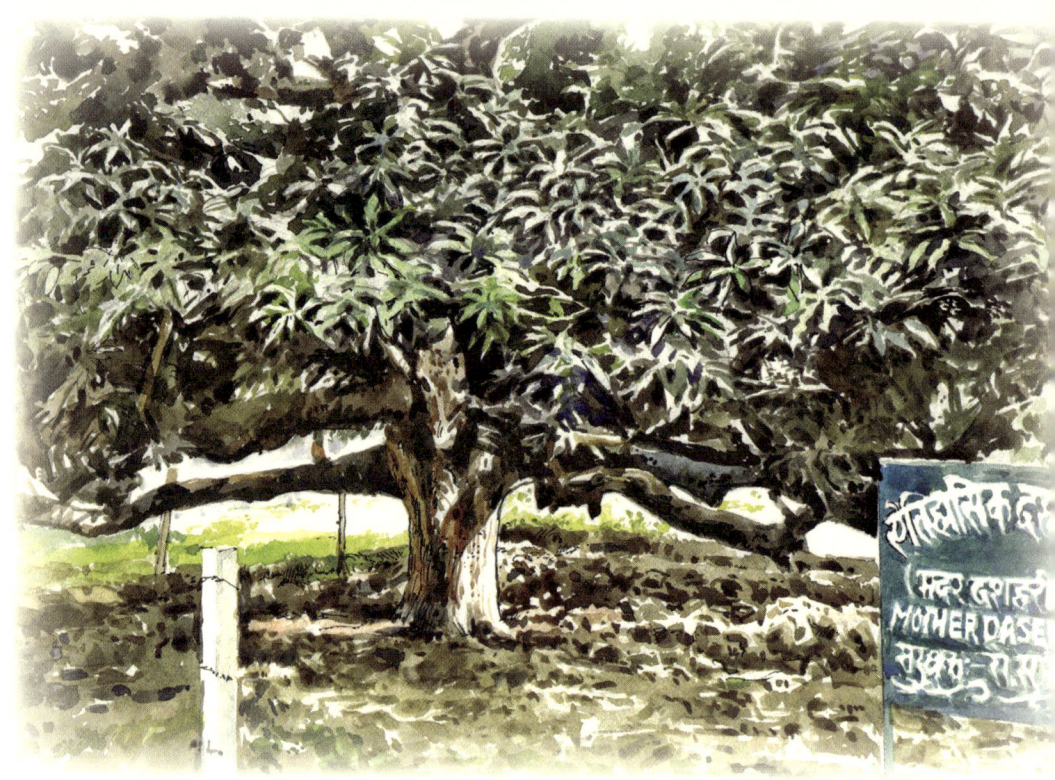

to the market through a checkpost near Dusseheri village. Toll in those days had to be paid in the form of a certain number of mangoes per basket. During the transaction, a dispute arose and the farmer, in a fit of rage and disgust, decided to throw his mangoes into a hole in the ground and left in a huff without paying the toll. In the ensuing monsoon season, a chance mango seed germinated from the discarded fruits and was duly cared for. Eventually, it matured into a tree and started producing fruits that were exquisitely different and unique. The new variety was named after the village in which it grew and came to be known as Dusseheri. The 'mother' tree became the exclusive property of the nawab who owned the village. In the years following its discovery, the tree was kept under strict and continuous watch to prevent pilferage. Despite this – so the story goes – an adventurous villager was able to obtain a seed and grow a tree. Soon grafts from the new tree were cultivated in orchards across the Malihabad mango belt. Eventually, the Dusseheri mango was available to common people.

In 2009–10, the Uttar Pradesh government wanted to set up a solid waste treatment plant in Dusseheri. It would have meant nearly 300 trucks coming in daily and dumping Lucknow's waste there, not to mention the noxious fumes from the exhausts of the vehicles. Worried by the scale of environmental degradation and its impact on their prized crop and the mother tree, the mango farmers were up in arms, forcing the government to shelve the project.

Baba Sheikh Taqi's Toothbrush?
The Ancient Baobab of Jhunsi

COMMON NAMES: **Baobab (English); Gorakshi (Sanskrit);**
Vilayti Imli, Baba ka Datun (Hindi)

SCIENTIFIC NAME: *Adansonia digitata* L.

FAMILY: **Hibiscus (Malvaceae, Bombacoideae)**

WHERE TO SEE: **Jhunsi, Prayagraj district, Uttar Pradesh**

LATITUDE: **25.2543° N;** LONGITUDE: **81.5398° E;**
ALTITUDE: **99 metres**

Standing at the Fort of Allahabad (now Prayagraj) and looking across towards the left bank of the Ganga as it flows past the Triveni Sangam you can see the suburb of Jhunsi. There, on a gradient stands a massive and ancient African baobab tree. About 14 metres tall, it has a squat hollow trunk with a circumference of approximately 18 metres,[67] partially damaged by fire. Before damage, its girth would have been approximately 21.2 metres. It forks out into six primary branches, three of them horizontal. The canopy is massive and spreads in all directions. Even now, the tree produces many flowers and fruit pods.

The baobab is barely five metres from the mausoleum of the Sufi Saint Saiyid Shah Sadrul Haq Taqiyuddin (1320–84), popularly known as Baba Sheikh Taqi. Born in Jhunsi in 1320, he renounced the world and left for Bukhara (now in Uzbekistan) seeking knowledge and returned after 12 years to spread the message of Islam. According to local belief, before he set off on his journey, Sheikh Taqi plunged his *datun* (a twig used as a toothbrush) upside down into the ground – and it grew into this tree.[68]

Scientific investigation has revealed that the tree is made of seven fused trunks and it has been standing here for close to 800 years.[69] It must have been born in the early thirteenth century, around 1220 CE, making it only the second of the two oldest baobabs outside Africa.

It is difficult to say how this African baobab tree came to be in Jhunsi. Its age does not support the Baba Taqi story; it was already there at least a century before Taqi's time. During the early nineteenth century there were three other ancient and enormous baobabs in Phaphamau, just about 17 kilometres from Jhunsi. Fanny Parks, an 'intrepid memsahib' and an independent traveller recorded in her journal[70] that one of them, reputed to be 1,100 years old with a trunk circumference of 10.67 metres, died in 1827 'on the day Lord Amherst (Governor-

General of India, 1823–28) arrived in Allahabad, on his return from the hills'. She sketched the other two baobabs. (These two are also dead now.) Fanny Parks also observed (but did not sketch) the Jhunsi tree, noting in her journal: 'Another of these trees which measured 11.28 metres in circumference is still in the grounds, which are on the banks of the Ganges.' During the previous century and a half, the tree's circumference has almost doubled.

During its silent journey of eight centuries, the baobab has seen many ups and downs. In recent times it seems to have been subjected to much neglect and public apathy. The fire damage to its trunk was sustained during a Kumbh Mela event in 2013 when its eastern side was accidentally set on fire by pilgrims who camped close to its base.[71] Its bark has been stripped and peeled off by locals for medicinal use, exposing the soft tissues to bacterial and fungal attacks. Continuous soil erosion over the years has partially exposed at least three big roots to the vagaries of nature. There is an additional worry: The Jhunsi baobab has an extremely low water content at 45.2 per cent, raising concerns that the tree is probably at the end of its life cycle and may topple over in the near future.

In 2016, someone from Mahoba, a small town in Uttar Pradesh, called attention to the pitiable condition of the Jhunsi baobab during Prime Minister Narendra Modi's monthly radio programme 'Mann ki Baat'.[72] The Prime Minister's Office promptly swung into action and directed the Botanical Survey of India to send a status report on its condition. The latter recruited the help of a local NGO, the Centre for Social Forestry and Eco-Rehabilitation for an inspection.

Three years later, in 2019, a boundary was built around the tree and soil was filled within it to protect its roots. While this might help it live a longer life, it is equally important to monitor its health at regular intervals. There is a need to raise awareness about the tree's status among India's natural monuments. Sprucing up its immediate ambience; providing interesting and accurate information on the iconic tree; cautioning against vandalism; as well as making access to the tree visitor-friendly are other steps that will help.

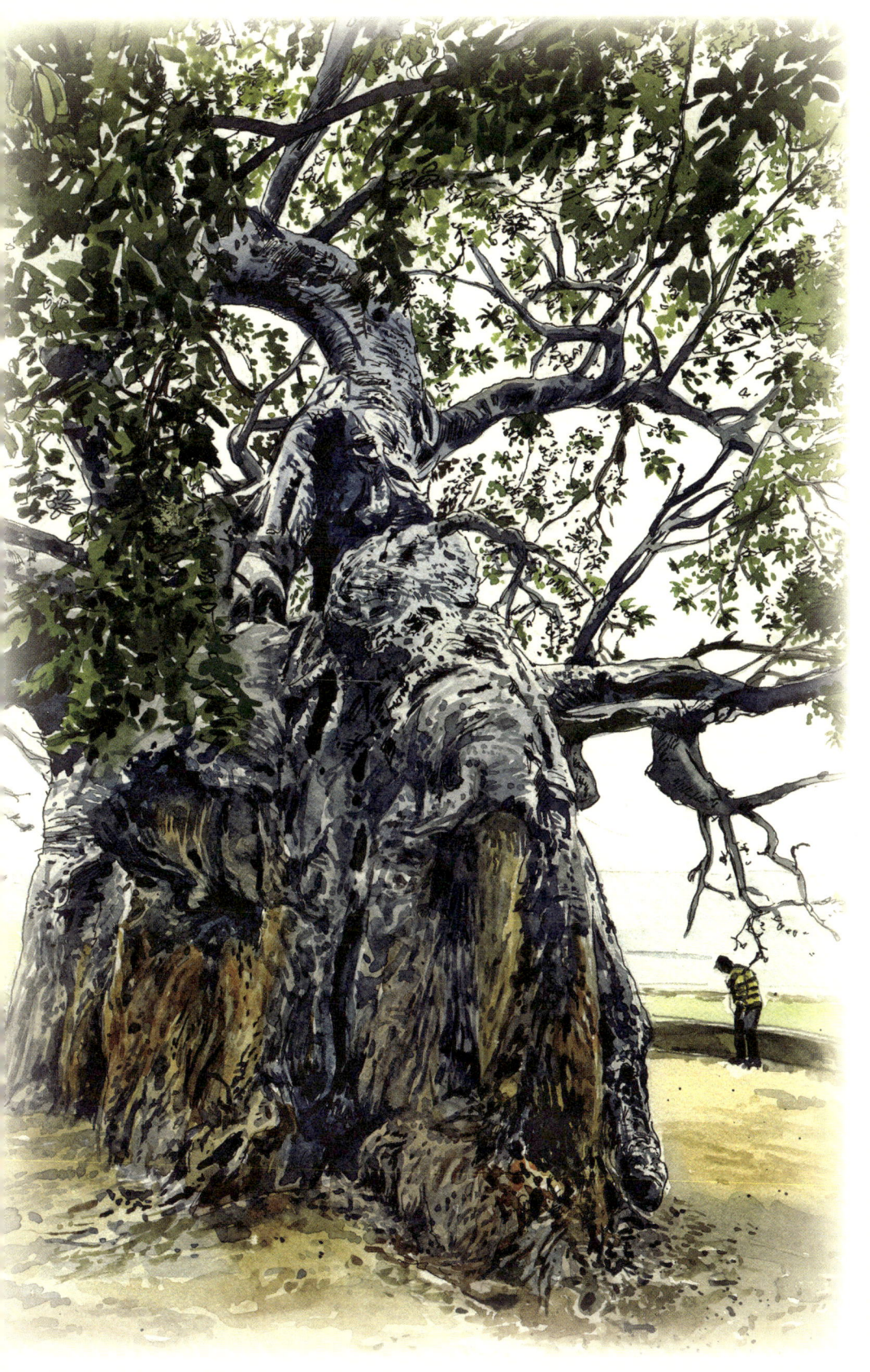

21

Khirni Tree of Chirag Dilli

COMMON NAMES: Ceylon Wood (English); Khirni (Hindi)

SSCIENTIFIC NAME: *Manilkara hexandra* (Roxb.) Dubard

FAMILY: Chikoo (Sapota) or Mahua (Sapotaceae)

WHERE TO SEE: Dargah of Hazrat Naseeruddin Chirag Dehlavi, in Chirag Dilli, New Delhi

LATITUDE: 28.539° N; LONGITUDE: 77.227° E; ALTITUDE: 209 metres

Once a fort with open grounds, greenery and water bodies, Chirag Dilli is now a crowded urban village in South Delhi. Its labyrinthine alleys and the dargah tucked away in a corner are perhaps among the few clues pointing to its past. The Dargah of Hazrat Naseeruddin Mahmud Chirag Dehlavi (approx. 1284–1356), the last Sufi saint of the Chisti order, still draws the faithful and a few tourists. The complex encloses a majestic khirni tree that many believe is Delhi's oldest.

Hazrat Naseeruddin was the disciple – and later successor – of the celebrated mystic and poet Hazrat Nizamuddin Aulia. Among the miracles attributed to him is the one where he humbles Ghiasuddin Tughlaq, who reigned over the Delhi Sultanate from 1320–1325. At the time Naseeruddin was supervising the construction of a *baoli* (stepwell) for his mentor in Nizamuddin, Tughlaq – who was building a fort in Tughlaqabad – had decreed that the labourers should concentrate work solely on the fort. However, filled with reverence for Nizamuddin, the labourers would toil on the *baoli* at night. An enraged Ghiyasuddin stopped the supply of oil to the area so that lamps could not be lit at night, effectively preventing any construction work. When the distraught masons went to Nizamuddin, he referred them to Naseeruddin, who told them to use the water from the *baoli* instead. The water lit the lamps and work on the *baoli* continued unhindered. An awestruck Tughlaq gave Naseeruddin the title *Roshan Chirag-e-Dilli*, which in Urdu means 'illuminated lamp of Delhi'. The khirni tree is said to have been planted by Hazrat Naseeruddin's own hands according to the *khadims* (caretaker) of the dargah.[73] Even if Naseeruddin planted it towards the end of his life, it should be over 660 years old now. Many of the local residents have fond memories of playing in the shade of the tree. Much has changed now. The narrow lane leading to the dargah is now abuzz with commercial activity, but the dargah continues to be a sanctuary of peace.

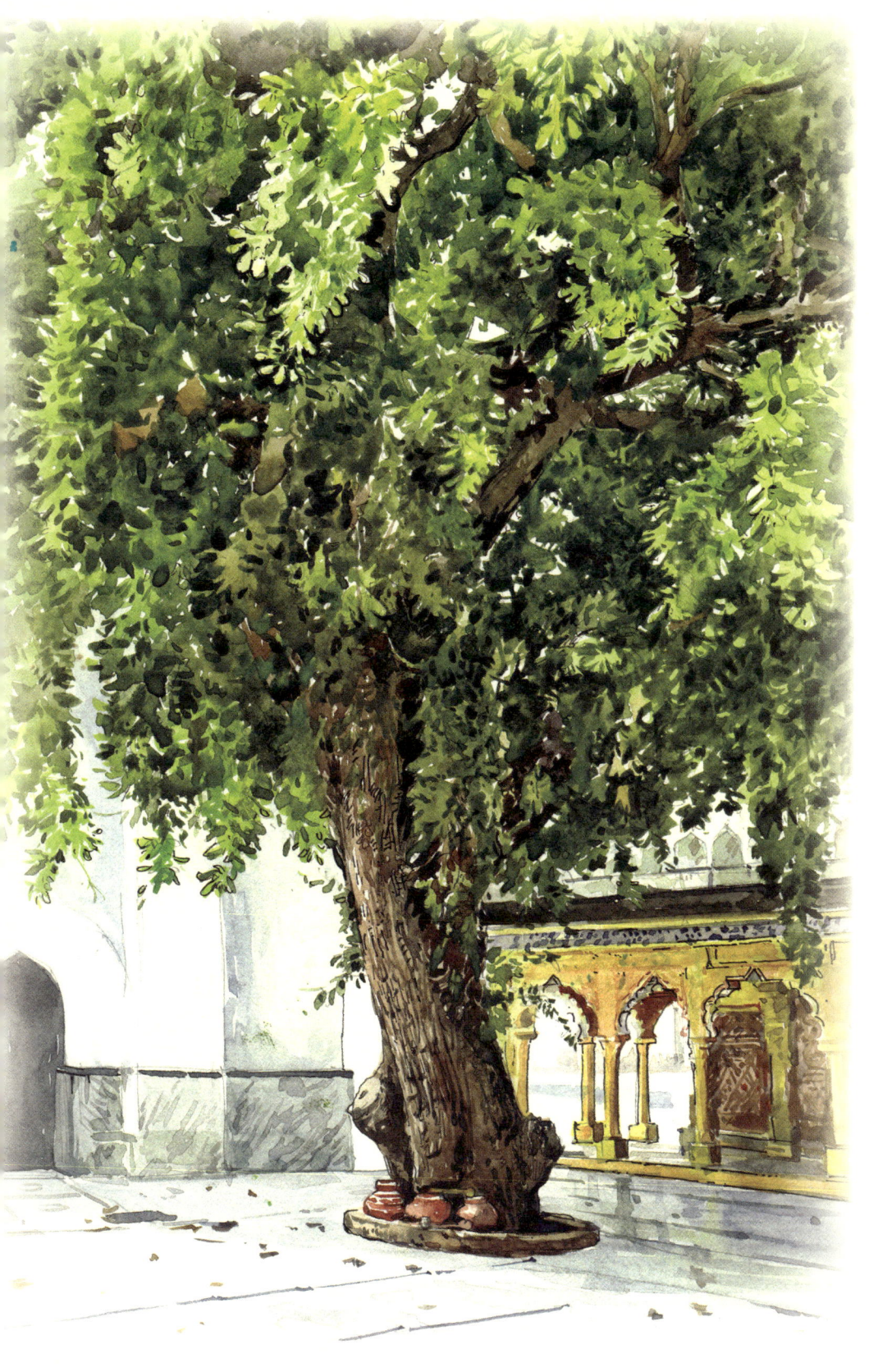

22

Tree of Enlightenment – Sri Mahabodhi

COMMON NAMES: **Bodhi, Sri Mahabodhi, Sacred Fig Tree (English); Ashwattha (Sanskrit); Sri Mahabodhi, Mahabodhi, Peepal Tree (Hindi)**

SCIENTIFIC NAME: *Ficus religiosa* L.

FAMILY: **Fig (Moraceae)**

WHERE TO SEE: **Mahabodhi Temple Complex in Bodh Gaya, Gaya, Bihar**

LATITUDE: **24.69602° N;** LONGITUDE: **84.99118° E;** ALTITUDE: **129.5 metres**

B uddhism took birth over 2,500 years ago under a sacred fig tree (peepal) in Bihar's Bodh Gaya. Meditating under that peepal, Siddhartha Gautama, received enlightenment and became the Buddha. Upon his enlightenment, the Buddha stood for seven days gazing at the tree, which came to be known as Bodhi, the tree of enlightenment. That hallowed spot at Bodh Gaya is a UNESCO World Heritage Site since 2002.

Buddhism might not have become a world religion but for Emperor Ashoka (died 238 BCE?), who helped spread it across India and neighbouring countries. He sent missionaries abroad to preach the *dharma*. As a goodwill gesture, the emperor sent a rooted branch of the sacred Bodhi tree, planted in a golden vessel by sea, under his daughter Sanghamitra's care to be planted at Anuradhapura, Sri Lanka. Ceremonially received at the royal gates, it was planted in 288 BCE in the presence of King Devanāmpiyatissa himself, at the spot where it stands today.[74] Known as Jaya Sri Mahabodhi, this is the only ancient tree in the world for which the exact date of planting is known.

The earliest available description of the sacred tree at Bodh Gaya is given by the Chinese Buddhist monk and fifth-century traveller Fa-Hien (Faxian),[75] who walked from China to India. He was followed two centuries later (637 CE) by the more illustrious Hiuen Tsang (Xuanzang), who described the Bodhi tree in his writings:

The Bodhi Tree at the Diamond seat is a pipal tree, which was several hundred feet tall at the time of the Buddha and although it has been cut down or damaged several times, it still remains forty or fifty feet high.... The trunk of the tree is yellowish white in colour and its branches and leaves are always green; they never wither away nor change their

100

lustre, whether in the winter or in the summer. Each year on the day of Tathāgata's Nirvana, the leaves fade and fall; but they grow out again very soon. On that day the monarchs of various countries and monks and laymen of different places, thousands and myriads in number, gather here by their own will to irrigate and bathe the tree with scented water and milk to the accompaniment of music; with arrays of fragrant flowers and lamps burning uninterruptedly the devotees vie with each other in making offerings to the tree.[76]

The tree in the temple complex today is not the original Bodhi, but a direct descendant. According to Hiuen Tsang, the first to cut down the tree was Emperor Ashoka himself, but it miraculously recovered.[77, 78] It was then destroyed by his Queen Tishya Rakshita employing a mandu thorn, as she was jealous of her husband devoting so much time to it.[79] This time too, the tree recovered miraculously.[80]

King Shashānka of Gauda (today's Bangladesh and West Bengal), a Shaivite king of sixth century CE, then felled it and dug up the roots. He then 'burnt it with fire and sprinkled it with the juice of sugar cane, desiring to destroy it entirely and not leave a trace of it behind'.[81] After a lapse of time, between 600 and 620 CE, King Purna Varma 'revived the roots of the tree with the milk of a thousand cows and in a single night, it sprang up again to a height of 10 feet (approximately 3.05 metres)'. This, of course, was a pious myth. Purna Varma planted in the original place a sapling of the Bodhi tree brought from Sri Lanka.[82] Thus, the sapling was the direct descendant of the original Bodhi tree.

In 1811, Francis Buchanan-Hamilton, a Scottish physician, noted that the tree was in full vigour and probably did not exceed 100 years in age. Examining the same tree in 1862, Major General Sir Alexander Cunningham, a British army engineer with the Bengal Engineer Group, who later took an interest in the history and archaeology of India, was disappointed to find it very much decayed,[83] one large stem on the western side was green, but other branches were bark-less and rotten. Between 1871 and 1875, the decay continued and in 1876, the only remaining portion of the tree fell over the wall during a storm and the old peepal tree was gone. The present tree was planted by Cunningham on the spot in 1881 from a seedling raised from the previous tree.[84] The much-venerated tree is now roughly 140 years old, with several spreading lateral branches. Some of its old and drooping branches require the support of iron pillars.

Today, the sacred tree is surrounded by a protective wall and kept under police guard as terrorists bombed the site in 2013. Fortunately, the Bodhi tree and the temple were undamaged.[85] The bombings did not discourage the inflow of pilgrims and tourists. They still gather under its sprawling canopy with varying sentiments – hope, curiosity, devotion, scepticisms and gratitude – as visitors must have done over many centuries. Unmindful of it all, the tree provides cool shade to all of them without discrimination.

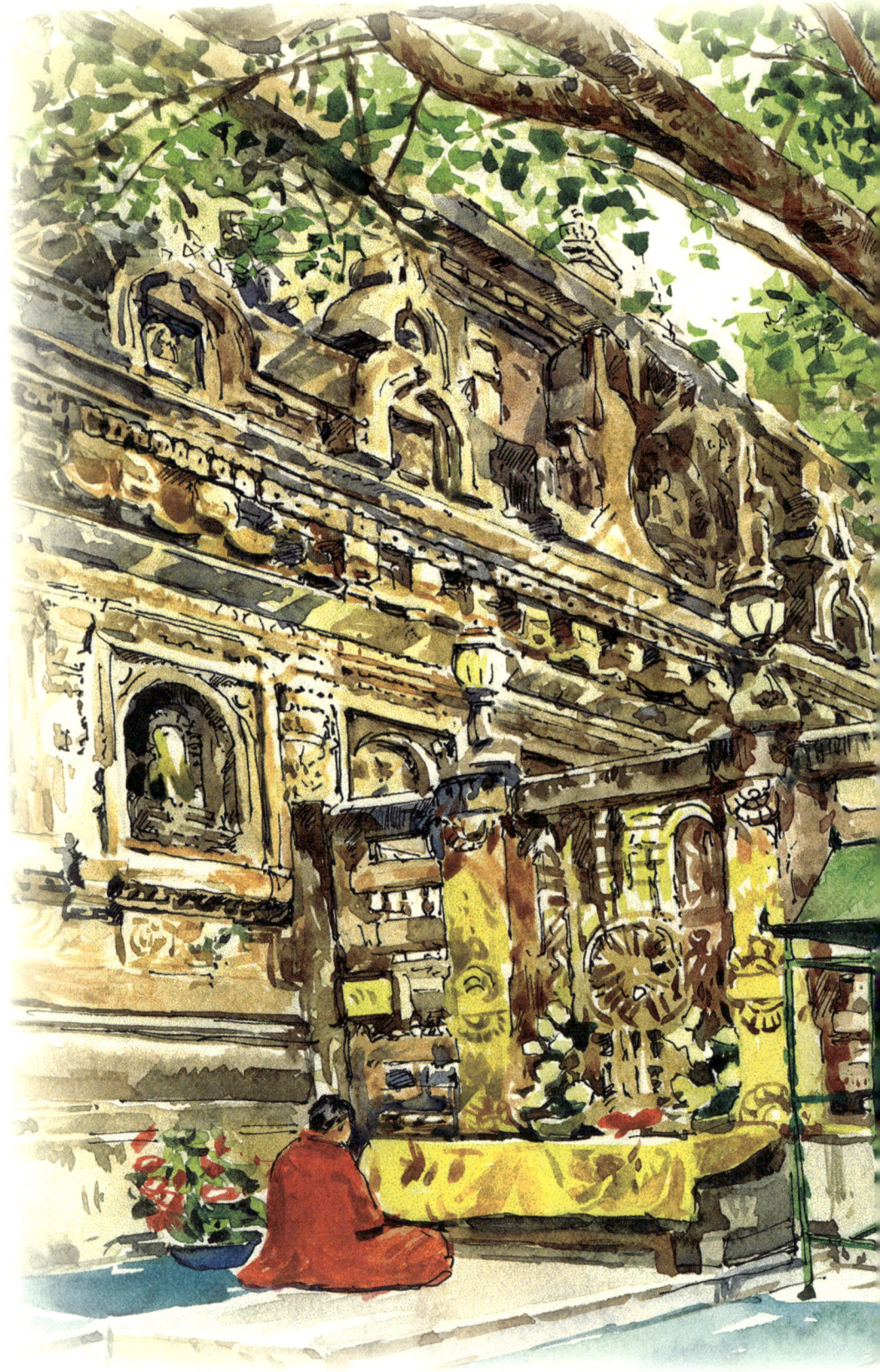

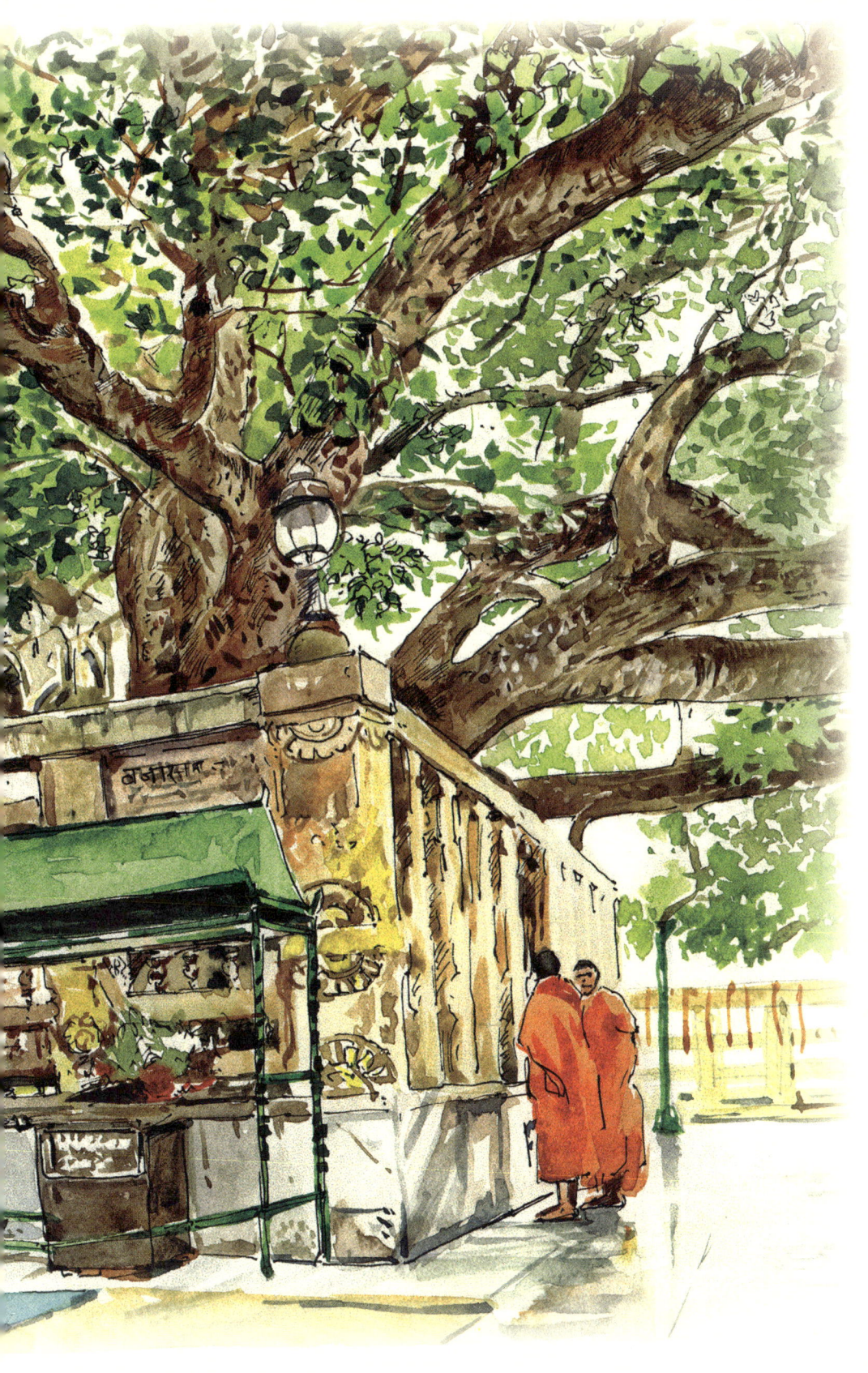

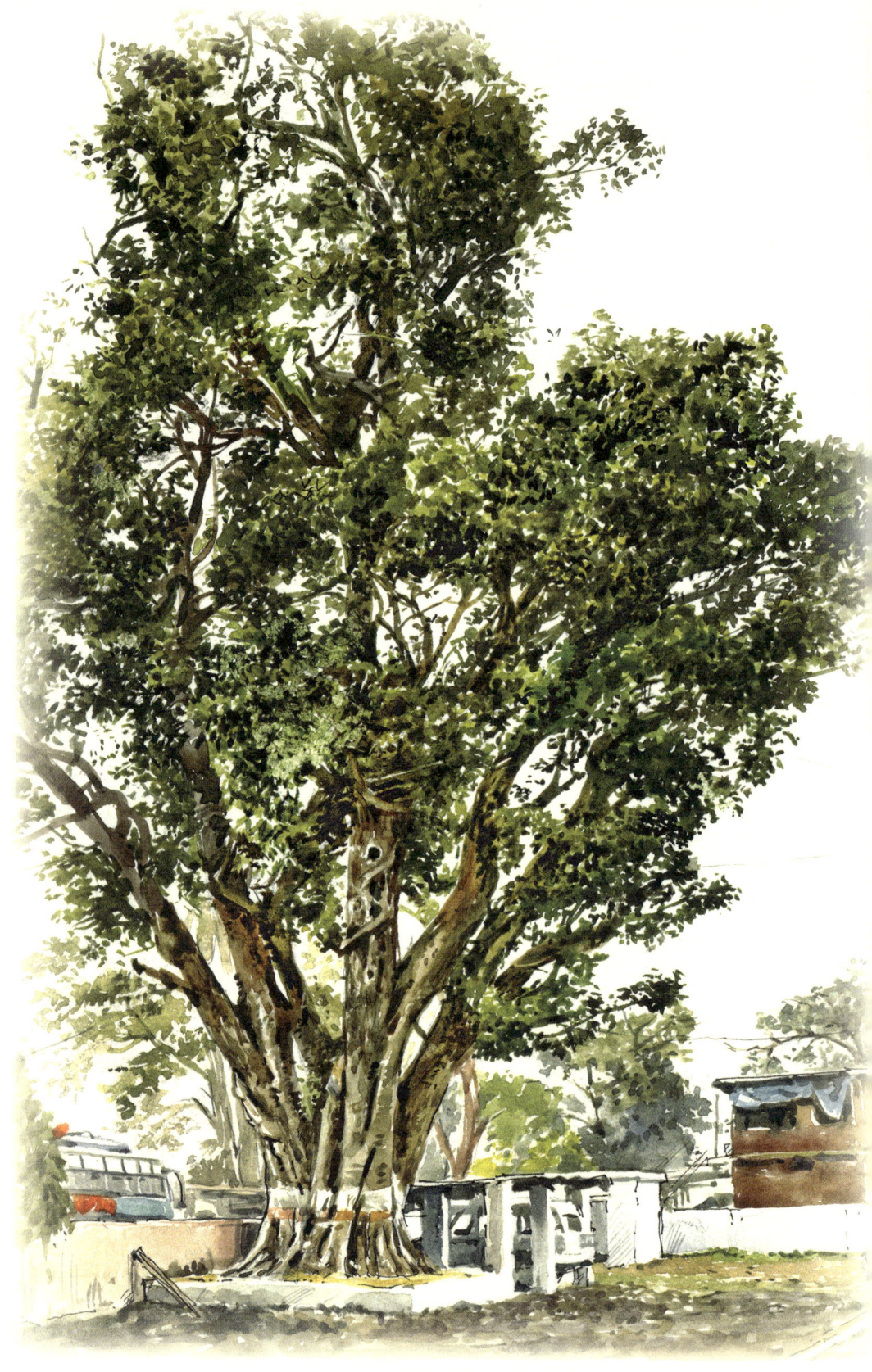

23

Sleeman's Tree for Hanging Thugs

COMMON NAMES: **Sacred fig (English); Ashwattha (Sanskrit); Peepal (Hindi)**

SCIENTIFIC NAME: *Ficus religiosa* L. | FAMILY: **Banyan (Moraceae)**

WHERE TO SEE: **Police Station, Sleemanabad, Katni district,**

Madhya Pradesh

LATITUDE: **23.3815° N**; LONGITUDE: **80.1526° E**;

ALTITUDE: **421.26 metres**

Northeast of Jabalpur, Madhya Pradesh, on NH 30, lies a village with a rather unusual name: Sleemanabad. Were you to drive through, you'd pass the local police station – located right on the highway – without knowing its walls contain a living piece of history. One that has aided and abetted the empire's atrocities. I am talking about a peepal tree that was used as gallows to hang scores of men sentenced to die as 'thugs' by the East India Company in the early 1800s. Both the peepal tree and the police station are of heritage importance, as many momentous events associated with thuggee are linked to them.

Thuggee, as their trade was called, was said to be rampant across vast tracts of the country, especially the Central and Northern Provinces. Early accounts of thugs were highly embellished. The contemporary fiction *Confessions of a Thug*[86] by the self-styled 'Captain' Philip Meadows Taylor, a British officer based in Hyderabad in the service of the nizam, changed the way thugs were perceived both in India and across the world. The three-volume book was one of the sensations of the 1839 London literary season.[87] Young Queen Victoria herself was so impatient to read its final chapters that running sheets were directly sent to her as they came off the press! This view persisted as late as 1984, with the film *Indiana Jones and the Temple of Doom* representing thuggee as first and foremost an evil cult.

The East India Company appointed Colonel William Henry Sleeman, a civilian with a military background, to address the thuggee menace. Sleeman was a natural choice; his *Ramaseeana* (1836), an exhaustive guide to the 'peculiar language of the thugs' had taken Britain by storm. He wasted no time in eradicating thuggee. Thus, between 1826 and 1848, some 4,500 men eventually stood trial for thuggee crimes. Of these, 504 were hanged. In Jabalpur and Sagar alone, there were 146 executions between 1830 and 1832. Some of the 'thugs' were hanged on the peepal tree in the police station but there is no record of the exact number. W. H. Sleeman became an 'eminent Victorian' whose name was widely recognized. [88] The British administration effectively used the image of thuggee as proof of the backwardness of India to justify colonization.

The Sleemanabad peepal stands close to the boundary wall at one end of the police station. A small square platform has been built around it. Although it is a fine specimen, it is neither grand nor imposing. At the time of the hangings, the tree must already have been quite big, probably 20–25 years old. Given that the hangings commenced around 1826, the current age of the tree could be estimated at around 220 years. The tall tree has several branches spreading out from the trunk in various directions, but the one used for the hangings apparently fell off years ago, leaving a prominent scar. Parakeets have made their home in several holes in the upper reaches of the branches.

Sleeman was a popular figure in this neighbourhood. He obtained 96 acres of land here for the poor and the people named the village 'Sleemanabad' in his honour. It is, however, pertinent to point out that post-colonial scholars have raised serious doubts about the authenticity of information contained in the corpus of colonial narrative about thugs and thuggee.[89] Did thugs – portrayed as religiously-motivated, secretly organized ritual killers – really exist? Or were they a convenient recognizable enemy? Did they kill so many people (two million according to one estimate)? More to the point, was the campaign against thugs no more than a witch-hunt drummed up by a hysterical and racist British? If yes, how many of the 504 hanged – largely on the testimony of other criminals – were innocent victims? The peepal tree may know the answers, but alas, we will not be told.

24

The Tree at the Heart of India

COMMON NAMES: **Belliric myrobalan, Belleric myrobalan, Bastard myrobalan, Bedda nut tree (English); Vibhitaki, Vibhitakah (Sanskrit); Baheda, Bhaira (Hindi)**

SCIENTIFIC NAME: *Terminalia bellirica* (Gaertn.) Roxb.

FAMILY: **Arjun (Combretaceae)**

WHERE TO SEE: **Karaundi village, Dheemerkheda tehsil, Katni district, Madhya Pradesh**

LATITUDE: **23.3048° N;** LONGITUDE: **80.1953° E;** ALTITUDE: **381.31 metres**

Before India's independence, Nagpur was the geographical central point of undivided India. To mark the location, the British erected the Zero Mile Stone, a sandstone pillar, in 1907. It still stands in Maharashtra's winter capital.

Independence from British rule and the accompanying partitioning of the country shifted this landmark approximately 343 kilometres northeast to Karaundi village in Madhya Pradesh's Katni district. It was Dr Ram Manohar Lohia, the socialist leader, who inspired the search for the geographical centre of modern India. A team from Jabalpur Engineering College headed by its founder Professor S.P. Chakravarti took up the task in 1956. Today, the area is surrounded by sprawling agricultural fields, framed by the picturesque Vindhyachal mountain range in the background. A monument was built in 1987 with a replica of the four lions of the Ashoka Pillar mounted on a short rectangular column at the centre of a three-tiered sandstone platform. Four larger lions made of cement are also seated at the edge of the platform, their faces pointing towards one of the four cardinal points.

Very close to the monument stands a tall and stately baheda tree. Native to south and southeast Asia, the baheda occurs naturally in the plains and lower hills all over India. This one must have sprouted naturally here probably 150 years ago. Today, it is a large presence that looms over the heart of India with a gradually tapering trunk and a massive dome-like crown. A large circular platform lined with sandstone tiles has been constructed at the base of the tree. The lower branches on the trunk have been cut (or fallen) off and burls are now growing in their place. A branch close to the tree base has been sawed off neatly and makes for a nice seat with the trunk as the backrest. In the higher branches, the large green leaves are typically crowded at the ends of the branchlets. Indeed,

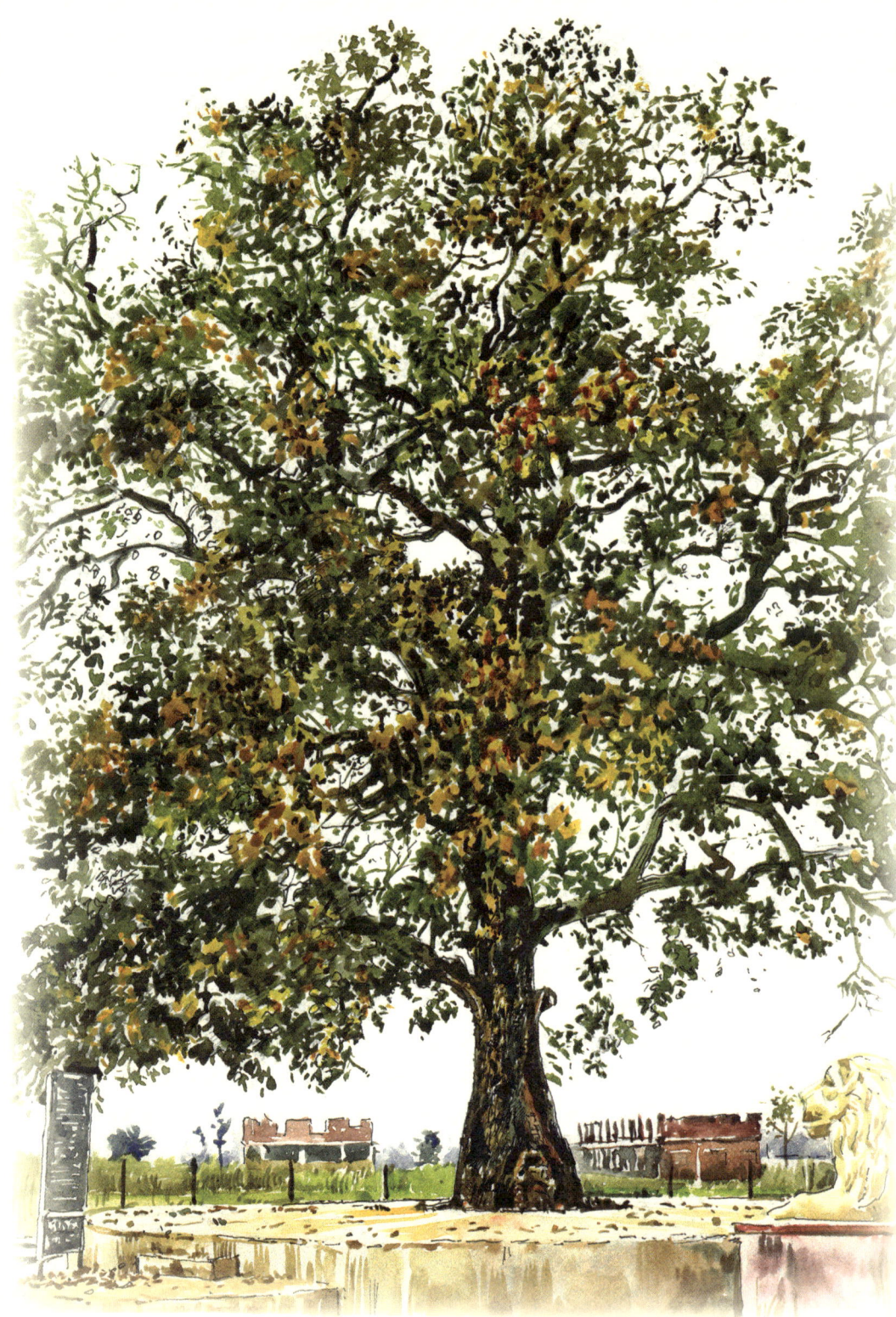

the generic name *Terminalia* is based on the Latin word 'terminus' or 'terminalis', referring to this trait of the leaves. During March and April, having just shed the old leaves, the tree dons new coppery-red foliage that later turns green. The new leaves are accompanied by small, fragrant, petalless flowers packed on long tassels at the ends of twigs.

Close to the monument and the tree, the sprawling campus of the Maharshi Mahesh Yogi Vedic Vishwavidyalaya attracts many overseas scholars. Sadly, few tourists have heard of Karaundi's importance and fewer still visit the remote village. Meanwhile, one can take heart in the fact that in addition to a cement-concrete monument, India's geographical heart is marked by a green living and pulsating being as well.

25

Weeping Cypresses of Dubdi Monastery

COMMON NAMES: Funeral Cypress, Chinese Weeping Cypress,
Mourning Cypress (English); Tchenden (Lepcha/Bhutia)

SCIENTIFIC NAME: *Cupressus pendula* Thunb. (= *C. funebris* Endl.)

FAMILY: Cypress (Cupressaceae)

WHERE TO SEE: Dubdi Monastery, near Yuksom, Sikkim

LATITUDE: 27.37° N; LONGITUDE: 88.23° E;
ALTITUDE: 2,011.68 metres

If you walk three kilometres from the bustle of Yuksom Bazar in the West Sikkim district you will reach Pauhungry hill. A broad but steep and winding pathway up the hill under the canopy of trees leads to a spectacular view of the surrounding mountainside. On the flat top, nearly 305 metres above Yuksom stands the Dubdi Monastery (sometimes referred to as Yuksom Monastery), considered by many as the oldest Buddhist monastery in all of Sikkim.[90] Established in 1701, during the time of Chakdor Namgyal, the third king of Sikkim, Dubdi literally means 'hermit's cell' after its founder Lhatsun Chembo (or Lhatsun Namkha Jigme) and belongs to the Nyingma sect – the oldest of four major schools of Tibetan Buddhism (the other three being Kagyu, Sakya and Gelug).

The Archaeological Survey of India has carried out considerable restoration work on the monastery and its surroundings. Surrounded by several ancient and beautiful weeping (or funeral) cypresses, the monastery's trees bear graceful, pale green conical crowns – their erect, cylindrical pine-like trunks holding the crowns aloft. The branches give rise to slender branchlets that are pendulous and weeping. Although the main trunks are not covered with moss, the upper branches are densely clothed with aerial plants such as lichens, mosses, ferns and orchids that wave in the breeze. Prayer flags flutter from many trees. One of the trees towering over the back of the monastery stands out from the rest with a girth of 6.5 metres, clearly exceeding by 1.4 metres the girth of the tree at the Norbugang coronation site (see page 113). This also seems to be the same tree that Sir Joseph D. Hooker, perhaps the most distinguished of botanical explorers to visit Sikkim during 1848–51, had described in the winter of 1849[91] in his journal as follows: '...One of these trees (perhaps the oldest in Sikkim) measured sixteen and a half feet (5.03 metres) in girth, at five from the ground and was ninety feet (27.4 metres) high: It was not pyramidal, the top branches being dead and broken and the lower limbs

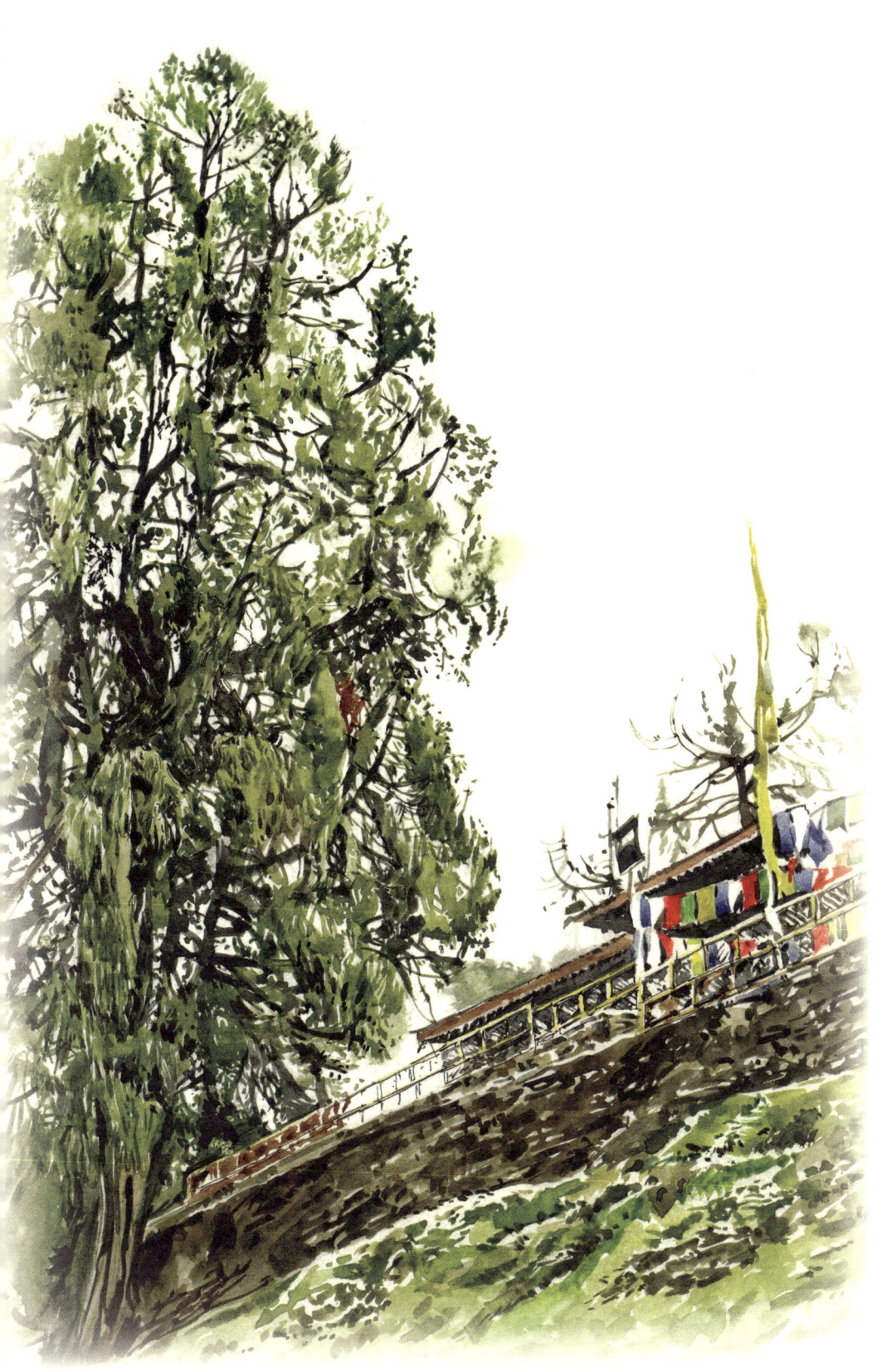

spreading; they were loaded with masses of white-flowered orchids.' However, its crown is now pyramidal contrary to what Hooker observed: possibly, one of the axillary branches took over apical function. Another beautiful tree, slightly lower down on the boundary, measures 6.25 metres in girth.

How old are these trees? Were they planted when the monastery was built or already existed earlier? Weeping cypresses are not wild in Sikkim and their seeds do not ripen here; they were most likely imported from Tibet, as the fragrant reddish wood of this tree is usually burnt in monasteries. Hence it is difficult to believe that the trees predate the monastery. In any case, they must be over 300 years.

Like the coronation site at Norbugang, the Dubdi monastery is part of the Khangchendzonga Biosphere Reserve, now included in the list of UNESCO's World Network of Biosphere Reserves.

26

The Coronation Cypress
of Norbugang

COMMON NAMES: Funeral Cypress, Chinese Weeping Cypress,
mourning Cypress (English); Tchenden (Lepcha/Bhutia)

SCIENTIFIC NAME: *Cupressus pendula* Thunb. (= *C. funebris* Endl.)

FAMILY: Cypress (Cupressaceae)

WHERE TO SEE: Coronation throne, Norbugang,
near Yuksom, Sikkim

LATITUDE: 27.2224° N; LONGITUDE: 88.1315° E;
ALTITUDE: 1,763 metres

Norbugang occupies a significant place in the history of Sikkim. A religious site for Buddhist pilgrims, it is believed to have been the meeting site for three great Buddhist monks who chose this place for the coronation of the first monarch of Sikkim in 1641. An ancient Chinese weeping cypress looming over the stone coronation throne is said to be contemporaneous with the coronation ceremony.

Legend connects the founding of the Sikkim kingdom to the ritualistic schism in the Tibetan Buddhist religion. It is said that in the middle of the seventeenth century, Lhatsun Chembo, a lama of the 'red hat' Nyingma sect, fled Tibet to escape persecution from reformers and came south to Denzong (Sikkim). After much wandering, he reached a place called Norbugang where he met two others who had also come from Tibet for the same reason.[92] After a brief parley the chief object of which was to establish a Buddhist monarchy in that place, they decided to search for a fourth individual who would help them in this mission. The place henceforth came to be known as Yuksom, or 'meeting place of three holy men'. The lamas' search culminated in identifying a certain Phuntsog Namgyal, an influential Tibetan then residing in the eastern part of Sikkim near today's Gangtok. An alliance was struck between them, the chief objective being the conversion of Lepchas to Buddhism and the installation of Namgyal as the monarch of the whole country. On an auspicious day in 1641, the lamas consecrated Phuntsog by sprinkling holy water from a nearby pond and declaring him the sovereign. The coronation ceremony was conducted on a four-seater rectangular stone platform facing east with a rubble enclosure wall behind the seats. The new monarch was given the title of *Chogyal* or 'religious and temporal ruler' and he made Yuksom the capital of his new kingdom. Twelve generations of Namgyals ruled Sikkim for the next 333 years.[93] To celebrate the historic coronation, Lhatsun Chembo is believed to have planted his wooden walking stick behind the coronation stone, which grew into the Chinese weeping cypress tree.

Only under cultivation in India, the species is native to central and south-western parts of China.[94] In contrast to the grey stones of the coronation throne and the surrounding rubble, its huge trunk – slightly over five metres in girth – has a smooth brown bark. While the branches are turned horizontal or slightly upwards, the branchlets clothed in small, light green, densely flattened leaves droop down, earning the species the sobriquet 'Chinese weeping cypress', 'funeral cypress' and even 'mourning cypress'. Colourful prayer flags adorn the site including the tree. Going by the legend, this tree ought to be approximately 380 years old. During its nearly four centuries of existence, it has seen several political and cultural upheavals.

The entire complex has been repaired and restored by the Archaeological Survey of India and is still venerated by the local people. Today, Yuksom is part of Khangchendzonga National Park (KNP). In 2016, the United Nations Educational, Scientific and Cultural Organization (UNESCO) recognized KNP as India's 35th World Heritage Site. With its diverse and scenic landscapes and unique ethnic rituals, this high-altitude park in Sikkim meets natural and cultural heritage criteria, making it India's first and only mixed heritage site.[95]

Trees do not mean timber alone. Trees also mean oxygen, soil and water. Those who believe that economy and ecology are at two different ends of the spectrum are mistaken. Ecology in fact is permanent economy.

Sunderlal Bahugana
Environmentalist and the leader of
the Chipko movement

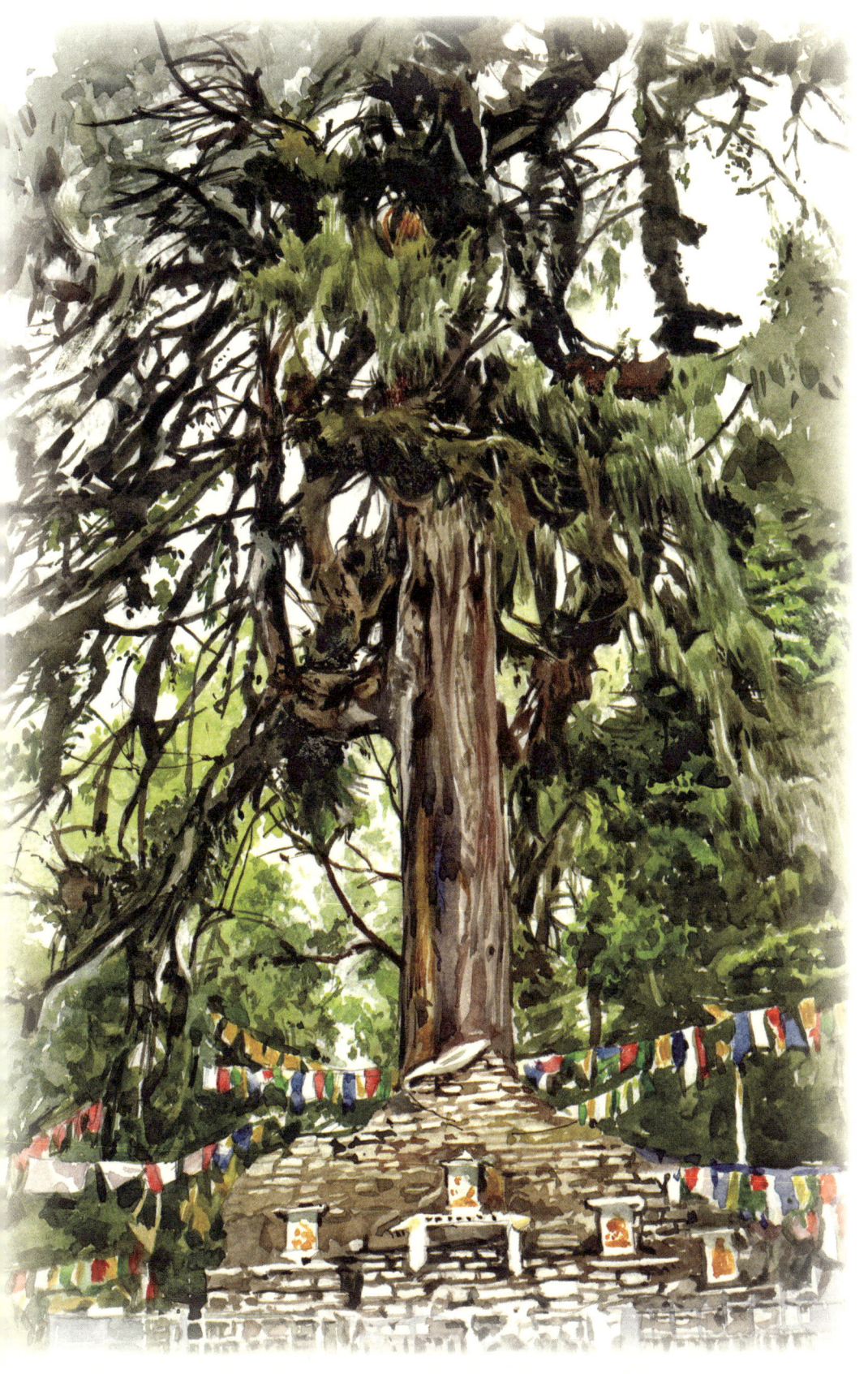

27

The Perils of Pregnancy – The Double Coconut Palm of Kolkata

COMMON NAMES: Coco de Mer, Double coconut (English)

SCIENTIFIC NAME: *Lodoicea maldivica* (J.F. Gmel.) Pers.

FAMILY: Coconut (Arecaceae)

WHERE TO SEE: Palm Conservatory, Acharya Jagadish Chandra Bose Indian Botanic Garden, Sibpur, Howrah, Kolkata

LATITUDE: 22.557247° N; LONGITUDE: 88.286135° E; ALTITUDE: 12 metres

O f the many living treasures growing in the Acharya Jagadish Chandra Bose Indian Botanic Garden, Kolkata, the double-coconut palm is by far the loneliest. Sadly, it is the only tree of double coconut in India. Its name is misleading, as the fruit usually has only one seed. Even the allusion to kinship with coconut is questionable, as the species is more closely related to palmyra palm. Its scientific name *Lodoicea maldivica* is inapt too. Its true home is 2,300 kilometres southwest in the Seychelles archipelago on just two neighbouring islands – Praslin (37 km²) and Curieuse (2.73 km²). This imposing 30-metres-tall, straight-trunked palm has huge fan-shaped, pleated leaves, just a few of which can thatch a roof. Even more noteworthy is its fruit – the largest wild fruit[96] ever, known weighing up to 45 kilograms (only the domesticated pumpkins and watermelons weigh more than that). Its seed is the world's heaviest by far at almost 25 kilograms – 10 more than the checked-in baggage allowance you get when flying economy on budget airlines! Two other records for which the palm is notable are the longest cotyledon (up to 4 metres) following seed germination and the largest female flower among all palms.

During the sixteenth century, the remarkable nuts were found all around the Indian Ocean, including Maldives, India and Sri Lanka and traded as far away as China. Since there was no evidence of their source, the nuts were initially believed to be produced by a plant species growing on the floor of the sea: thus, the common name *coco de mer* (coconut of the sea). The fresh seeds of the palm are too dense to float or retain viability in seawater unlike those of coconut and seeds washed up on distant beaches are merely empty shells. Therefore dispersal by sea route is ruled out making its common name coco de mer a misnomer too.[97]

A single nut of double coconut has a pair of large, round lobes which bear a striking resemblance to female buttocks (hence its other name coco fesses).

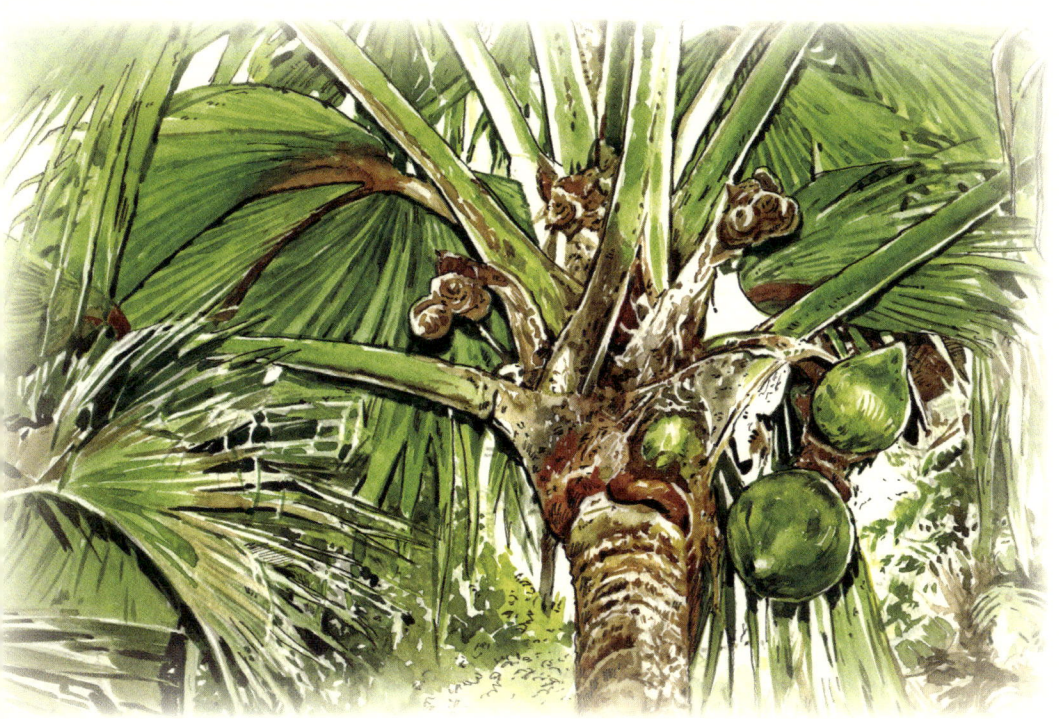

This led people to believe that the seed has aphrodisiac and medicinal properties and a thriving business began in the sixteenth and seventeenth centuries, with a single seed in London fetching £400 (roughly £70,000 in today's times). Even today, nuts are highly valued, making them a prime target for poachers. The high esteem in which they are held may be illustrated by the fact that the Government of Seychelles presented the UK heir to the throne and his wife, the Duke and Duchess of Cambridge, Prince William and Kate Middleton, a coco de mer nut as a wedding gift when they were spending their honeymoon on the islands in 2011.

The garden in Kolkata successfully raised a single plant of double coconut in 1894 using seeds obtained from Seychelles and planted it in the central part of its large palm house. In this species, male and female flowers are borne on separate trees, so until it flowers there is no way of telling the tree's sex. In their natural habitat the trees are notoriously slow growers and take up to 25 years to attain reproductive maturity. This one kept everyone guessing, just gaining in height and putting out leaves for 94 long years! Then in October 1988, it put forth a bunch of flowers announcing itself to be a female. Although the metre-long inflorescence persisted on the tree for nearly two years, fruit set could not occur in the absence of male palms. By 2006, the number of inflorescences had doubled and thereafter, two to four inflorescences appeared regularly on the tree from March to September each year.

In its natural state, the double coconut is known to survive for up to 350 years but the garden authorities were unsure how long the lone palm would last in alien soil and climatic conditions; it would be wonderful if they could pollinate

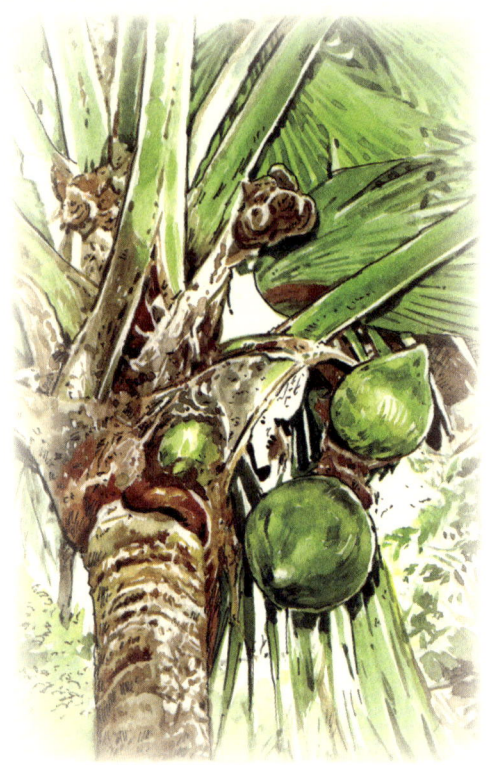

the flowers artificially to produce a few fruits, ensuring that the country would continue to have specimens of the curious tree. They started looking for suitable 'grooms' – fertile male palms – that could act as pollen donors for their palm 'bride'. In 2006, a team of scientists attempted to fertilize the tree with pollen obtained from the Royal Botanical Gardens, Peradeniya, Sri Lanka. However, the mating was unsuccessful.[98] In August 2013, the team tried again, this time with pollen from the Nong Nooch Tropical Garden, Thailand, hand-pollinating seven fresh female flowers. Weeks of anxious vigil followed before it could be confirmed that two of these had resulted in fruits – one slightly larger than the other. By 2015 both young fruits showed encouraging growth. They continued to grow on the mother tree for over six years until 2019, kindling hopes of a healthy 'induced pregnancy'. Back in Seychelles, fruits are known to take six to seven years to ripen on the tree. If all went well, they thought, by the end of 2020, there could be two mature double coconut fruits in India.

However, there was now a new worry regarding the health of the mother tree. Usually, each year, the palm produced a new 'spear leaf' at the tip of the tree, which slowly unfurled like a Japanese hand fan as the blade expands to its full size. Its health was a major indicator of the well-being of the tree as a whole: If the spear was infected or otherwise damaged, no new growth would occur on the plant. In the preceding eighteen months, the bizarre palm had not produced a spear leaf and the older leaves were slowly turning yellow. The question was: Would the iconic tree survive at least until the fruits matured?

Disaster struck in early February 2020 as the smaller of the two fruits (weighing 8.5 kg) fell off the tree. Just eight or ten days later, the larger fruit (18 kg) also fell off. After maintaining them in the dark for nearly eight months to induce germination, they have been planted in two large wooden potting boxes. In the normal course, germination takes 24 to 36 months. At the time of going to press, there is no encouraging sign yet. Will the seeds germinate successfully? They say prayer is a powerful thing. Let us put our hands together in prayer and hope they do.

28

Kolkata's Oldest Resident – The Great Banyan

COMMON NAMES: **Banyan (English); Nyagrodha (down-growing) and Bahupada (many-footed) (Sanskrit); Bot Gacha (Bengali)**

SCIENTIFIC NAME: *Ficus benghalensis* **L.**

FAMILY: **Mulberry (Moraceae)**

WHERE TO SEE: **Acharya Jagadish Chandra Bose Indian Botanic Garden, Sibpur, Howrah, West Bengal**

LATITUDE: **22.56075278° N;** LONGITUDE: **88.28677778° E;** ALTITUDE: **14.6 metres**

U ntil late March 2006, Advaita (Adwaita) was the oldest resident of Kolkata. Following his death, that mantle of honour has passed on to the 240-year-old Giant Banyan Tree in the Acharya Jagadish Chandra Bose Indian Botanic Garden.

Advaita was an Aldabra giant tortoise,[99] said to have been a pet of Robert Clive, whose triumph of arms at Plassey in 1757 is considered by many as the defining moment in gaining Britain her jewel in the crown. The great banyan tree too is the crown jewel of the famous botanical garden, drawing more crowds than all the garden's other treasures put together. Although neither the oldest nor the largest, it is undoubtedly, the star among the Indian banyans and certainly, the best-known and the best-looked-after banyan tree in the country.

No one seems to know when the tree sprouted to life, but it was already there when the garden was set up in 1786.[100] Banyans usually originate from seeds contained in droppings of fruit-eating birds on old walls or trees and this one is believed to have commenced its historic journey in the crevices of a wild date palm tree.[101] Its tiny roots started moving towards the earth in search of water and nutrients, meanwhile surrounding and eventually crushing, the host in a lethal embrace. Once entrenched in the soil, branches started pushing upwards and outwards. As is typical of banyans, secondary and tertiary branches emerged from these, securing them firmly to the ground. In due course, they became additional trunks.

The tree soon assumed breathtaking proportions with hundreds of large and tangled column-like roots in all directions. Early travellers from Europe to Kolkata could hardly contain their wonder and amazement at this arboreal marvel. One of them was Viscount Valentia, a British peer and politician who, calling

119

at the botanic garden in early February 1803, described it as the finest object in the garden, spectacularly festooned in orchids and ferns.[102] Maria Graham, a British writer, traveller and accomplished illustrator who visited the garden on 30 November 1810, could not stop gushing about the banyan tree whose rough bark was adorned with plants of the 'gayest colours and most elegant forms'.[103]

The tree has had more than its share of struggles and misfortunes though. In 1864, a cyclone devastated the garden when massive tides inundated the grounds, uprooting or severely damaging over a thousand trees. Three years later, another cyclone almost completed the obliteration of the arboretum, but the Great Banyan miraculously survived. In 1925, the tree was struck by lightning and its main trunk – then around 15 metres in girth – was severely infected by an opportunistic fungal growth and had to be amputated to save the rest of the tree. The spot where the main trunk stood is marked with a tablet now. Today, the tree stands only on its pillar-like roots. Despite this, it has continued to grow and expand with a whopping 3,772 prop roots and occupying an area of more than 1.89 hectares.[104] In 1985, fencing was built around the tree by the authorities to keep vandals away,

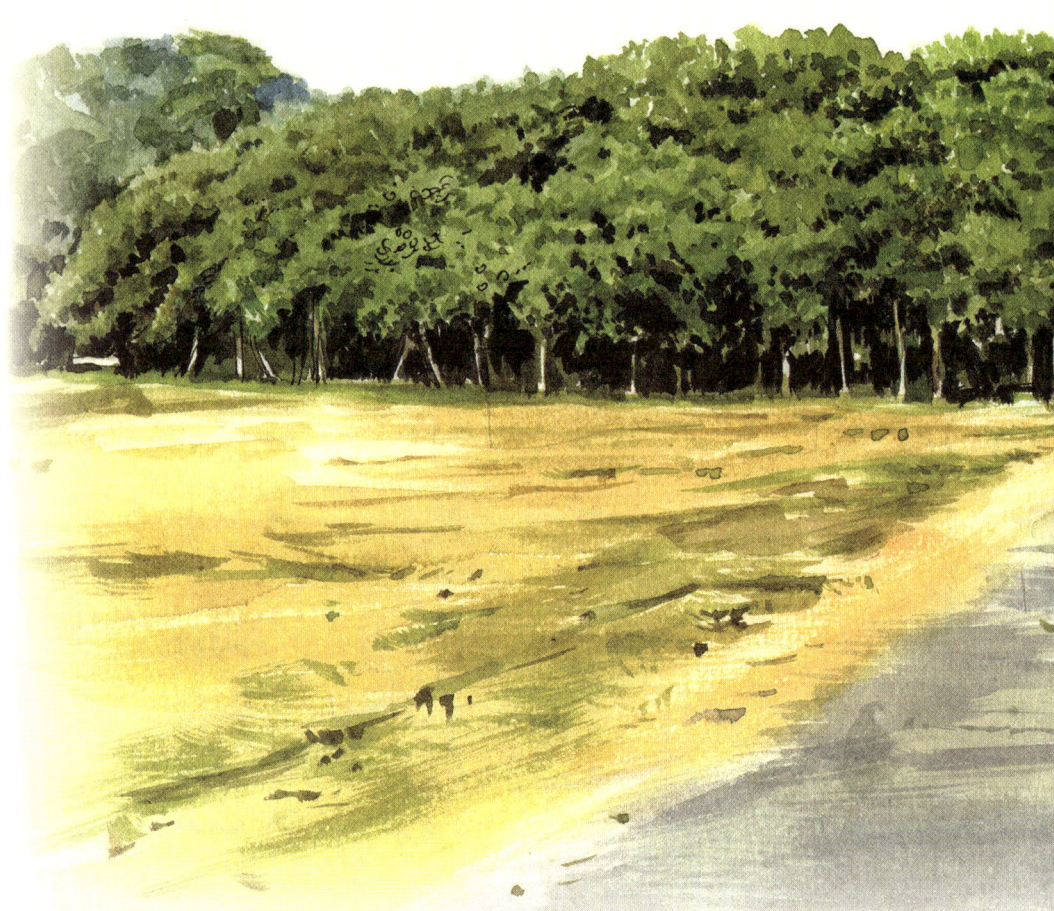

but it continued to grow over and beyond it. Interestingly, growth does not occur towards the western boundary of the garden along which there are residences and a busy road with traffic; instead, the tree is 'walking' mostly eastwards, towards the rising sun. Thus, it is probably the world's 'widest' tree.

Using net canopy coverage area as the primary measure of size, the Great Banyan Tree of Kolkata ranks fourth[105] with a net area of 15,531 m^2. The Botanical Survey of India, established on 13 February 1890 under the direction of Sir George King, the then superintendent of the garden, chose to adopt the Great Banyan on its logo.

Cyclone Amphan – which struck Kolkata on 20 May 2020 – wrought much havoc on the big tree, uprooting 44 prop roots. The hard work of the garden staff for a year could save 24 of these, giving reason to hope that, in time, the banyan will recover completely. It is not for nothing that this arboreal champion is celebrated as a proud symbol of majesty, endurance and permanence. Surely, Advaita would have approved.

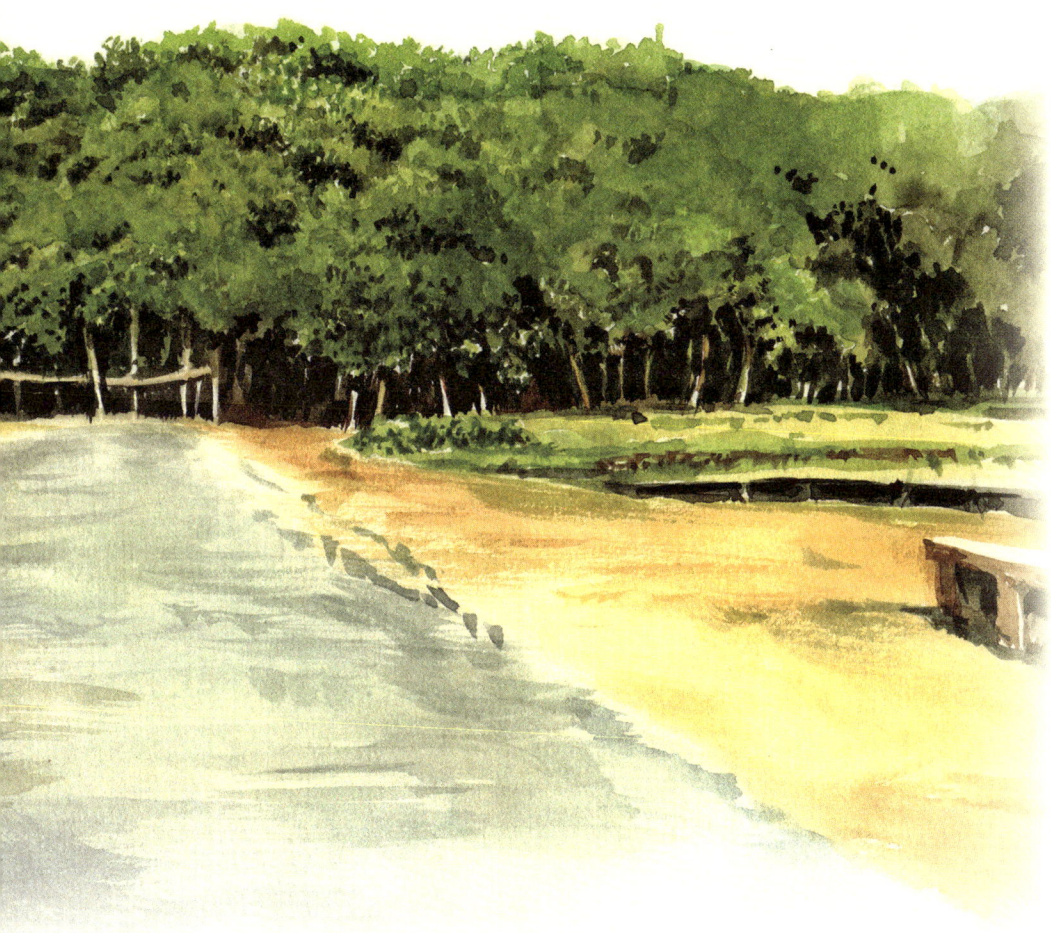

29

The Wishing Tree of Assam – Bakhor Bēngana

COMMON NAMES: Divine Jasmine Tree (English); Pindelu (Sanskrit); Bakhor Bēngana, Bakhar Bēngana, Bana Bēngana, Bon Bēngana (Assamese)

SCIENTIFIC NAME: *Tamilnadia uliginosa* (Retz.) Tirveng & Sastre

FAMILY: Coffee (Rubiaceae)

WHERE TO SEE: Bokata Mauza/Bokata Bakhor Bēngana, Nemuguri-Khamun Road, Demow tehsil, Sivasagar district, Assam

LATITUDE: 27.02385° N; LONGITUDE: 94.75552° E; ALTITUDE: 92 metres

Sivasagar (formerly Sibsagar) district in Assam is popularly referred to as the 'land of kings' and has great importance in Assam's history. It was from here that the late medieval Ahom kings ruled for over 600 years (1228–1826) until the Burmese routed them. Known for its scenic beauty as well as historical monuments, the district is also home to what is probably the oldest known living tree in Assam. The tree is locally known as Bakhor Bēngana, literally the 'divine brinjal' in Assamese because of its rounded fruit and 'divine jasmine' on account of its white, highly fragrant flowers. People believe it can grant wishes and seek its blessings to ensure the longevity of near and dear ones.

Bokata Mauza, the village where it stands, was surrounded by dense forests and not much was known about the tree except some local legends. The tree was thought to be the oldest in the world and one of its kind. With the publication of *Flora of Assam* by Upendra Nath Kanjilal and co-workers,[106] its identity was established as *Randia uliginosa* and was later changed to *Tamilnadia uliginosa*, its currently accepted scientific name.[107] With *uliginosa* (Latin) meaning 'growing in moist/wet conditions', the species is native to moist deciduous forests of India, Bangladesh, Cambodia, Laos, Sri Lanka, Thailand and Vietnam. It is also a popular garden plant.

During the eighties and the nineties, when the terrain came under the grip of the separatist outfit, the United Liberation Front of Assam (ULFA), the wishing tree was practically forgotten. Restoration of peace and normalcy took several years and it was only about a decade ago that interest in the tree was rekindled. In 2012, a team from BSIP, Lucknow was studying pollen sediments from this region and carried out some work on the soil surrounding the tree as well. They found evidence of pollen grains of the species in the top 80 cm of sediment. The age of the sediment as analysed through radiocarbon dating was 580 years at the time of the study. Thus, by inference, the tree has been existing here for at least 580 years and possibly longer.[108]

This new information tied in nicely with the legends circulating around the tree and the local people began taking a keen interest in their unique heritage. Although no historical evidence exists to support them, the legends have been popularized through the local media.[109] According to one, the tree was a friendly gift from the Chinese Emperor to the 14th–15th century Bārāhi-Kachari King Mahamanikya (or Manikpha) of Jayantipur, who planted it here. According to another legend, once there was a long-drawn battle between King Mahamanikya and the Chutiya King (14th–16th CE) and with no side likely to win decisively, they decided to bury the hatchet and called for a truce. Mahamanikya then planted the tree as a token of the new friendship between the two dynasties.

The tree is approximately seven metres tall, armed with sharp spines in pairs. Several low branches have surrounded the main trunk with shiny leaves arranged in opposite pairs on the branchlets. White fragrant flowers appear on the tree during April and May and small, round, many-seeded fruits are seen during September.

The renewed local interest has led to the establishment of a Bakhor Bēngana Rakshya Samiti. The trust erected a boundary wall with an entry gate to protect the tree and placed a board nearby informing people of the tree's importance. The village itself came to be named Bokata Bakhor Bēngana. Today, several institutions within it are also named after the tree: Bakhor Bēngana Gram Panchayat, Bakhor Bēngana Branch Post Office, Bakhor Bēngana High School and so on. Unfortunately, the tree's fame has not spread beyond the borders of the Sivasagar district and few know of it even within Assam.

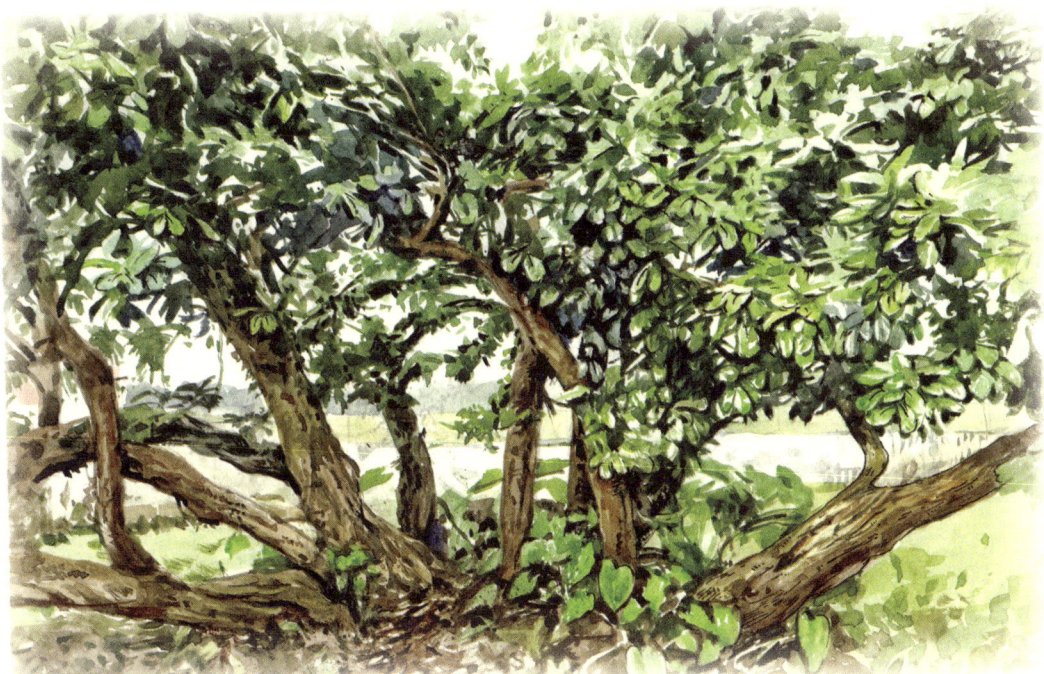

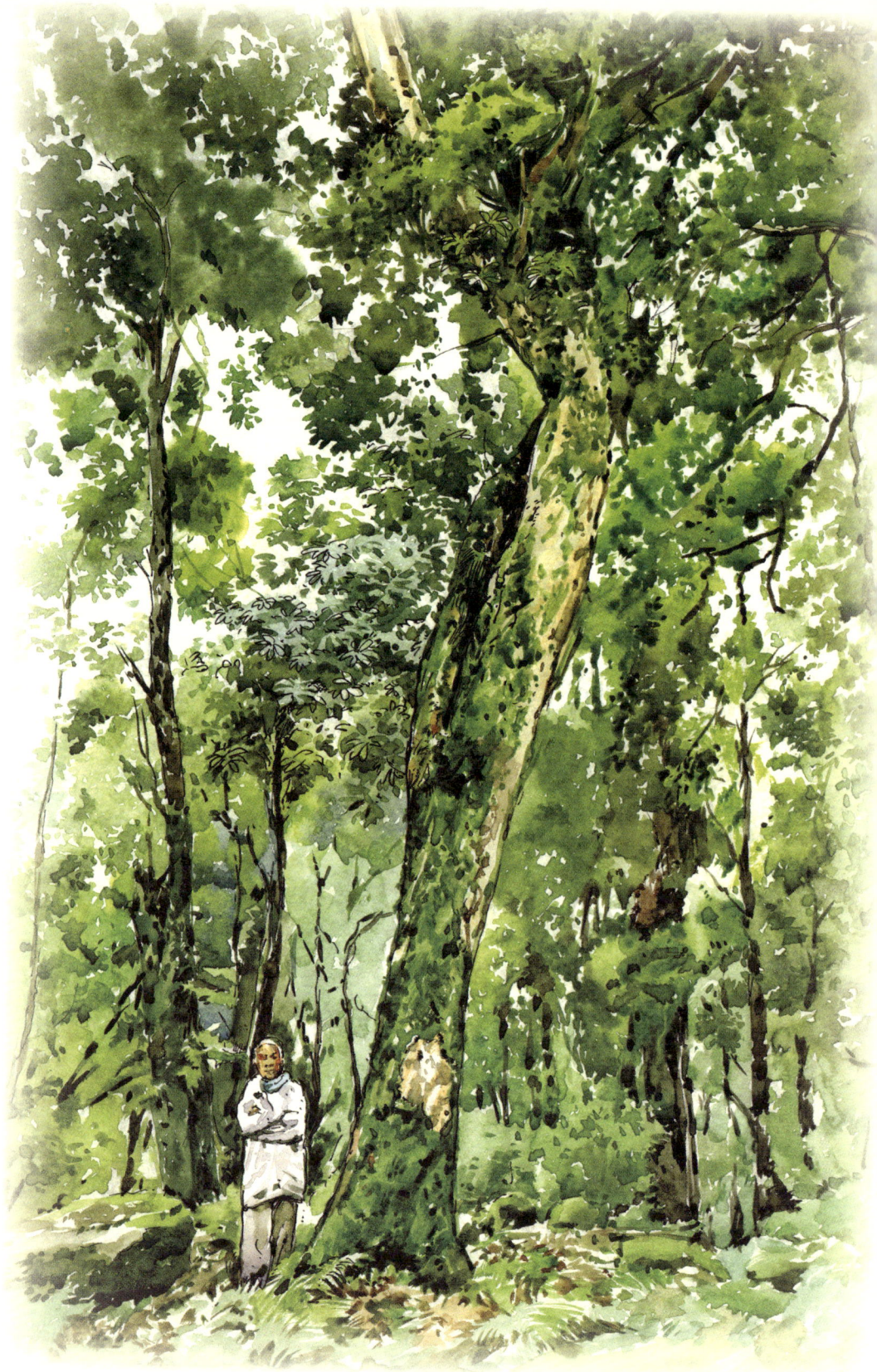

30

The World's Tallest Rhododendron

COMMON NAMES: **Rose Tree, Rhododendron Tree (English);**
Nithu (Angami Naga); Lindai (Mao Naga)

SCIENTIFIC NAME: *Rhododendron delavayi* **Franch.**[110]

FAMILY: **Rhododendron or Azalea (Ericaceae)**

WHERE TO SEE: **Japfü Mountain, near Kohima, Nagaland**

LATITUDE: **25.3634° N;** LONGITUDE: **94.04007° E;**
ALTITUDE: **2,500 metres**

Every seasoned outdoor adventure lover knows that in Nagaland climbing the Japfü Peak signifies one of the most challenging tasks. Its captivating beauty and thrilling trails attract trekkers and nature lovers from afar. However, the fact that Japfü is also home to the world's tallest rhododendron tree is, alas, not publicized enough. Consequently, few, if any, end up admiring this arboreal wonder.

Mount Japfü is Nagaland's second highest mountain (3,048 metres) after Mount Saramati (3,826 metres). Depending on your stamina, it might take anywhere between five and eight hours to reach the peak. The climb is tough with dense shrubs and climbers and lush green undergrowth that hides the ground beneath. Many gurgling streams rush down the steep gradient increasing the beauty of the landscape.

Between March and May, several species of rhododendrons set the Japfü alight with their purple, white, pink, yellow and scarlet blooms. All of them are shrubs barring two tree species – the white-flowered McCabe's rhododendron (*R. macabeanum* G. Watt ex Balf. f.), which is endemic to Manipur and Nagaland and the more widespread scarlet-blossomed rhododendron (*R. delavayi*). The latter, known to the Angami Nagas as Nithu, is an eye-catching stately tree that rarely exceeds 15 metres in height. It is readily recognizable as much by its pink to deep-crimson flowers held in clusters of 10 to 20, as by thick, leathery leaves crowded at the ends of branches with a green glossy upper surface and fawn to rusty-brown hairy lower surface.

One day in April 1985, Doneipa Khale and Viyale Tachü[111] – two Angami Naga friends from the nearby Phesama village – were camping on the Japfü range for game hunting. At the height of 2,500 metres, they came upon a lone tree of nithu that was spectacularly taller than any they had seen before. It took considerable effort to get the slow government machinery moving. Finally, some officials in the Department of forestry helped confirm the height as 32.9 metres (108 ft.) – as tall as a nine-story building – and the tree got included in the *Guinness Book of World Records* in 1993.[112]

No recent measurements of height are available but unconfirmed reports claim that in 2002 it was 38.1 metres (125 feet). And the tree is still growing! The trunk is unbranched until about three metres above the ground. At the time of our visit (16 July 2019), its girth was 2.59 metres. As is common for this species, the bark is rough and clothed with moss, ferns and small flowering plants. Come spring, it produces a riot of scarlet flowers.

Sadly, this priceless arborescent treasure – which could be the flagship for Nagaland's conservation and tourism programmes – is largely unheralded and in a state of neglect. Neither the local community nor the state government has taken any visible measures to promote the tree or protect it from vandals. No signboards lead the average tourist to the tree. Nor is there an authentic route map with information on distances and time required to locate it on Japfü. It will also greatly help if the immediate neighbourhood of the champion tree is at least partially cleared to enable visitors to admire it from various angles. Anyone listening?

31

Bridges that Breathe – Living Roots that Connect Meghalaya

COMMON NAMES: **India Rubber (English)**

SCIENTIFIC NAME: *Ficus elastica* L.

FAMILY: **Fig (Moraceae)**

WHERE TO SEE: **Nohwet village, close to Mawlynnong in the Cherrapunjee area, East Khasi Hills district, Meghalaya**

LATITUDE: **25.207° N;** LONGITUDE: **91.898° E;**
ALTITUDE: **Approximately 1,490 metres**

The people of Cherrapunjee (native name Sohra) in southern Meghalaya are used to rain – no, barrels of rain. Indeed, the region is one of the wettest places on earth[113] with an average annual rainfall of nearly 12 metres. During the rainy season, the many streams and rivers here swell and overflow in the remote hilly terrains all over southern Khasi and Jaintia hills. Conventional bridges using bamboo or wood do not last, as they rot away under warm and humid conditions or are washed away by the surging waters. Construction of steel and concrete bridges is unaffordable and difficult due to the constraints of transporting the building material through the poor network of roads. The War-Khasi and War-Jaintia people living in these parts hence do not build bridges, but simply grow them.[114]

On the slopes of these mountains, a species of fig known as the India rubber tree occurs naturally. With the help of its incredibly strong secondary roots arising from parts of the trunk, the tree can comfortably perch on boulders along river banks or even in the middle of rivers or streams themselves. The local population has used different strategies to employ the roots to grow their bridges across streams and rivers. The simplest way is to pull, tie and twist the roots by hand and encourage them to fuse, forming natural grafts so that over time they form the desired architectural structure. Often bamboo or wood scaffolds are used to train the young roots and direct them to grow in the desired manner. In some places, longitudinally split and hollowed-out betel nut trunks are employed instead of wood or bamboo to guide the roots. The betel nut trunks not only prevent the young roots of the rubber tree from fanning out but also protect them and provide additional nutrients as they decay. The scaffolds may have to be replaced several times over the years. When the roots reach the other side of the river, they are encouraged to anchor themselves in the soil. In time, a sturdy bridge

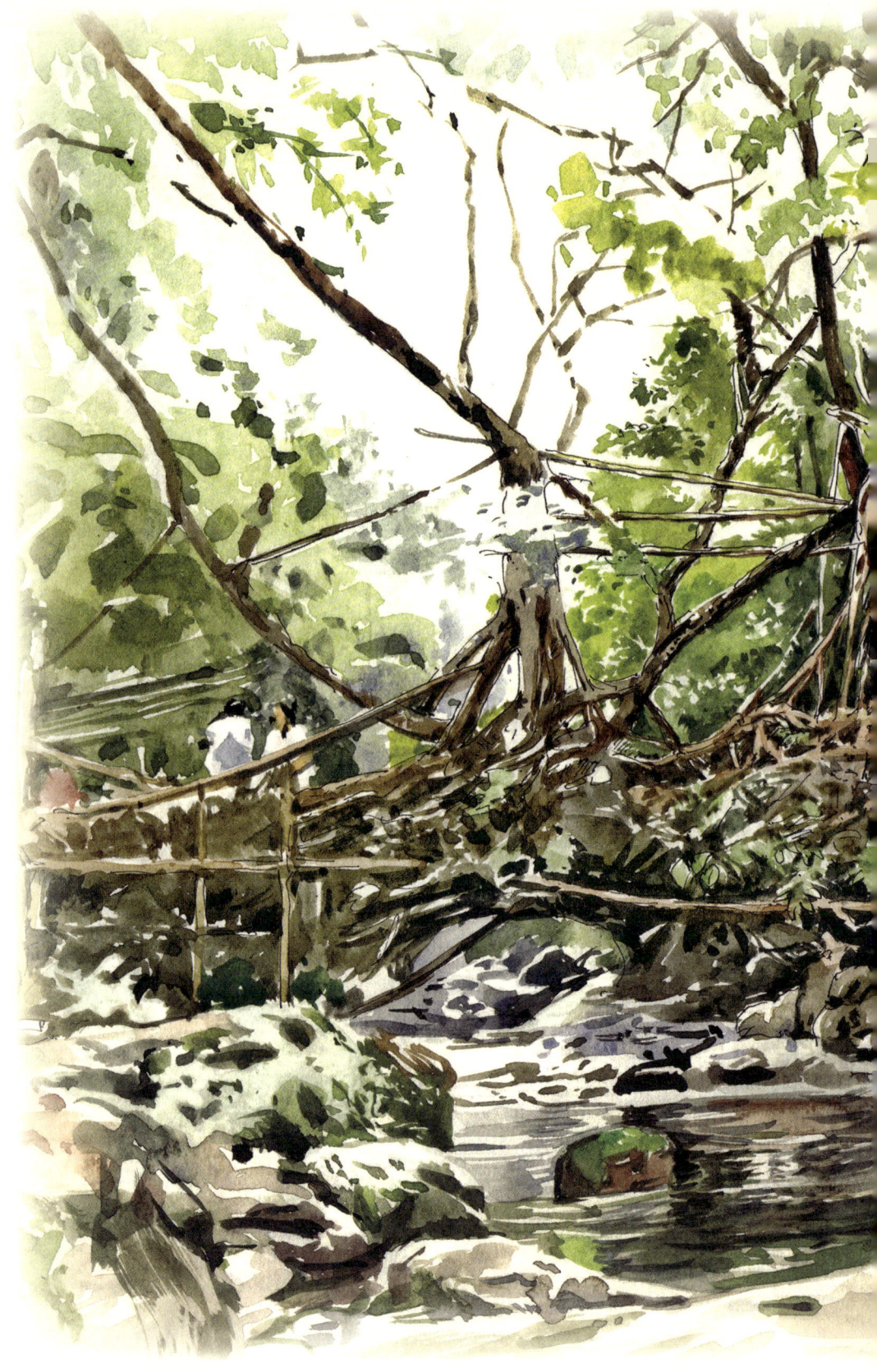

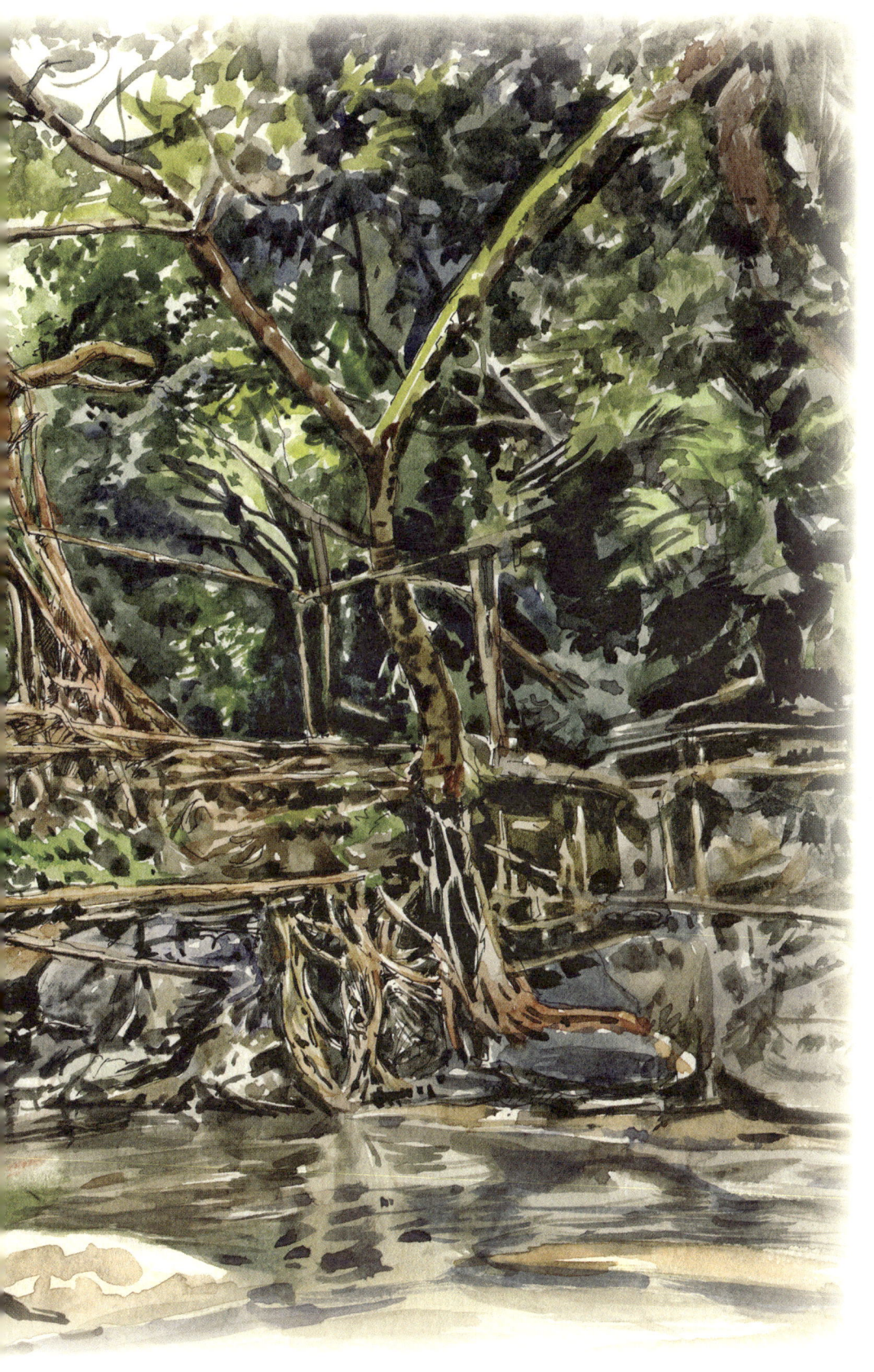

of intertwining roots is formed. No two bridges are exactly alike as each one is uniquely handcrafted with sustained community care and patience. Villagers constantly keep an eye on their upkeep and maintenance over decades or even centuries and weave a new root into the mesh whenever they chance upon one to further strengthen the lattice.

This is why root bridges come in various shapes and sizes. Some are as short as two metres and others as long as 52 m;[115] some are within touching distance from the river water, while some soar 24 m above the streams they straddle. Most bridges have a single span, but some have more than one – arranged as two parallel spans, or stacked one above the other such as the famous 'double-decker' root bridge in Nongriat, built across the Umshiang river.

The time taken to form a root bridge varies depending on several factors including the health of the mother tree, availability of nutrients in the soil and the manner of growing the bridge, but generally, it takes at least 15–20 years before it can be used. Once established, it gains in strength as it constitutes a living and growing system – strong enough to bear the weight of 50 or more people simultaneously. The lifespan also varies: Some of the bridges in the villages near Cherrapunjee are well over 500 years old. One such root bridge in Nohwet village is near Mawlynnong village, reputed to be Asia's cleanest village,[116] and close to the Bangladesh border.

No one seems to know when exactly the practice of growing and maintaining root bridges over centuries across rivers and streams started in the region. Sadly, the practice seems to be on its way out as steel and concrete bridges have started taking over. Since 2004, at least some of the bridges have been attracting considerable tourist attention. Recently, the Living Bridge Foundation has added strength to this endeavour.[117] Encouragingly, a few new bridges are under construction. For example, in the village of Rangthylliang, a new bridge is being made with bamboo and wood scaffolding. In Nongriat, a new third span is currently being grown above the existing two.

There is a need to carry out a more in-depth scientific study of the living root bridges in Meghalaya. Rough estimates indicate a tally of nearly a hundred. Dozens of such bridges can be seen in the West Jaintia Hills district. They are also known to exist in Nagaland and in Java and Sumatra. Recently, the district administration of West Garo Hills has shown interest in promoting such bridges in their region.[118] Uniquely characteristic of the southern mountainous regions of Meghalaya, the bridges represent human ingenuity and skill at their best and a living cultural heritage that needs to be preserved and nurtured. Fittingly, these green engineering marvels have now been placed on the tentative list of UNESCO's World Heritage Sites.

32

The Sacred Fig Tree of the Nagas

COMMON NAMES: **White fig (English); Charasü Marabo Kaji (Mao language); Plaksha (Sanskrit); Pilkhan (Hindi)**

SCIENTIFIC NAME: *Ficus virens* **Aiton**

FAMILY: **Fig (Moraceae)**

WHERE TO SEE: **Makhel village, Mao Maram tehsil, Senapati district, Manipur**

LATITUDE: **25.4294° N;** LONGITUDE: **94.1294° E;** ALTITUDE: **Approximately 250 metres**

Makhel has a special place in Naga history, mythology and folklore. It is one of the locations to which many tribes historically trace back the original settlement from where they dispersed over the Naga Hills. Several interesting monuments significant to both Naga tribes and others (Ahom and Meitei) make Makhel a very attractive spot for local tourists. Among these is an ancient fig tree associated with the primaeval Naga mother and hence considered highly sacred.

The origin of Nagas is shrouded in mystery. The fact that there are no written historical records either on their origin or routes of migration to the present places of habitation leaves ample room for various interpretations and is far from settled. Lately, consensus seems to be emerging that they came in waves from Southeast Asia, probably Thailand and Indonesia, through the Indo-Myanmar corridor.[119] One group – comprising the present-day tribes such as Angami, Ao, Chakhesang, Lotha, Mao, Poumai, Seema and Tangkhul – originally settled in Makhel and nearby places in today's Senapati district of Manipur before dispersing to the states of Nagaland, Assam and Arunachal Pradesh, and the contiguous western part of Myanmar. Other tribes trace their dispersal point to other legendary places. It was only after their dispersal from these locations that different tribes came to be known by separate and distinct names.

Locally known as charasu marabo kaji, the tree is wrongly identified as either peepal or banyan in most popular narratives. It is in fact a pilkhan tree (botanically *Ficus virens* Aiton). Towering over the village chief's house with its woodcraft motifs, the tree constitutes a prominent landmark of Makhel. Locals believe it is several centuries old and have cordoned it off in a cemented enclosure paved with flat stones, with seats along its circumference. As is typical for this species, aerial roots arise from the upper part, but unlike in banyan, they tightly hug the main trunk. In this age-wizened tree, the fused aerial roots impart an irregularly furrowed look. At some point in history, one of its main branches must have fallen off after being singed by lightning, leaving a dark hollow in its place. The branches and crevices on the trunk are home to a variety of luxuriously growing lichens, ferns and flowering plants, adding to the general vibrancy of the tree. Stone tablets under the tree explain its significance.

According to a popular Mao Naga legend,[120, 121] the Naga ancestral mother Dziilimosiiro from whom the entire human race descended was resting under this tree. She was suddenly enveloped by clouds and was impregnated with three offspring: her firstborn was Okhe (tiger, representing the animal world); the second Orah (god, representing the supernatural world); and the third and youngest Omei (man, representing the human race). Only with Omei, her youngest and favourite son, was she happy as he nursed her with tenderness and love. After her death, a period of bickering followed over who should inherit the native home. Eventually, it was settled that god would occupy the open spaces and the tiger the forest. Man thus came to retain the native home as per his mother's dying wish. That is why the youngest son gets to inherit the ancestral home in Naga culture.

In the olden days, a dead branch of the Marabo tree was an indication of the direction in which calamity would happen. To avert the calamity, the entire village would observe *Genna*, meaning 'it is forbidden'. Normal life came to a standstill; it was forbidden to work in the fields, travel, have sexual intercourse or even eat certain types of food.

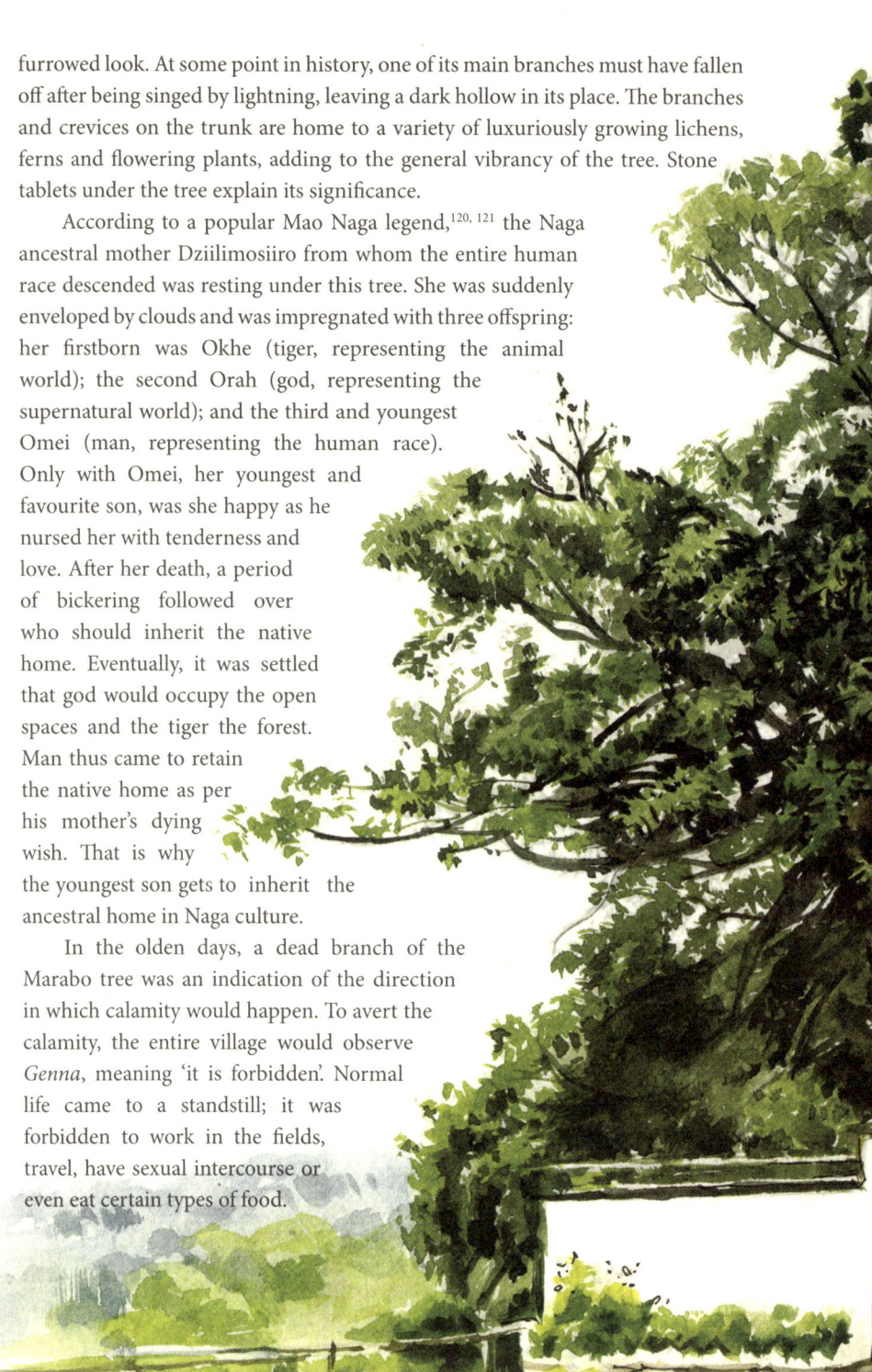

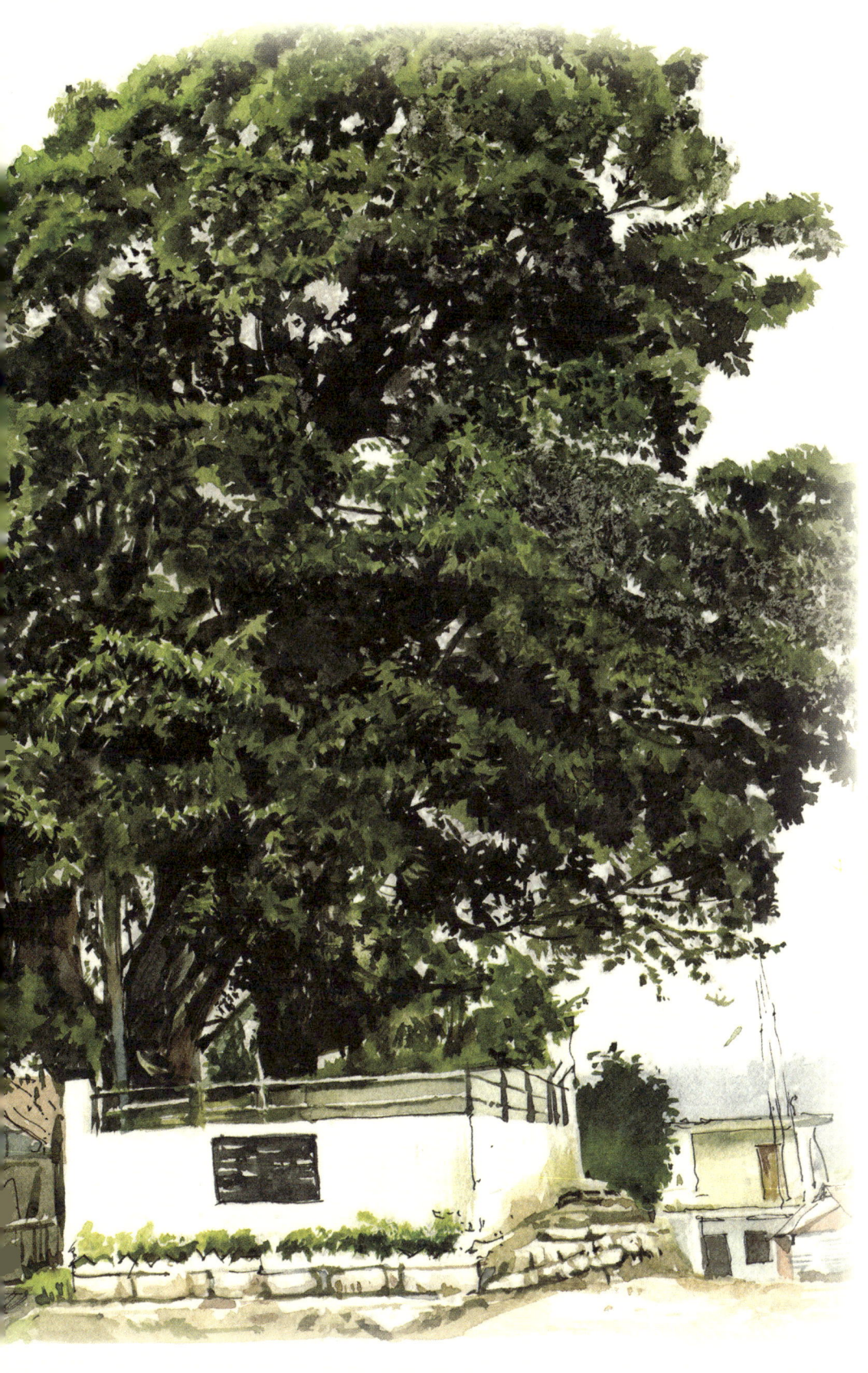

33

The Sacred Wild Pear
Tree of Shajouba

COMMON NAMES: Wild pear (English); Chetibo Kaji /
Chitebu Kaji/Cheti-Bu Kaji (Mao Naga)

SCIENTIFIC NAME: *Pyrus pashia* Buch.-Ham. Ex D. Don.

FAMILY: Rose (Rosaceae)

WHERE TO SEE: Shajouba village, Mao Maram tehsil,
Senapati district, Manipur

LATITUDE: 25.478° N; LONGITUDE: 94.147° E;
ALTITUDE: Approximately 1,290 metres

Excepting the sacred fig tree at Makhel (see page 131), there is no tree more significant to the Nagas than the wild pear tree in the small village of Shajouba in the Senapati district of Manipur. Many Naga tribes know it as chitebu kaji. They hold it in high reverence for it is the historical landmark from where their ancestors migrated to their present areas of habitation. Hence they also refer to it as the 'departure tree'.

I have already mentioned the Naga legend in which the primordial mother Dziilimosiia was impregnated by a white cloud under a fig tree in Makhel and bore three sons. The youngest of these, Omei, representing humankind produced three sons. The youngest of those was the ancestor of the Nagas.[122]

The Naga community multiplied and prospered in the native hilly terrain. Finally, a point was reached when resources dwindled and could scarcely support the burgeoning population. The worried community held an assembly and came to the sad conclusion that members must now branch out of Makhel in search of new fertile lands to find new settlements. Before the imminent parting of ways, the Nagas congregated at Shajouba for a farewell feast and took a solemn pledge that sometime in the future they would all reassemble at this very spot. To commemorate the event, they planted a chitebu kaji or wild pear tree at the place.

According to a Poumai Naga legend, Pou, one of the elders, dug his walking stick in the soil before his departure and it sprouted into the wild pear tree. Thus, the tree is believed to represent the time when the migration of Nagas began from Makhel.

It is believed to be among the oldest trees in all of Naga history. The people who migrated from Makhel/Shajouba settled down in other villages

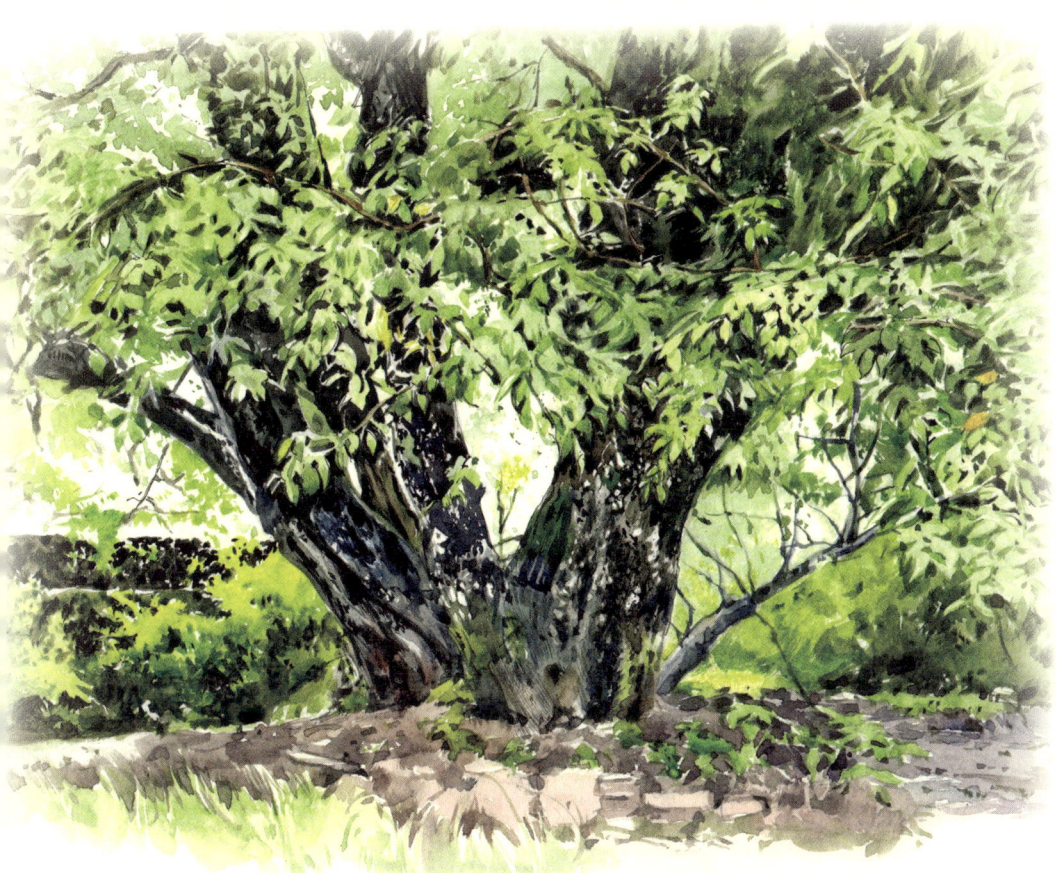

assuming new tribal names. These include Angami, Lotha, Rengma, Chakhesang and Sumi.

The Shajouba tree is strikingly good-looking. Its main trunk is short and hollow and branches arise close to the ground. All around the main trunk, new shoots have shot up from the roots and together they present a grand look. Every year in February and March, its leaves fall off and beautiful white flowers with a slight pink tinge begin to appear on the branches. At that time, the tree is a sight for sore eyes. Small roundish fruits follow in due course. They are initially green but turn a darker shade of brown with white and yellow dots.

Although the fruits are edible, they are not consumed by the locals due to the tree's sacredness. Even the plucking of a single leaf or the falling of a branch in a storm signifies a bad omen. The entire community observes *Genna* (community worship of God). Today, the ancient wild pear tree is sacred to several Naga tribes and stands in Shajouba as a symbol of the oneness and unity of the Naga people.

34

Hātiyan ka Jhad of Golconda Fort

COMMON NAMES: **African Baobab** (English); **Hātiyan ka Jhad** (Urdu); **Gorakh Imli** (Hindi)

SCIENTIFIC NAME: *Adansonia digitata* L.

FAMILY: **Hibiscus (Malvaceae, Bombacoideae)**

WHERE TO SEE: **Naya Qila, Golconda Fort, Hyderabad, Telangana**

LATITUDE: **17.2357° N**; LONGITUDE: **78.2466° E**; ALTITUDE: **518 metres**

Naya Qila is an extended portion of the Golconda Fort, a fortified citadel in Hyderabad. Built in 1656 by Sultan Abdullah Qutb Shah (1626–1672), the seventh ruler of the Qutb Shahi dynasty, it was meant to add strength to the fort against Mughal attacks. That the Mughals under Aurangzeb still succeeded in annexing the Shahi Kingdom in 1687 is another story. Close to the ruins of the Naya Qila is an enormous baobab tree – locally known as hātiyan ka jhad (literally elephantine tree) – now considered the largest such tree outside Africa.[123]

Baobabs belong to the botanical genus *Adansonia* of which there are eight species – six[124] being endemic to Madagascar, one native to mainland Africa and part of the Arabian Peninsula and another to north-western Australia. Of these, the African baobab is the best known and has been introduced throughout the tropics.

The tree's current height exceeds 19 metres and the trunk has a circumference of 25.48 metres, making it by far the biggest baobab tree outside Africa. A popular local legend claims that 40 infamous thieves had made the tree's commodious and cavernous trunk their home; they hid there by the day and carried out their odious activities by night. Systematic studies by Professor Adrian Patrut and his team show that all large African baobabs with a girth exceeding 15 metres are multistemmed. Although baobabs usually start life as single-stemmed, they generate new stems periodically. In large trees, the stems fuse to form a ring leaving a big space in the middle. The fusing stems are of different ages and hence different generations. The space is a false cavity, having never been filled with wood. If the stems fuse perfectly over the years, the cavity may be 'closed' but in others, it is 'open'. Some large baobabs have two false cavities. This enables African baobabs to survive for many years and become large. Because of this complex structure, baobabs do not lend themselves to age determination through the counting of annual growth rings. Using AMS radiocarbon dating, Patrut's group found that

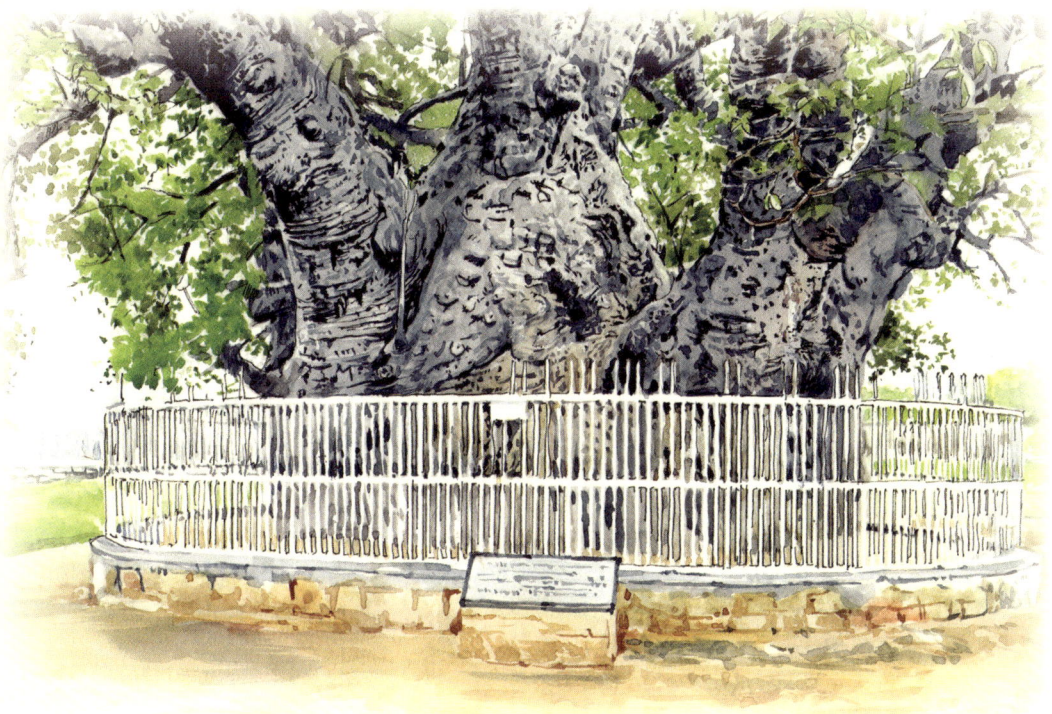

the oldest African baobabs could reach ages greater than 2,000 years,[125] making them the longest-living angiosperms.

It turns out that the Golconda baobab has eight stems – six of them fused to form a ring in such a way that there are two false cavities of unequal size and two stems free outside the ring structure. Between 4 and 4.5 metres of height in the trunk, there are two openings corresponding to the two false cavities; the opening of the larger cavity is wide enough to permit an adult human to enter and descend a staircase fashioned out of the bark inside. That of the smaller cavity is too narrow to permit entry. However, the two cavities are interconnected below.

The age of this tree giant has been estimated to be 475±50 years.[126] Thus, it must have been planted around 1540 CE. The Naya Qila itself was built a century later in 1656! Let us hope this arboreal monument of Hyderabad lasts for centuries to come.

35

Thimmamma's Gigantic Banyan

COMMON NAMES: **Banyan (English); Nyagrodha (down-growing) and
Bahupada (many-footed) (Sanskrit); Thimmamma Marrimanu (Telugu);
Vat Vriksh, Bargad (Hindi)**

SCIENTIFIC NAME: *Ficus benghalensis* L.

FAMILY: **Fig (Moraceae)**

WHERE TO SEE: **Gootibylu village, near Kadiri, Anantapur district,
Andhra Pradesh**

LATITUDE: **14.0282° N**; LONGITUDE: **78.3256° E;**
ALTITUDE: **576.8 metres**

The dusty town of Kadiri is likely to experience scorching temperatures of
40 degrees during the summer. The hillocks that border the town on its
northern and eastern sides look parched and brown with scrubby growth.
Unmindful of the heat and the dust, the visitors – almost invariably pilgrims –
come here to offer prayers at the fourteenth-century temple of Lakshmi Narasimha
(the fourth incarnation of Vishnu), or to visit the tomb of the Telugu poet-saint
Yogi Vēmana nearby. Most of them are unaware that just 25 kilometres away
to the southeast of the town, there is an awe-inspiring living green monument
that entered the *Guinness Book of World Records* in 1989 because it is the largest
plant on this planet. The record holder is a banyan tree locally referred to as
Thimmamma marrimānu or Thimmamma's banyan tree.

The tree derives its name from a local legend: Thimmamma, a woman said
to have lived here during the first half of the fourteenth century was so devoted
to her husband that when he died in 1434, she jumped into his funeral pyre to
commit *sati*. To the amazement of the assembled people, one of the banyan poles
on the pyre started sprouting buds and leaves and grew into a banyan tree. Since
then, Thimmamma is revered as a goddess and a shrine has been built for her
under the tree. Locals believe that childless couples who pray at the shrine will
be blessed with a baby. An annual *jatra* on Maha Shivaratri attracts hundreds
of people.

Banyans are the largest of 881 accepted species of figs characterized
by branches that are spreading and almost horizontal, sending down aerial
roots to the ground below. Having obtained a foothold in the soil, these roots
develop into additional trunks, enabling the tree to spread out in all directions.
supporting a large canopy. Over the years, Thimmamma's banyan has attained
unmatched proportions: Its canopy covers 19,107 m^2 (close to two hectares)

– an area equalling 2.70 football fields that can probably accommodate 20,000 people under it, making it the largest tree by two-dimensional canopy coverage area.[127] It is also the world's largest tree with a perimetre length of 846 metres.

Standing under its shade is like being in a mini-forest. The temperature is a couple of degrees cooler than the ambient. The floor is littered with a thick carpet of fallen leaves that crunch under your feet. Since the tree produces reddish, fleshy figs, insects, birds, squirrels and other mammals chirp and chatter on it continuously. Marrimanu is a green oasis in the reddish dusty desert of Gootibylu. Indeed, there is no tree more awesome to behold.

Going by the legend, it is claimed to be 600–700 years old. That said, Marrimanu's fame has not spread far and visitors are not many except during the Maha Shivaratri *jatra*. With such a living treasure in hand, it is surprising that the state government has neither created better awareness about the tree nor good access. Since 1990, the local forest department and concerned village officials have joined hands to protect and preserve the tree by fencing off the tree's periphery. This might keep stray cases of vandalism at bay for the time being. Nevertheless, the arrival of more tourists will likely bring additional problems for the tree's safety. But for now, the tree continues to grow, unaware of these problems or despite them.

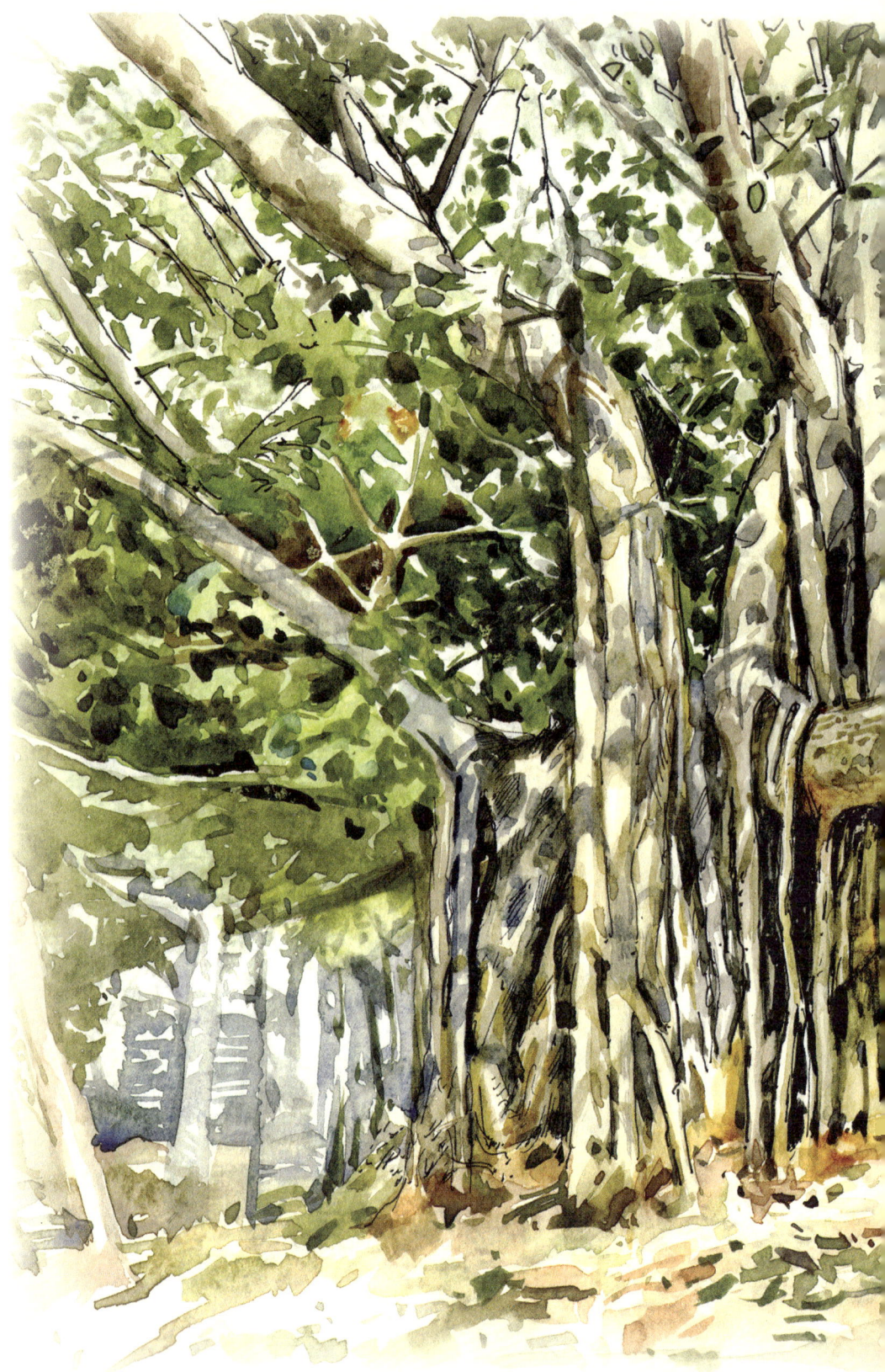

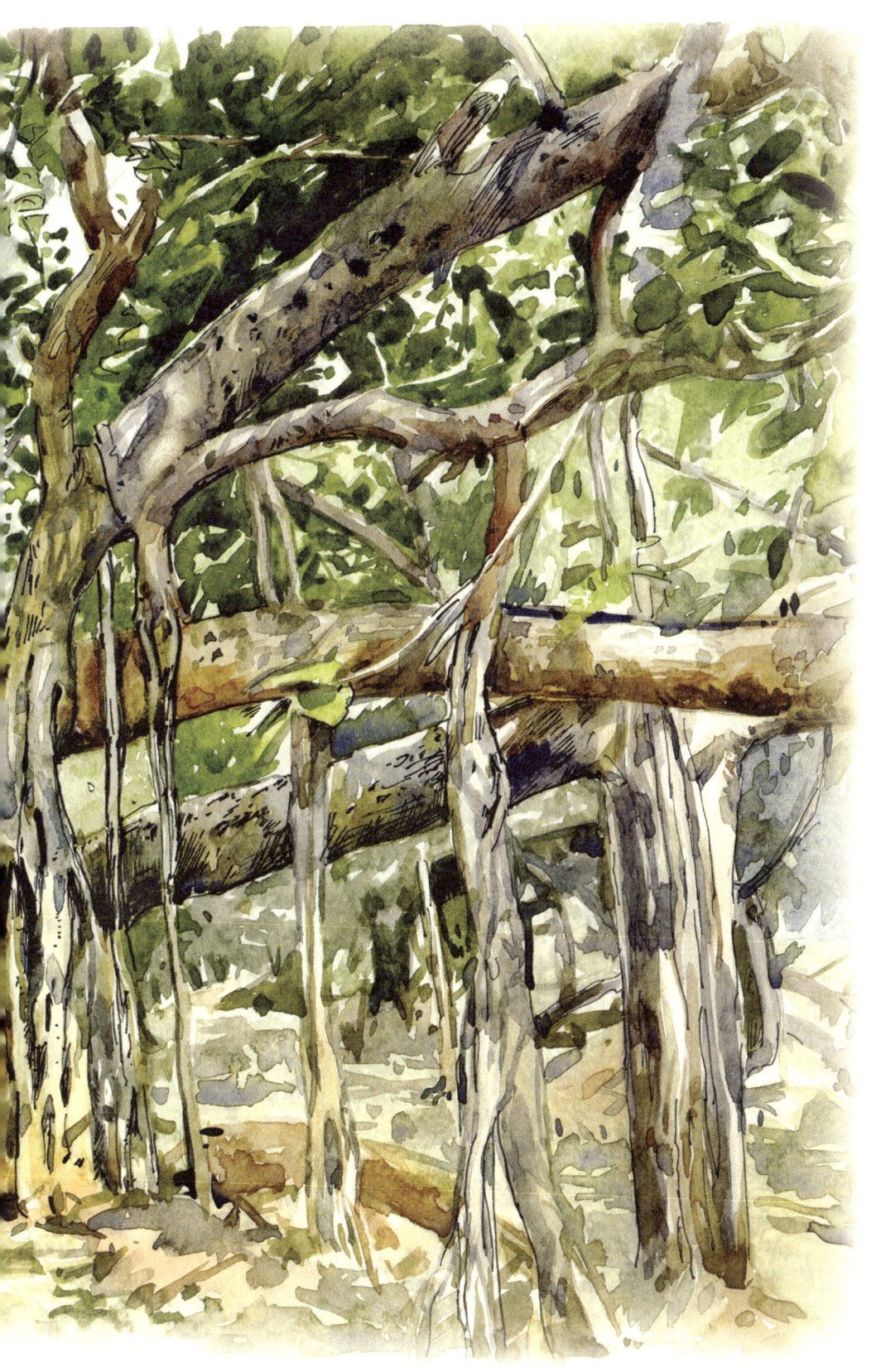

36

Horsley's Eucalyptus

COMMON NAMES: **Blue Gum, Southern Blue Gum, Tasmanian Blue Gum (English); Neelagiri Chettu (Telugu); Safeda (Hindi)**

SCIENTIFIC NAME: *Eucalyptus globulus* Labill.

FAMILY: **Guava (Myrtaceae)**

WHERE TO SEE: **Next to W. Horsley's bungalow at Horsley Hills, Chittoor district andhra Pradesh**

LATITUDE: **13.39° N**; LONGITUDE: **78.25° E**;
ALTITUDE: **1,290 metres**

An old eucalyptus tree stands at the summit of Horsleykonda or Horsley Hills, a small hill station in the Chittoor district of Andhra Pradesh. A board at its base declares that the tree was planted by William D. Horsley, the collector of the Cuddappah district in the erstwhile Madras Presidency in 1859.

During one of his travels on horseback in these parts in 1840, Horsley decided to explore the hills – the tallest of which was not bare like the others, but full of trees. The view was excellent from the height and the climate was delightful. Most importantly, the place was free from the fever that was prevalent everywhere in the district. Charmed by its beauty and climate, Horsley decided to build his summer house there – something that was easier said than done for the hill was for a long time supposed to be haunted by demons. The ravines were inhabited by wild animals such as hogs, bears and even an occasional tiger. It was with the greatest difficulty that workmen could be persuaded to go up.[128] Soon the building began taking shape.

Horsley must have planted the blue gum adjacent to the house even before it was completed. There are no records that explain why he chose this particular species or from where he obtained the seed or seedling. By that time blue gums were already in India for nearly two decades, having been introduced by the British themselves in the Nilgiri Hills as a source of fuel wood, starting from the 1840s.[129] By 1856, regular plantations were already available. It is, therefore, likely that Horsley obtained it from the Nilgiris.

Today, Horsley's tiny sapling has grown to a stately tree of 40 metres in height with a trunk girth of 4.7 metres. The bark on the trunk and branches can be seen peeling characteristically in large strips. Incidentally, eucalyptus means 'well covered' and refers to the typical and highly distinctive flowers in which the petals are modified into a cap-like structure technically known as the operculum. The operculum falls off as the flower matures, exposing numerous showy stamens that take over a decorative role. The fruits are woody, cone-shaped and release seeds through valves at the tip. Horsley's Tasmanian blue gum is healthy and produces flowers and fruit copiously. Today, Horsley's bungalow is used as a guest house by the state forestry department. The tree has been decorated with the Mahavriksha Puraskar in 1995 by the Government of India.

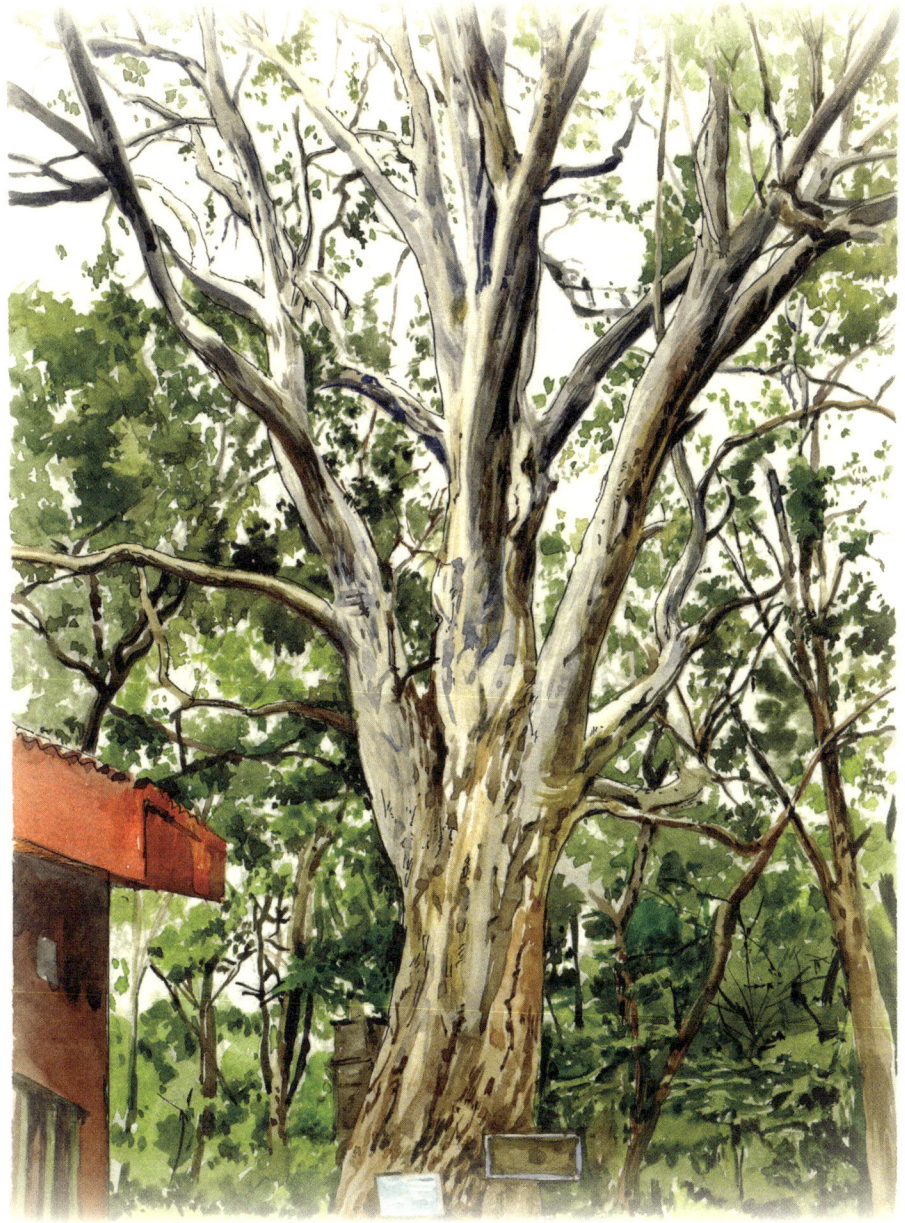

37

The Sentinel Baobabs of Yogapur

COMMON NAMES: Upside-down Tree, Monkey Bread Tree (English); Gorakshi (Sanskrit); Devva Hunasé Mara, Huccha Hunasé Mara (Kannada); Gorakh Imli (Hindi)

SCIENTIFIC NAME: *Adansonia digitata* L.

FAMILY: Hibiscus (Malvaceae, Bombacoidae)

WHERE TO SEE: Dargah at Yogapur on Sindagi road, to the northeast of Vijayapura, (formerly Bijapur), Bijapur district, Karnataka

LATITUDE: 16.4947° N; LONGITUDE: 75.4510° E;
ALTITUDE: Approximately 592 metres

On the outskirts of the historic city of Vijayapura – formerly Bijapur – is a village called Yogapur. It has a dargah close to a mosque and, nearby, an exceptional but contrasting pair of African baobabs. Two tombs – dedicated to Sufi saints Hazrat Syed Shah Imamuddin Quadri and Hazrat Syed Shah Abdul Gafoor Quadri – are housed in the small dargah. The mosque stands to the north of the dargah. Towards the south of the dargah in an open field, is a stunningly beautiful baobab that rises to about seven metres in height. It stands out from a distance because of its squat trunk (over nine metres in girth) and beautiful, umbrella-shaped crown. One can see the great Gol Gumbaz framing the background of the tree. Its bark is greyish brown and folded and pitted from years of growth. The tree regularly bears pendulous white blossoms on long drooping stalks.

The second baobab stands further north behind the mosque and is not visible unless you walk a little distance. It could be as old as the first tree but looks more like its poorer cousin. Located next to a row of houses, its swollen trunk is too close to the wall. Its spreading branches have been chopped off more than once as they overarch the roof of a house, imparting the crown a distorted, mutilated, even forlorn look. Nonetheless, it seems to be holding on and, at the time of my visit, gamely bearing many fruits. Baobabs being native to – and even a defining icon of – the African bushland, you would be curious to know how these trees came to be here. Local people call them devva hunasé or huccha hunasé (devil is tamarind or mad tamarind in Kannada). Nobody is sure when these trees were planted or by whom, but they are believed to be about 350 years old.

Bijapur was under the Adil Shahi rulers for over two centuries (1480–1686). Interestingly, the distribution of baobabs in India coincides remarkably with areas under Muslim control[130] and the Adil Shahi rulers, a Shia Muslim dynasty, did have contact with Arab and other traders as well as maintained an army of African slaves.[131] Rather than Arab traders, it must have been the African migrants that brought the seeds and fruits of the tree that has great economic, cultural and artisanal significance to them in their homeland. In the absence of scientific investigation, there are different stories regarding their presence here. According to one version, the Adil Shah sultans, being fond of things natural, obtained these trees from Turkey. Another even more fascinating legend[132] claims that around 400 years ago, an ogre tyrannized the people of Yogapur. The tormented citizens clandestinely petitioned a Persian Sufi fakir, Syed Shah Imamuddin Quadri, who was camping nearby. He duly arrived with his wooden staff and challenged the ogre to a duel. The ogre pounced on the holy man and a pitched battle ensued at the end of which the fakir managed to kill the ogre. The grateful people urged the fakir to stay with them permanently. Gently refusing their request, the holy man broke his staff into two and planted the stumps on the northern and southern sides of the mosque. Prophesizing that each stump would grow into a baobab tree, he stated that as long as the two trees stood guard, no evil would befall the people of Yogapur. And so the two trees stand there like alert sentinels protecting Yogapur even today. The saint's tomb – along with that of his brother – is still worshipped here.

After the sudden demise of the majestic (and more famous) baobab near Ibrahim Roza in 2015, the Yogapur baobabs are probably the only relics of the Adil Shahi era in Bijapur.

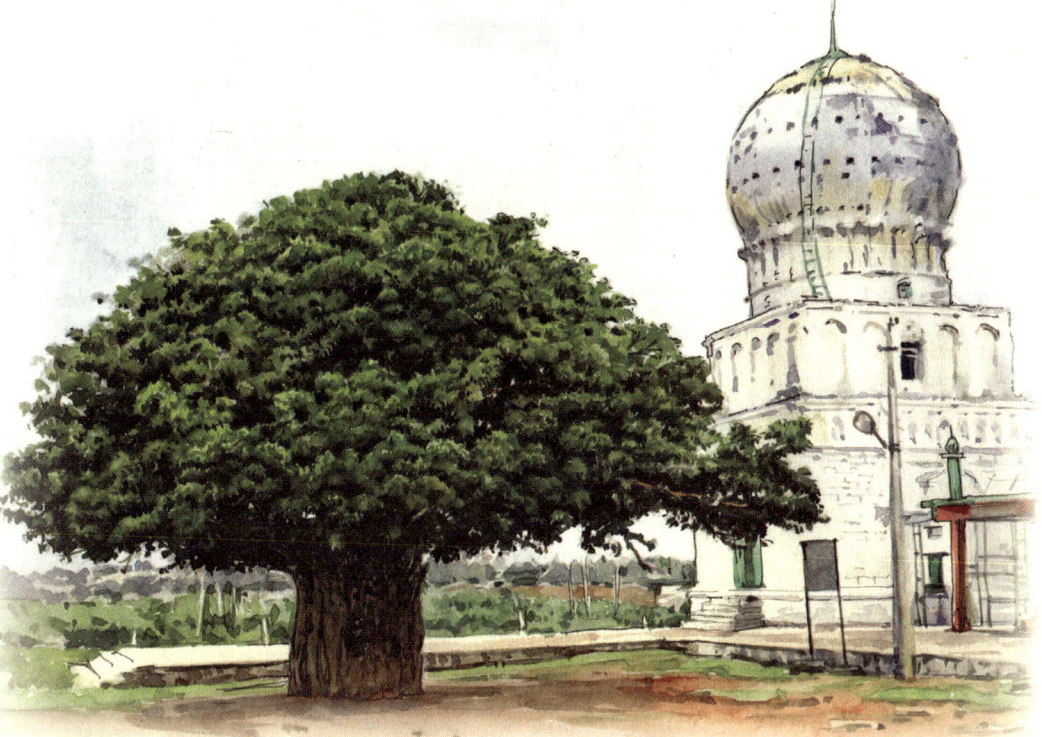

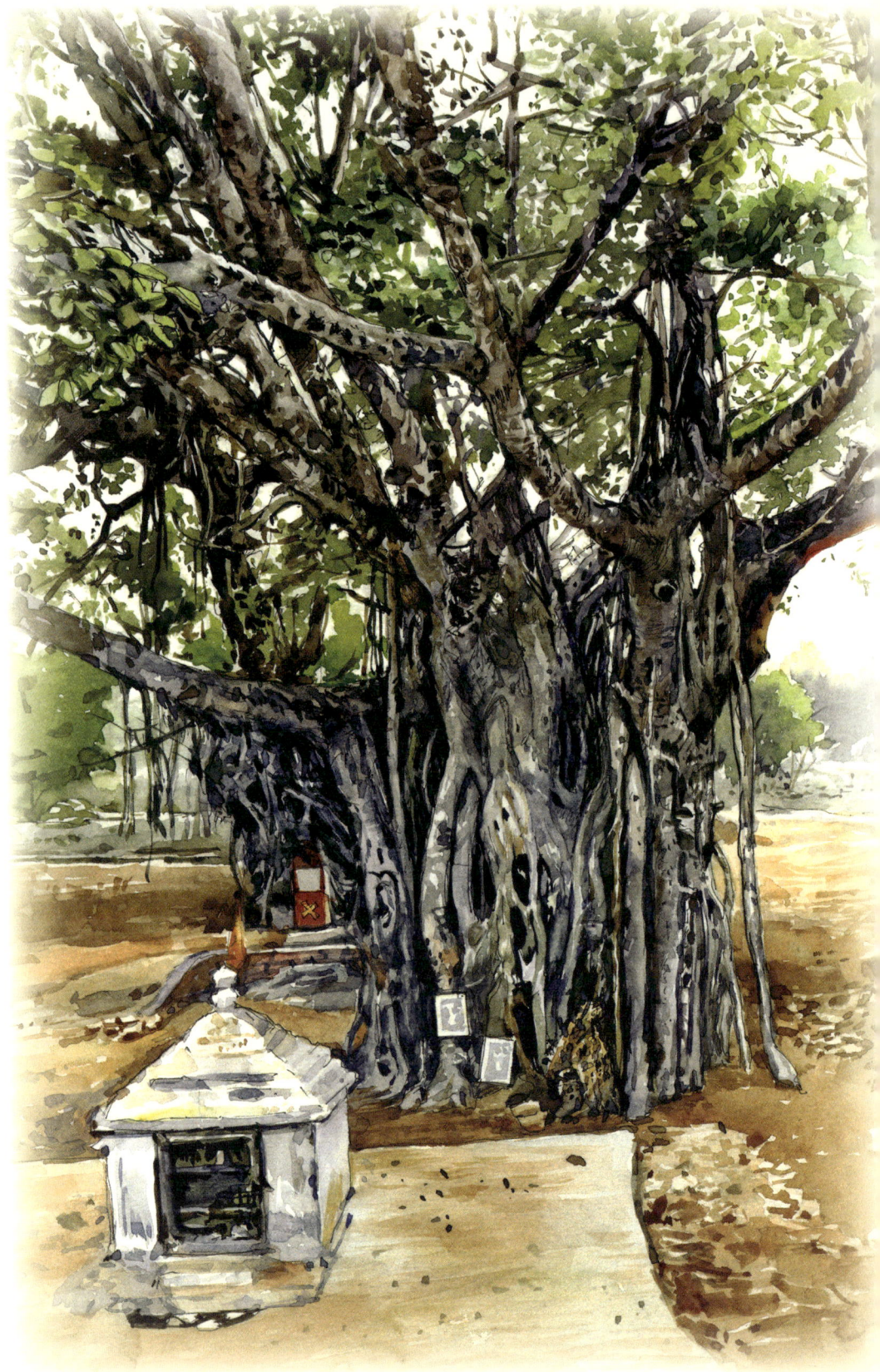

38

Sangolli Rayanna's Banyan Trees

COMMON NAMES: **Banyan (English); Nyagrodha (down-growing) and bahupada (many-footed) (Sanskrit); Ālada Mara (Kannada); Vat Vriksh, Bargad (Hindi)**

SCIENTIFIC NAME: *Ficus benghalensis* L. | FAMILY: Fig (Moraceae)

WHERE TO SEE: **Nandgad village in Khanapur taluk, Belagavi district, Karnataka (The samadhi and the place of hanging are roughly 4 km apart.)**

TREE AT SITE OF HANGING: LATITUDE: **15.041° N**; LONGITUDE: **74.577° E**; ALTITUDE: **Approximately 1,171.3 metres**

TREE AT THE SAMADHI: LATITUDE: **15.578° N**; LONGITUDE: **74.567° E**; ALTITUDE: **Approximately 1,171.3 metres**

The samadhi of Sangolli Rayanna, Karnataka's icon of anti-imperialist rebellion, who rose in revolt against the British in the nineteenth century, as well as the place where he was publicly hanged by the East India Company in 1830, have banyan trees that are central to the sites.

Kittur, a small principality in today's northern Karnataka, had more powerful kingdoms of the Mughals and the Peshwas in its neighbourhood. Even so, Raja Mallasarja led it with impeccable integrity and distinction, managing to escape the clutches of the Peshwas. His son Shivalingarudra Sarja who followed him to the throne was not an astute administrator and during his reign, Kittur became a vassal state of the British colonial powers. Since he was childless, Shivalingarudra decided to adopt a boy, but unfortunately died in 1824 before the formal adoption could take place. Therefore, Rani Chennamma, Shivalingarudra's stepmother organized the adoption ceremony just before performing his final rites and pressed the claims of the adopted boy, wishing to administer the territory as his regent until he came of age. The British, of course, did not recognize the adoption and recommended the annexation of the principality as there was no successor. Chennamma, however, refused to surrender. Wearing battle gear and leading the charge, she rose in revolt and succeeded in routing the British forces that laid siege to the fort. Within a month, however, the British struck back. With a larger army and a bit of treachery, they took Chennamma and many of her loyal followers captive. Kittur was annexed to the Bombay Presidency and the rani was imprisoned in Bailahongal Fort where she died in confinement in 1829.[133, 134,]

An armed retainer of Shivalingarudra, Sangolli Rayanna was born into a family of shepherds in 1798. Having rushed to Kittur in response to the call of his queen, he was among the people arrested during the uprising and later released with a warning on grounds of general clemency in 1826. A witness to Chennamma's imprisonment, Rayanna was greatly influenced by her passion and longing for the repossession of Kittur. Anger against the colonial masters increased manifold on his return to Sangolli. New land regulations under the British regime had resulted in half of his property being confiscated and a heavy tax being imposed on the rest. A fight with the local kulkarni (an official who maintains tax and village records) forced him to leave his native village. Rayanna then organized a band of rebels with others who were similarly marginalized, with the avowed intention of restoring Kittur to Chennamma's grandson. In 1829, his group – experts at guerrilla warfare – started targeting government offices and the treasury and destroying tax records. Their daring acts against the mighty British struck an emotional chord with the locals and soon became the stuff of legends. From a minor irritant to the Company, Rayanna had suddenly turned into a festering sore that had to be surgically removed quickly. In the end, betrayed by a local landlord, the British caught and hanged him from a banyan tree on the outskirts of Nandgad village on 26 January 1831. As per his last wishes, he was buried on the eastern side of Nandgad village, about four kilometres from where he was hanged.[135] According to local legend, one of his close friends planted a banyan tree that stands on his samadhi today.

The tree on which Rayanna is believed to have been hanged is large, with many well-entrenched columnar roots and could be close to 250 years old. Someone has built a small shrine below the main trunk. The district administration has acquired 4.85 hectares of land surrounding the tree and work seems to be in progress to make it a tourist spot. At present, the number of people visiting the spot is small. Rayanna's story is an integral part of Karnataka's folklore. Many ballads celebrate his devotion to Kittur and its queen Chennamma as well as his heroic exploits against the British. The banyan trees that stand at the two sites are living and tangible reminders of his indomitable patriotic spirit that refused to be cowed down by colonial powers.

39

Maulsari Mahavriksha of Dudhgali

COMMON NAMES: **Bulletwood, Indian Medlar, Spanish Cherry (English);**
Bakula (Kannada); Omval (Konkani); Maulsari (Hindi)

SCIENTIFIC NAME: *Mimusops elengi* L.

FAMILY: **Chikoo (Sapotaceae)**

WHERE TO SEE: **Adjacent to Vitthal Rakhumai temple, Dudhgali, Nujji, Joida**
(or Supa) taluk, Uttara Kannada district, Karnataka

LATITUDE: **15.02275° N;** LONGITUDE: **74.21234° E;**
ALTITUDE: **592 metres**

D udhgāli gets its name from a delicious combination of two languages – *dudh* (milk in Marathi) and *gāli* (breeze in Kannada). Part of Joida taluk in the Uttara Kannada district of Karnataka state, the entire area is surrounded by dense forests constituting the Kali Tiger Reserve, earlier known as Dandeli-Anshi Tiger Reserve. The focal point of the hamlet is the Vitthal-Rakhumai temple, built on land owned by the Desai family with contributions from the local community. The Desais are also the hereditary priests here. Right in front of the temple is a tall and stately bakula tree that commands the visitor's immediate attention. This tree bagged the Mahavriksha Puraskar of the Government of India in 1997.

Native to South Asia (Western Ghats, Sri Lanka) Southeast Asia and northern Australia, maulsari is also widely cultivated along avenues and as an ornamental evergreen tree all over India. The generic name *Mimusops* interestingly means 'monkey-faced' in what is probably an imaginary reference of the flower to a monkey. The specific epithet *elengi* refers to the common name of the plant in Malayalam. Not a fast-growing species, its height rarely exceeds 15 to 16 metres. This majestic specimen, however, appears to be over 18 metres tall and is reputed to be the tallest such tree in India.[136] Devoid of branches up to seven or eight metres from the base, the trunk has dark brown or greyish, deeply fissured and cracked bark, as is typical for the species. At 4.51 metres, the tree's circumference catches the eye. The somewhat dome-shaped, densely foliated crown seems to be slightly tilted to one side.

During March, April and May, the tree is full of fragrant white to slightly off-white flowers that are offered to the temple deities. The flowering season also happens to coincide with the wedding season around here. Flowers left over after worship are collected by the local people to make garlands for brides and grooms. The fragrance of the flowers can be retained for years if they are sun-dried and

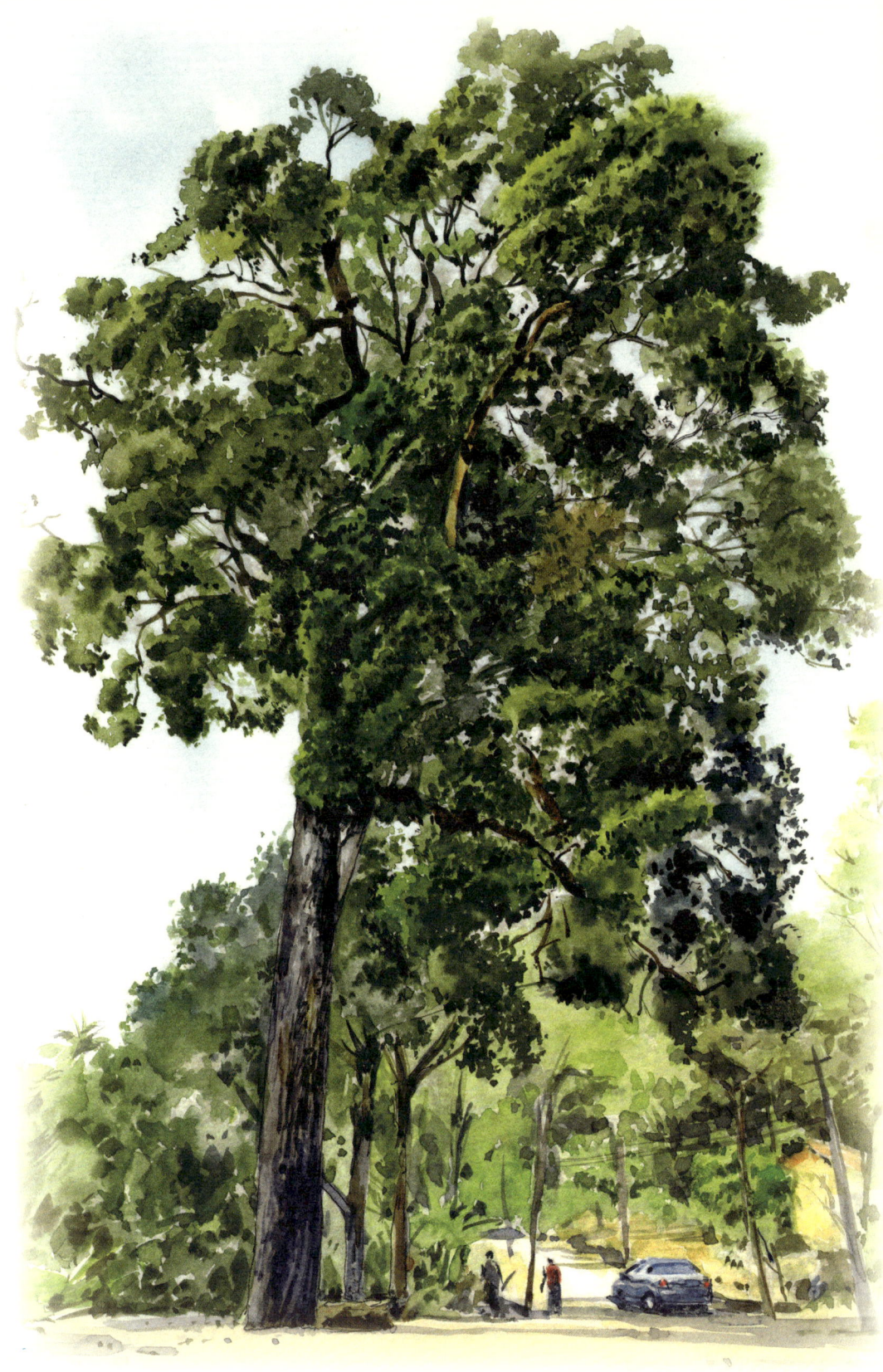

preserved properly. A deep-rooted popular belief among the Desais, Kunbis and Brahmins here holds that a wedding is not solemnized until the bride and groom have exchanged garlands of bakula flowers. So much so, that weddings held during the seasons when fresh flowers are unavailable use dried garlands. The fruiting season follows the flowers but the tree does not produce very many fruits. Although the timber of the species is highly valued, no one has harmed the tree over the years.

At the base of the tree, there is a short stone column symbolizing the *grām devta* or village deity known as *khuti*. Enclosed by a roofless circular enclosure open on one side, local people offer their first harvest to the deity and seek its blessings to ward off evil spirits, protecting people, crops and livestock from pests and harmful diseases.

Originally from Goa, the Desais came here a long time ago. It was in 1942 that an ancestor felt the presence of God here. The images of Vitthal and Rakhumai were ceremoniously installed in 1957 by the father of the present priests. By 1962, a small temple was built. Soon its fame spread in the neighbourhood through word of mouth. The present expanded-and-renovated structure was completed in 2014 with contributions from devotees from the surrounding villages. The bakula tree has always been there for at least three generations of Desais, standing near the temple like a benediction from the deities. It would be nice to see a plaque providing some information on the tree including the fact that it is a recipient of the prestigious Mahavriksha Puraskar.

Let them know we will not allow the felling of
a single tree. When their men raise their axes,
we will embrace the trees to protect them.

Chandi Prasad Bhatt
Environmentalist and social activist

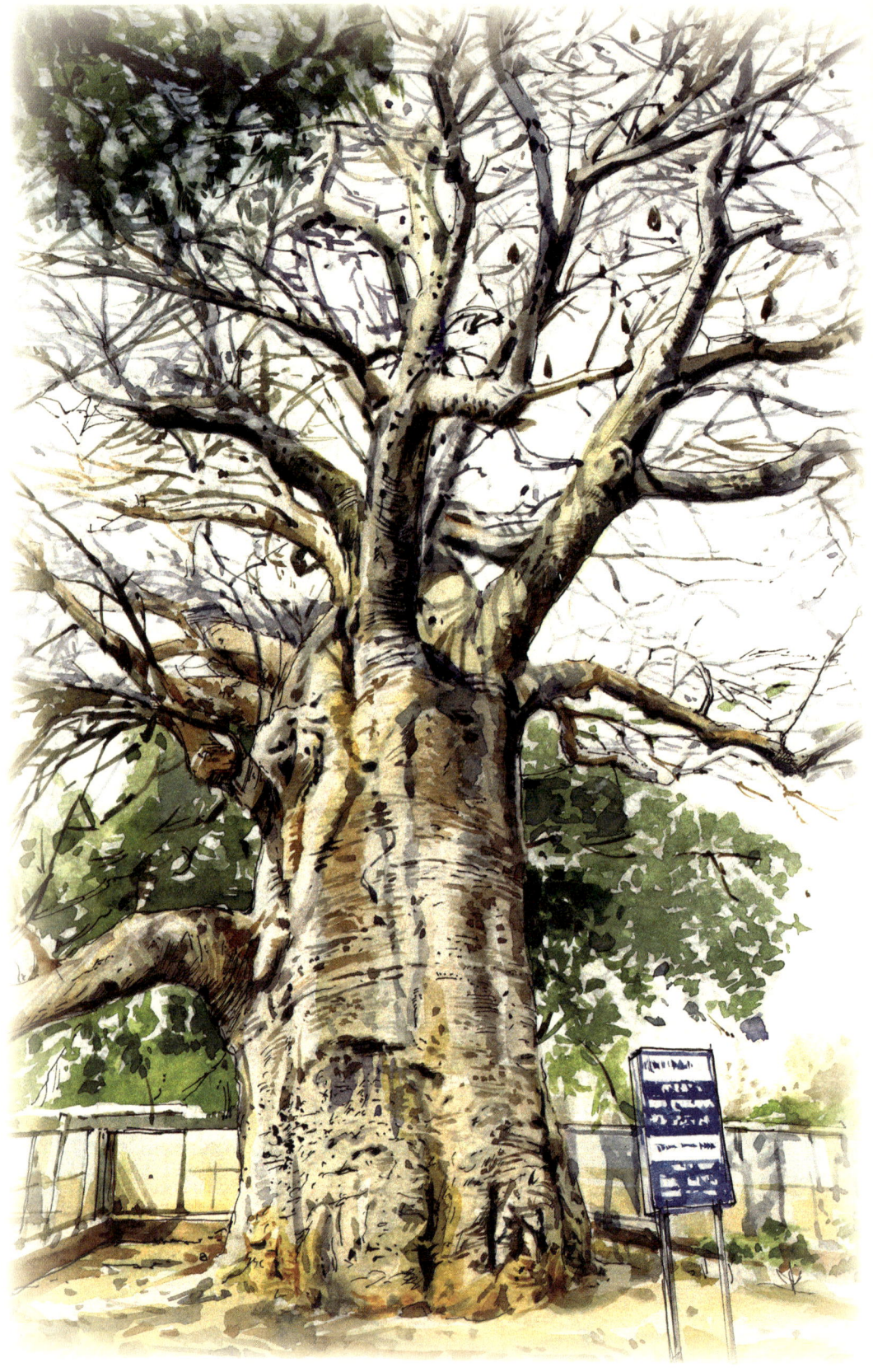

40

Three Baobabs of Savanūr

COMMON NAMES: Baobab (English); *Gorakshi* (Sanskrit); *Dodda Hunasé Mara* (Kannada); Gorakh Imli (Hindi)

SCIENTIFIC NAME: *Adansonia digitata* L.

FAMILY: Hibiscus (Malvaceae, Bombacoideae)

WHERE TO SEE: Dodda Hunasé Kalmatha, Savanūr, Savanūr taluk, Haveri district, Karnataka

LATITUDE: 14.9791° N; LONGITUDE: 75.3332° E; ALTITUDE: Approximately 573 metres

S avanūr offers two things that delight visiting gastronomes: a home-grown variety of paan (betel leaves) popular all over the Haveri district and beyond and a type of savoury locally known as khāra. Its real treasure, however, comprises three massive baobabs that are tucked away in a little monastery known as Dodda Hunasé Kalmatha on the edge of the town.

The monastery has been running a school for the last 20 years. As you enter the property, the baobabs can be seen at the far-left corner planted in a triangular formation – two in front, with the third behind them. Enclosed in a metal fence – presumably to prevent people from carving graffiti on the trunks or otherwise damaging the trees – their size and shape make an immediate impact. The trunks are huge, bottle-shaped and rather stubby, tapering suddenly as is typical for baobabs. Their grey bark is wrinkled in places, reminiscent of an elephant's skin. The large crowns regularly bear flowers and fruit and support a colony of fruit bats. The stoutest tree has a girth of approximately 16.20 metres, followed by the other two at approximately 13.70 metres and 12.40 metres.

There are no historical records to explain how these strange African trees came to Savanūr. Legend has it that Lord Krishna brought the seeds from Africa and planted them here. Barely 40 kilometres away is the historic town of Hānagal, once part of the ancient kingdom of the Kadamba dynasty. Medieval inscriptions reveal that it was known as Virātakoté and Virātanagari (i.e., the fort and the city of Virāta).[137] *Virāta* of *Mahabharata* fame was the king at whose court the Pandavas spent their thirteenth year of exile incognito. According to lore, Krishna had come to visit his friends there. Local people believe that the trees are over 2,000 years old.

As fascinating as this lore is, a more reasonable estimation of their age would be 350-400 years. You would recall that baobabs have been repeatedly introduced into the Indian subcontinent through the Indian Ocean trade routes by African travellers who carried dried baobab pods as a source of nourishment. In due course, these people came to be collectively known as Habshis or Sidis (Siddis/Sheedis). It is no coincidence that baobab trees in India are predominantly found in locations where the Sidi people settled.

Savanūr has been in existence as a separate province for nearly 350 years. Founded by an Afghan from Kabul in the service of the Adil Shahi regime, it was later taken over by the Mughals. The Marathas chiselled away much of its territory and the remainder was annexed by Tipu Sultan of Mysore. After the collapse of the Maratha empire in 1818, Savanūr province became a protectorate of the British East India Company with its original territories restored.

Interestingly, there has always been a concentration of African Indians in the neighbourhood of Savanūr. Even today, Sidi settlements are seen in places such as Yellapura, Haliyal, Ankola, Joida, Mundagod, Khanpur and Kalghatgi[138, 139] – all within a 150 kilometre radius. Some of the Sidis had occupied prominent positions in governance and polity. The first Nawab of Savanūr, Abdu'l Karim Khan I (r. 1672–1686), had married the daughter of a Sidi chief Masaud Khan, a general in the Adil Shahi sultanate. Sidi Masaud's services were requested in 1678 to defend the Bijapur kingdom when it was under attack by the Mughals,[140] and he obtained Bagalkot as dowry when he became Sidi Jauhar's son-in-law.[141] Seen in this light, the presence of three grand African baobabs in Savanūr is hardly surprising. It would, of course, be important to investigate their genetic history and the correct age to validate these inferences.

It is because of the religious connotation attributed to baobabs, the Savanūr baobabs are safe within Kalmatha. Much has been written in the popular media about them, resulting in the visit of curious people. Even so, according to the current chief of Kalmatha, there has been neither recognition nor support from the state government.

41

A Sultan's Legacy – The Gums
of Nandi Hills

COMMON NAMES: **Gums (English); Nilagiri (Kannada); Safeda (Hindi)**

SCIENTIFIC NAME: *Eucalyptus* L'Her.

FAMILY: **Guava (Myrtaceae)**

WHERE TO SEE: **Nandi Hills, Chikkaballapur district, Karnataka**

LATITUDE: **13.39° N;** LONGITUDE: **77.70° E;**
ALTITUDE: **1,478 metres**

A little less than an hour's ride from the bustle of Bengaluru is the well-known hill fort of Nandi Hills. Once known as Nandidurga – the British called it Nandidrug or Nandidroog – the name celebrates the ancient temple atop the hill with a 1,000-year-old stone sculpture of Nandi, the sacred bull of Lord Shiva. Among the many attractions here are the imperial-looking gum trees dotting the hilltop. Standing straight and tall, they are not only the largest trees in the entire area but also represent the first-ever introduction of gums or eucalyptus into India. Displaying wide variation in bark, flower and leaf characters, these trees have played a historic role in popularizing gums in several parts of the country.

During the second half of the eighteenth century, the province of Mysuru was ruled by the Muslim adventurer Hyder Ali until 1782 and later his more illustrious son Tipu Sultan, popularly known as the 'Tiger of Mysore'. Tipu built a fort on Nandidurga, considered impregnable by many. Located at an elevation of 1,478 metres with an annual rainfall of 750 mm, the climate on the hill was very congenial. The sultan also constructed a summer residence where he would occasionally stay. A keen horticulturist, he had imported several exotic plants from distant parts of the globe into his dominion. One such introduction – sometime between 1782 and 1799, most probably around 1790[142] – was that of several species of gums.[143] He had them planted in his garden atop the Nandi hill, including some (especially blue gums) in front of his summer resort. However, no official records are available to show exactly when these plants arrived here or to what species of gums they belonged. Regrettably, we shall probably never know why Tipu Sultan obtained them or whether they were sourced directly from Australia or through Mauritius, or even if they were gifted to him by his French allies or Dutch traders.[144] Unfortunately, he got embroiled in successive Anglo-Mysore wars and was killed in 1799.

Gums have had a history of fluctuating fortunes in India and have been the subject of long and often contentious debate as to their negative impact on the environment.[145] Thus, one way or another, Tipu's introduction can be viewed as having had a long-term implication on India's ecological history.

For nearly 150 years, the gums kept growing quietly on Nandi Hills without the species even being scientifically identified (it was not until 1954–55 that herbarium specimens of the mother trees were sent to Australia and 16 species were identified[146]).

As early as 1937, small samples of seeds collected from these trees were used for sporadic cultivation as ornamentals in rest house premises, public gardens and roadsides all over Mysuru province. It turns out that these were hybrid seeds involving a mixture of four species.[147] The early 1950s saw the Mysore hybrid/ Mysore gum finding its way into forestry practices in Karnataka. The hybrid gum soon became popular not only in Karnataka but also in other parts of the country due to its unique abilities to grow fast, tolerate drought and thrive under widely varying geographic conditions. By the 1960s, Karnataka was supplying nearly five tonnes of Mysore gum seed each year to the rest of the country.

As mentioned earlier, one of the spots where the sultan had planted blue gum trees was in front of his summer residence. Now close to 250 years old, most of them with girths over 3.7 metres tower over the building. In the winter of 2018, some overzealous officials from the State Horticulture Department discovered that three such trees were leaning towards the summer house and had them cut down immediately to protect the building. Tipu's summer house is dilapidated and in a state of neglect and disrepair for a long time.[148] Whether it was apathy or a shocking lack of awareness of the trees' historical importance is not clear. This case typically illustrates how built heritage in India always takes precedence over living, breathing heritage!

There is still a lone blue gum tree with a girth of 3.72 metres standing tall and erect close to the summer residence, though I can't help feeling that its days are numbered. Perhaps it is only a matter of time before the official axe will come crashing down on another piece of Indian history.

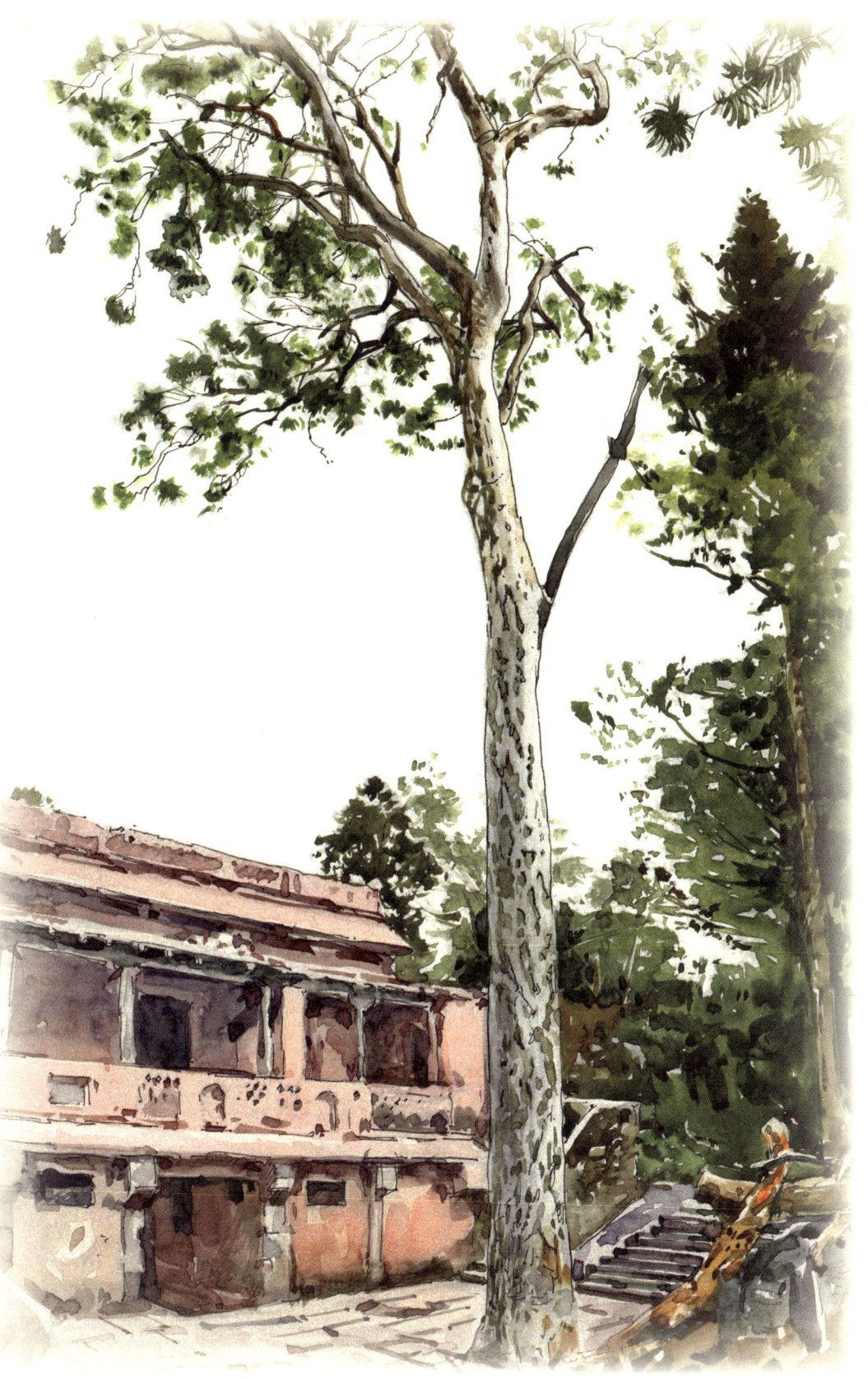

42

The Tree that Hosts
Millions of Bees

COMMON NAMES: Banyan (English); Nyagrodha (down-growing)
and bahupada (many-footed) (Sanskrit); Ālada Mara (Kannada);
Vat Vriksh, Bargad (Hindi)

SCIENTIFIC NAME: *Ficus benghalensis* L.

FAMILY: Fig (Moraceae)

WHERE TO SEE: Ramagovindapura, Hoskote tehsil,
Bengaluru rural district, Karnataka

LATITUDE: 13.22° N; LONGITUDE: 77.91° E; ALTITUDE: 875 metres

Red letters on a white-painted background announce in Kannada, 'Ramagovindapura – the place for observing honeycombs'. Behind that board is a noticeably large banyan tree, standing tall on a traffic island around which the road bifurcates and rejoins to accommodate its huge girth of over 24 metres. Its branches arch over 18–20 metres on all sides supporting a crown that resembles a massive green umbrella. Meet the 'bee tree' on which up to 630 honeycombs of the Asiatic giant honeybee have been documented, setting a record for the highest concentration of nesting bees.[149] The number of bee colonies on the tree has been increasing steadily since 1998.

The giant honeybee, scientifically *Apis dorsata*, is a species of wild social bees locally known as hejjēnu. Its massive colonies are characterized by large combs, usually attached to a tree branch or cliff overhang, but sometimes also to the eves of urban buildings. Colonies usually occur singly or in aggregates of two or three, rarely reaching 200 at a single site. Hence, abuzz with bees, the banyan tree of Ramagovindapura is truly exceptional!

The number of hives, however, does not remain constant throughout the year. Typically, the majority of the bees start arriving by October of each year and the population swells by January. Honey is sometimes harvested in March and the bees generally depart in April leaving a few colonies behind. Each colony could comprise up to 100,000 individuals, but usually, the number ranges between 20,000 and 75,000 depending on the age of the hive and the season. At the height of the season, the big tree plays host to between 45 million and 63 million honeybees!

The giant honeybee is very ferocious. Intruders, if any, are chased for long distances and stung. But this has not prevented local communities throughout its range from exploiting the species as a source of honey and beeswax. Villagers of

Ramagovindapura used to earn ₹30,100 on average through the regular harvest of honey from the colonies until 2008. In 2009, extraction of honey was prohibited, as large swarms of bees used to invade the neighbourhood for several days following.

Fortunately, members of the local community are aware of an even more valuable service provided by the bees through pollination. In its rural setting, the tree is surrounded by cultivated fields that grow grain, flowers, fruits and vegetables providing ample forage for the bees. The villagers claim that they get five harvests per year (as against four) and attribute it to the concentration of bees in their area. Animal pollination is responsible for 35 per cent of all food production.[150] Insects, particularly bees, provide the majority of this pollination. The Food and Agriculture Organization has estimated that three out of four crops across the globe producing fruits or seeds for human use as food depend, at least in part, on pollinators.[151] Pollination appears to be economically much more important than previously recognized and needs better support through agricultural management and policy.[152] A new study by researchers at the University of Maryland has found that the median lifespan of caged worker honeybees has been declining since the 1970s, from an average of 34.3 days to 17.7 days.[153] In the real world, shorter honeybee lifespans translate to less foraging time, less honey production and less pollination services. These are compelling grounds for protecting bee trees such as this one as heritage sites.[154]

Why the bees favour this particular banyan tree out of several available on the road is hard to explain. The area in and around Ramagovindapura has a high concentration of nesting giant bees – as many as 2,000 colonies have been detected here on eleven trees within a 5.7 kilometre radius. To the human eye though several sites in each area may appear perfectly suitable, giant bees are very fastidious in choosing their nesting sites. They are inclined to build their nests on conspicuous landmarks,[155] perhaps in association with other colonies or the remains of combs of previous years. They also reoccupy nesting sites year after year over periods of several decades or more. Amazingly, some returning colonies are known to find their way back to the same tree – or other sites – they occupied the previous season. Such precise homing instincts are also characteristic of migratory salmon, toads, turtles, storks, seals and certain other vertebrates.

The Karnataka government planned to set up a special economic zone in the nearby Nandagudi area but was forced to back off following strong opposition from the farming community. A recent three-year study in Bengaluru has shown that even mildly dirty air could kill 80 per cent of giant Asian honeybees.[156] Without bees and other insects – and the trees that host them – domestic production of fruits, vegetables, nuts and legumes could be at risk.

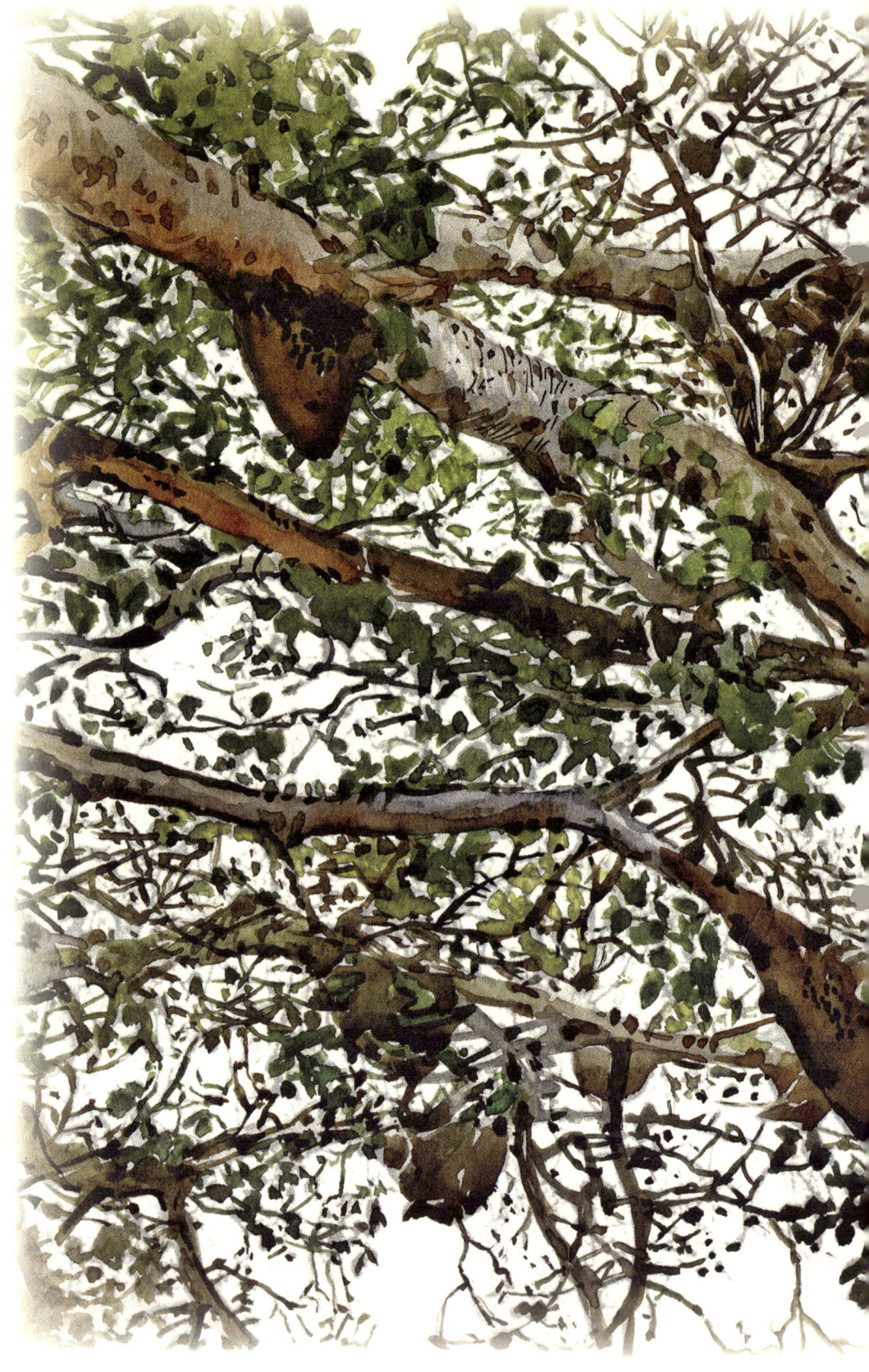

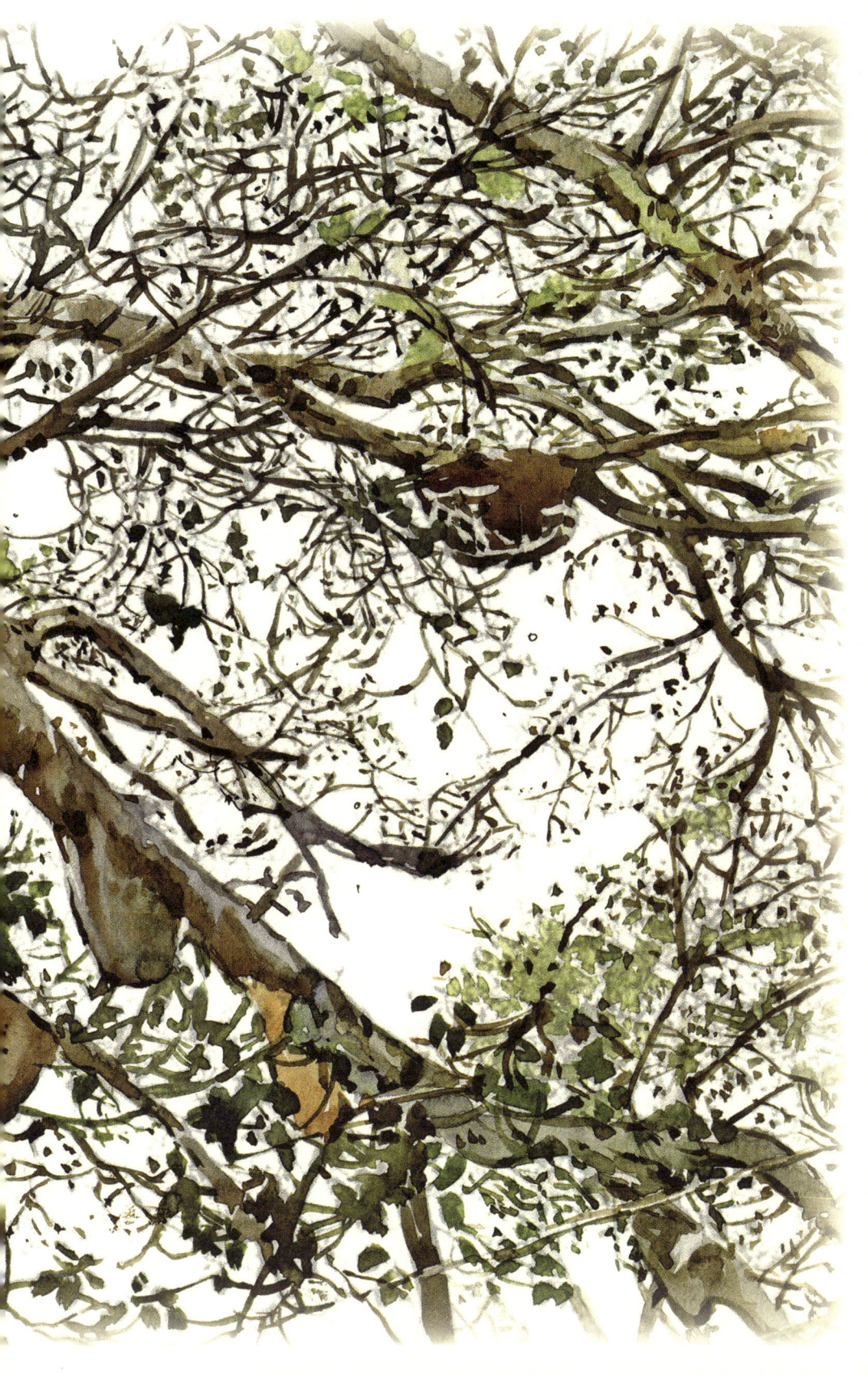

43

A King's Scattered Treasure – Nallur Tamarind Grove

COMMON NAMES: **Tamarind (English); Amlika, Tintiri (Sanskrit); Hunasé, (Kannada); Imli (Hindi)**

SCIENTIFIC NAME: *Tamarindus indica* L.

FAMILY: **Legume or bean (Fabaceae)**

WHERE TO SEE: **Nallur village, Devanahalli taluk, Bengaluru rural district, Karnataka**

LATITUDE: **13.1155° N;** LONGITUDE: **77.4601° E;**
ALTITUDE: **892 metres**

A medium-sized village in Devanahalli taluk of Bengaluru is home to an exceptionally well-preserved grove on the outskirts of the city. Locally known as amaroy topu, the Nallur village grove covers an area of 21.46 hectares (53.02 acres) and contains about 300 tamarind trees, some of which are believed to be centuries old. The grove is reputed to have originated during the early thirteenth century when the imperial Chola dynasty ruled over this region. Within the tamarind grove an ancient temple dedicated to Gopalaswamy (Lord Krishna) – no longer worshipped and now in ruins – testifies to the antiquity of the site.

The Nallur grove is a scrub jungle interspersed with tamarind trees. Eighteen of the 300 tamarind trees are very old, the rest being middle-aged, or young. Carbon dating of wood samples showed that one of the trees (Tree No. 155) in the grove is over 400 years old[157] and several others are half that age. The old trees are the most interesting: some have large trunks whose circumference measures between eight and nine metres. The surface of the trunk varies from smooth to very rugged, carved into ridges and furrows or strikingly twisted and gnarled. The majority of trunks are hollowed out into cavities of varying shapes and sizes due to the degeneration of the core tissues (this is a common feature associated with tamarind trees). Interestingly, many of the trees have developed prop roots from the inner side of the cavities. Most prop roots get arrested at different points before reaching the ground. However, some do hit the ground and anchor the mother tree. In another first for tamarind trees, the prop roots even produce suckers that develop into young plants that bear flowers and fruits. Flowering generally takes place during April and May and even the very old trees do bear fruits. The yield from the grove is auctioned off annually by the forest department.

162

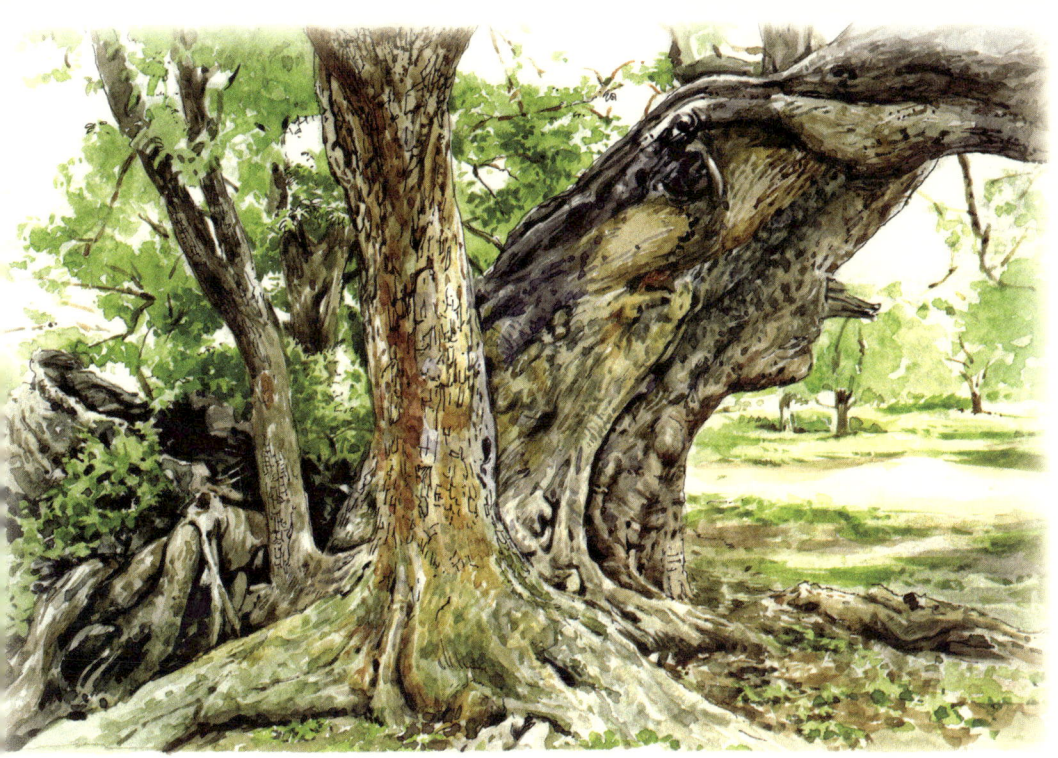

Tamarind trees grow slowly but are usually long-lived. A local legend pushes back the age of the grove to 800–900 years. According to the legend,[158] Nallur in the olden days was a small kingdom surrounded by a fort. The royal couple had a beautiful daughter. Smitten by their daughter's beauty, the prince of the neighbouring state Dalikoté sought her hand in marriage. The king, however, rejected the alliance as unsuitable. Enraged at this insult to his honour, the prince laid siege to Nallur with his entire army. Meanwhile, unbeknownst to the king, the princess secretly loved the prince and sent him clandestine messages informing him of an underground entrance to the fort, through which the fort was easily conquered. The vanquished king hurriedly had all his gold and jewellery buried along with tamarind seeds all over the kingdom to recover them later on. The tamarind grove you see now sprouted from those seeds.

Sadly, having braved the seasons over centuries, this historic grove is in a state of neglect today. The Karnataka Forest Department has done very little beyond declaring the place as a bio-heritage site. The fence around the grove is incomplete, leaving it open to the collection of firewood and grazing by cattle of the nearby villages. The popular media keeps highlighting threats to the grove from time to time, but so far, the warnings seem to have fallen on deaf ears.

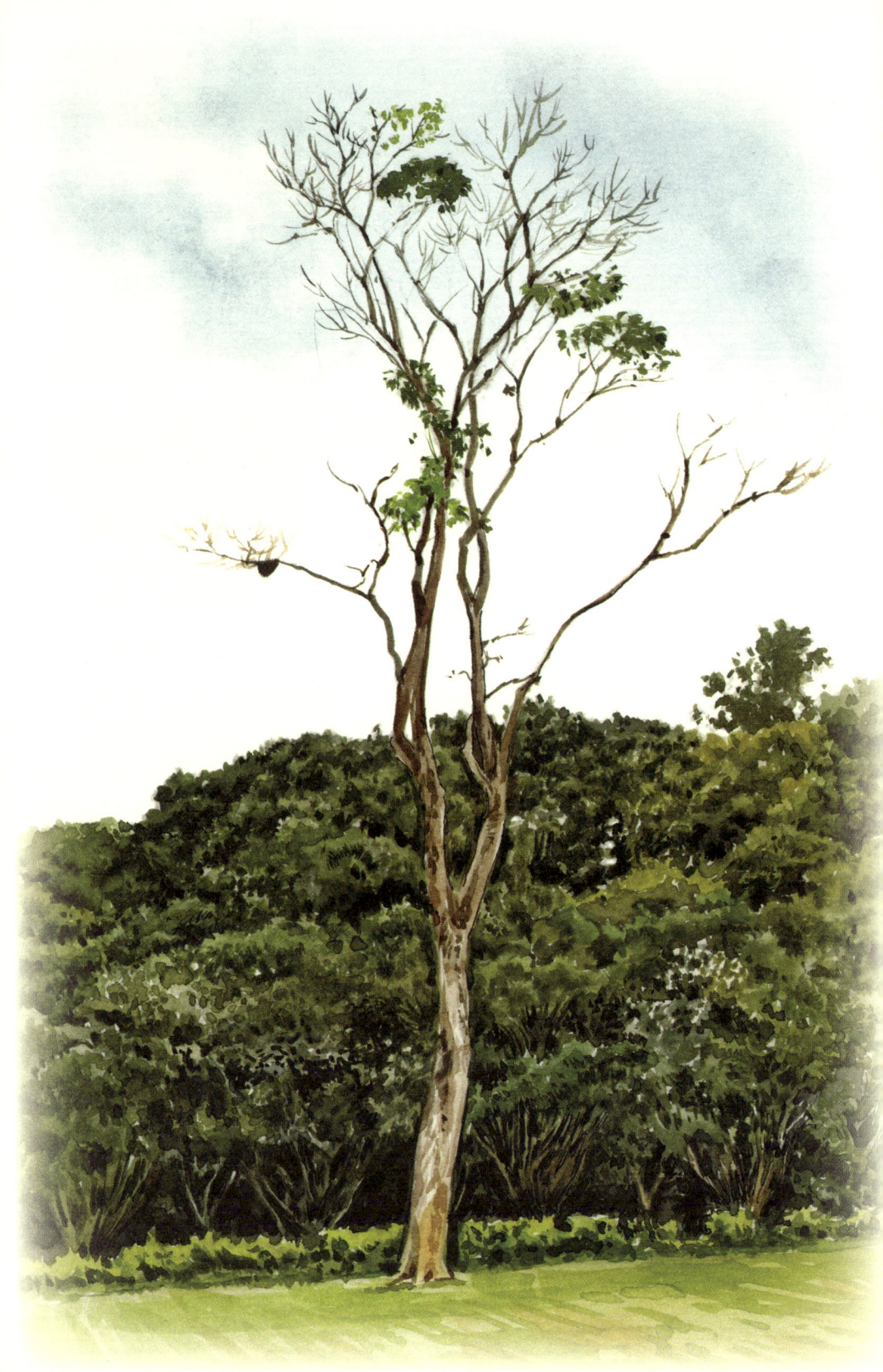

44

Sir C.V. Raman's Primavera

COMMON NAMES: **Primavera, Gold Tree (English)**

SCIENTIFIC NAME: *Roseodendron donnell-smithii* **(Rose) Miranda**

FAMILY: **Jacaranda (Bignoniaceae)**

WHERE TO SEE: **Raman Research Institute,
C. V. Raman Avenue, Sadashiva Nagar, Bengaluru**

LATITUDE: **13.0132° N**; LONGITUDE: **77.5806° E;**
ALTITUDE: **Approximately 940 metres**

A solitary but stunningly beautiful primavera tree stands out prominently on the green lawn in front of the main building of the Raman Research Institute. This rare tree[159] was planted as a memorial to the great scientist Professor C.V. Raman, winner of the 1930 Nobel Prize in Physics.

The institute was created by Professor Raman in 1948 to provide an ambience conducive to the conduct of pure science research. It is well known that the physicist took a keen interest in landscaping the institute's garden. He planned, planted and supervised hundreds of flowering and avenue trees as well as pretty bushes and climbers. He tended with care 168 exquisite rose plants at the institute. One of the great joys of his life was to gaze at the trees and flowers in his garden. He also delighted in showing his institute to school and college children, teaching them science and encouraging them to study the nature around them.[160]

Shortly before his 82nd birthday, Raman took ill and had to be hospitalized. Doctors gave him four hours to live but he survived. After a few days, he refused to stay in the hospital as he preferred to die at the institute he had lovingly built, surrounded by flowers and trees. He was moved back to his residence on the institute campus. Though back in his favourite surroundings, he had to lie down in his bed. The window was high and, to his distress, he could not catch a glimpse of his beloved garden. 'If I had known I would die here,' grumbled a disappointed Raman, 'I would have arranged for the windows to be lower'. His bed was raised by his caregivers so that the garden came into view.[161] The end came peacefully in the early hours of 21 November 1970.

The Karnataka government gave special permission to cremate his mortal remains within the institute premises amidst the surroundings he loved and enjoyed. The institute decided that there was no better way to perpetuate his memory than by planting a beautiful flowering tree on the spot. A committee was set up, both to scout for and source a suitable tree. One of the members on that committee was Dr Nalini Dhawan, wife of Professor Satish Dhawan – the renowned space scientist and the then director of the Indian Institute of Science (IISc), Bengaluru. Nalini was very knowledgeable about plants. Her horticulturist father, Mr Bhavani Shankar Nirody, was well-known and had a significant share in transforming the IISc campus into what it is today.[162] It was Nirody who recommended a Primavera tree to commemorate the great man. Although not native to India, its saplings were available in the gardens of the plants in Jamshedpur and Pune. Three saplings of primavera arrived from Pune and were under Nalini's loving care at IISc, getting acclimatized to Bengaluru, over the next few months. It was in July/August of 1974 that the tallest of the three saplings was planted at the Raman Institute.[163] At that time primavera's scientific name was *Cybistax donnell-smithii* (Rose) Seibert but a taxonomic revision changed it to *Roseodendron donnell-smithii* (Rose) Miranda, which is currently accepted.

Primavera in old Spanish signifies the spring season, a name justly earned due to its habit of flowering early as nature's most magnificent harbinger of tropical spring in the land of its nativity. When the rainforests of Central and South America come alive with a bedazzling show of exquisitely bright yellow primavera blossoms, one can be sure that spring is not far behind. The spectacular floral display occurs in clusters at the ends of bare branches when the deciduous tree has shed leaves. In its home, primavera is a source of valued timber and can grow to heights of up to 61 metres (200 ft.). However, coming from a continent some 17,000 to 20,000 kilometres away as the crow flies, no one knew how it would fare in the alien climate of Bengaluru.

The next few years were filled with suspense. The tree did survive and grow tall but did not flower. Finally, in 1988, it timed its flowering to the year of Raman's birth centenary. Flowering thereafter has not occurred every year and fruit set has never occurred so far. Even so, when the primavera flowers it is a riot of yellow! It is nice to see that the Indian giant honeybee (*Apis dorsata*) has made a honeycomb on one of its high branches – surely a sign of acceptance of the tree in its adopted home. Nevertheless, in 2020, it suffered from fungal infection, along with termite and root grub infestations. Treated by an expert from the nearby University of Agricultural Sciences, the iconic tree responded well and was nursed back to health.[164] Today, the commemorative tree is very tall and stately, rising above others in its background, just as Raman was the tallest Indian scientist of his time. Is that the 'Raman Effect'?

Note: Sadly, the iconic Primavera tree collapsed on 7 September 2022 following unprecedented heavy rains in Bengaluru. Efforts to revive it were unsuccessful. An Ashok tree [*Saraca asoca* (Roxb.) W. J. de Wilde] has been planted in its place.

45

The Gentle Giant of Lalbagh

COMMON NAMES: **White Kapok, White Silk Cotton Tree (English);
Shweta shalmali, kuta shalmali (Sanskrit); Bili būruga (Kannada);
Safed Semal (Hindi)**

SCIENTIFIC NAME: *Ceiba pentandra* **(L) Gaertn.**

FAMILY: **Hibiscus (Malvaceae, Bombacoideae)**

WHERE TO SEE: **Near the West Gate, Lalbagh, Bengaluru, Karnataka**

LATITUDE: **12.95° N;** LONGITUDE: **77.58° E;**
ALTITUDE: **Approximately 912.8 metres**

Near the west gate of Bengaluru's iconic Lalbagh Botanical Garden, a giant white kapok tree of jaw-dropping proportions is a huge draw for visitors. Said to be over 200 years old, the tree rises close to 26 metres above ground dominating the immediate landscape. The trunk is massive with a girth of 23 metres at breast height. Enormous buttresses issuing from the trunk anchor the tree, fusing with their neighbours and creating cavities, some of which are large enough for an adult person to hide in. Several branches arise from the trunk at an angle, collectively upholding a colossal umbrella-like crown that has a spread of some 6,100 m², with a circumference exceeding 170 metres. The leaves, typically palmately divided with five to seven leaflets, shed just before the flowers appear. Although called white kapok or white silk cotton on account of the flowers being usually off-white or creamy, the petal colour in this species can vary. The tree produces pinkish flowers that are pollinated commonly by two smaller bat species – the fulvous fruit bat (*Rousettus leschenaultii*), the short-nosed fruit bat (*Cynopterus sphinx*) and, less frequently, the larger Indian flying fox (*Pteropus giganteus*).[165] However, fruit set has not been observed.

According to the website www.monumentaltrees.com/en/ the Lalbagh kapok holds the world record for girth (23 metres) among eight other contenders from Cape Verde, Costa Rica, Cuba, Guinea-Bissau, Mexico, Senegal, Singapore, and the USA.[166] The website claims the tree's age in 2012 as 216 ± 20 years, third in position after a 420-year-old tree in Cuba and a 250-year-old tree in Cape Verde.

Kapoks are native to the rainforests of tropical America but must have been introduced into India a long time ago. How and when did the tree get included in Lalbagh? Lalbagh was established by Hyder Ali in 1760 and later nurtured, enlarged and enriched by his son Tipu Sultan.[167] The popular belief is that the kapok must have been introduced during Tipu's time. The British took over Lalbagh in

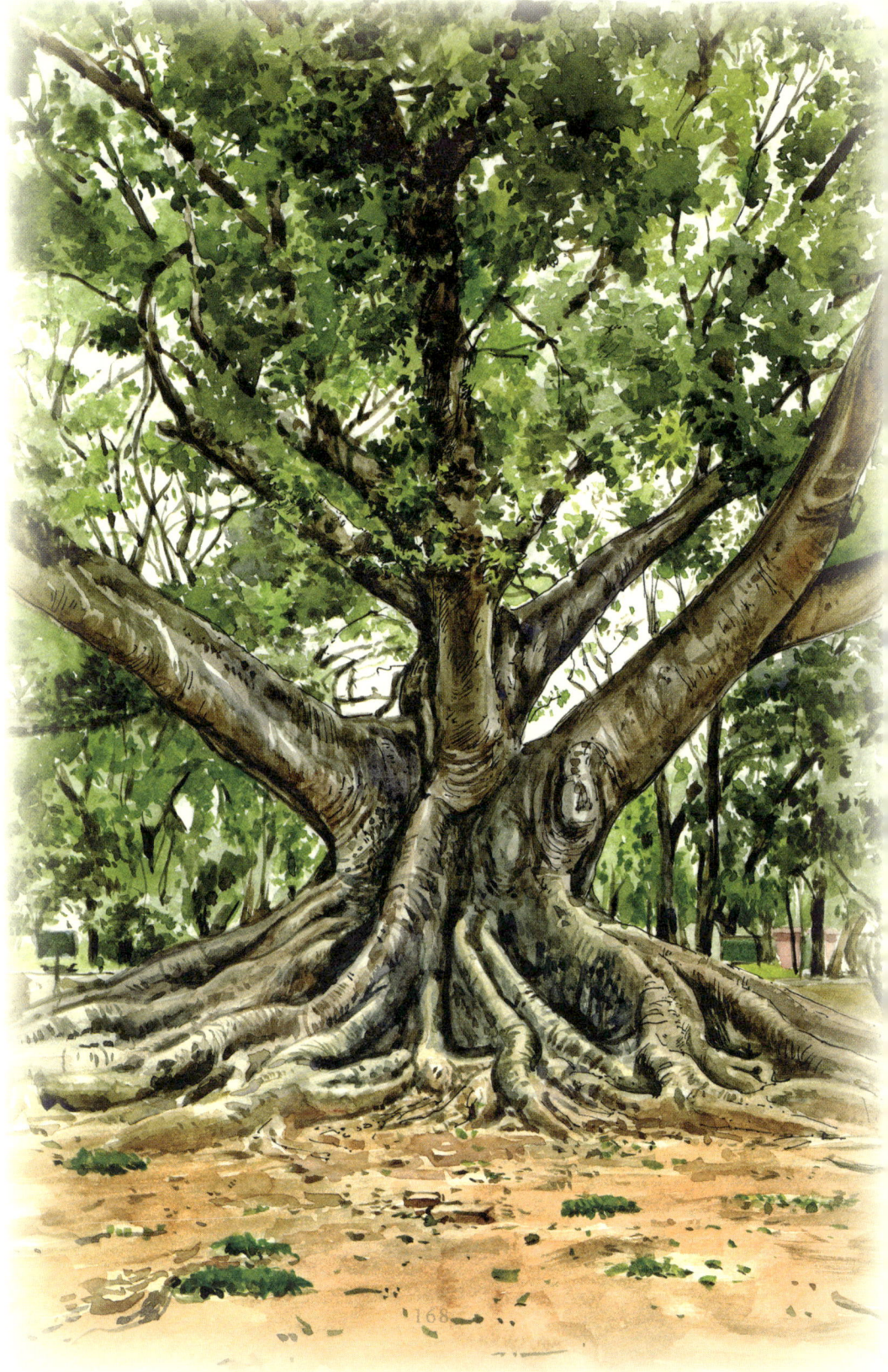

1799 after Tipu's death in the fourth Anglo-Mysore war. Over the following five decades, it kept changing hands within the British colonial bureaucracy. During much of that period, the garden received little attention. It was only in 1856 that a decision was taken to make it the Government Botanical Garden.

The Karnataka government, through its Department of Horticulture, officially depicts the original Lalbagh as a single, contiguous chunk of 16.19 hectares (40 acres). However, a 2012 study using a combination of historical maps, old paintings and current remote sensing images[168] revealed the actual boundaries of the garden as they existed during 1760–1856 – that there were as many as five distinct garden patches that collectively formed the area then called the Cypress Gardens. Only one of these – 'Tipu's Garden' – lies within the boundaries of today's Lalbagh, the remaining (Hyder Ali's) patches having been lost to the growth of the city outside the garden. More pertinent to our story, the study also revealed that the giant kapok tree was not within the original garden patch attributable to Tipu, but located at a distance of 195 metres to the southwest corner of this patch, on land that was apparently acquired between 1865 and 1950.

When then did the tree get incorporated into Lalbagh? We can only guess as there are no records to establish the date. In 1861, William New, the then superintendent of Lalbagh, published a catalogue of the garden's plants in the *Transactions of the Botanical Society of Edinburgh*,[169] and included the kapok tree under the name *Eriodendron anfractuosum* Dec., a synonym of its currently accepted name *Ceiba pentandra* (L) Gaertn. The tree was acquired sometime between 1856 and 1861, six decades after Tipu Sultan's death. It is not clear if there was any land acquisition during that period.

Today, this gentle giant of Lalbagh is an object of wonder and awe for visitors. Not all callers seem to accord respect due to this gentle giant, for despite a metal fence built around the tree, one can unfortunately see graffiti defacing some of the branches.

46

Gurudev's Weeping Fig Tree

COMMON NAMES: Java Fig Tree, Benjamin's Fig Tree, Weeping Fig Tree
(English); Java Atti, Jeevi (Kannada); Kabra (Hindi)

SCIENTIFIC NAME: *Ficus benjamina* L.

FAMILY: Fig (Moraceae)

WHERE TO SEE: Near the main gate, Lalbagh, Bengaluru

LATITUDE: 12.953° N; LONGITUDE: 77.585° E;
ALTITUDE: Approximately 912.8 metres

As soon as you enter the famed Lalbagh Botanical Garden in Bengaluru through the main entrance (earlier known as Cameron Gate) and move straight ahead towards the oval garden, a large and historic weeping fig tree beckons you toward it. Under its canopy, Rabindranath Tagore was accorded a grand civic reception in 1919 during his maiden visit to Bengaluru.

Bengaluru has several grand specimens of the weeping fig – also known as the Java fig – but the grandest are the ones here in Lalbagh. The earliest specimens were brought into Lalbagh sometime between 1860 and 1880 from Java[170] (hence its name). Trees of this species can grow to 30 metres in height under natural conditions but this one is about 24 metres. Branches start low and the dense canopy bears smooth, leathery leaves with pointed tips typical of the species. Aerial roots are not prominent on the tree. Now close to 150 years, the historic tree must have been nearly 50 years old when Gurudev came calling in 1919.

Tagore was awarded the Nobel Prize in Literature in 1913. He undertook his first tour of south India in 1919, arriving in Bengaluru from Madras (now Chennai) by train. Tagore preferred to have the planned reception in the open under the trees. Hence the event was shifted to the sprawling canopy of weeping figs near the main entrance. Tagore, perhaps in harmony with his surroundings, wore a green gown to the meeting and paid fulsome praise to the beauty of Lalbagh.[171] The programme was a great success and the poet won the hearts of the thousands who had gathered there with his simplicity and charming manner. Among the audience was Māsti Venkatesha Iyengar, the well-known Kannada writer and Jnanpith awardee, who wrote about the function[172] as did several other dignitaries.[173]

At the time of Tagore's visit in 1919, there were eight trees at the site, four on either side at the entry point to the oval garden. By the 1980s, three weeping fig trees were gone and now only this one survives, reminding us of a bygone era.

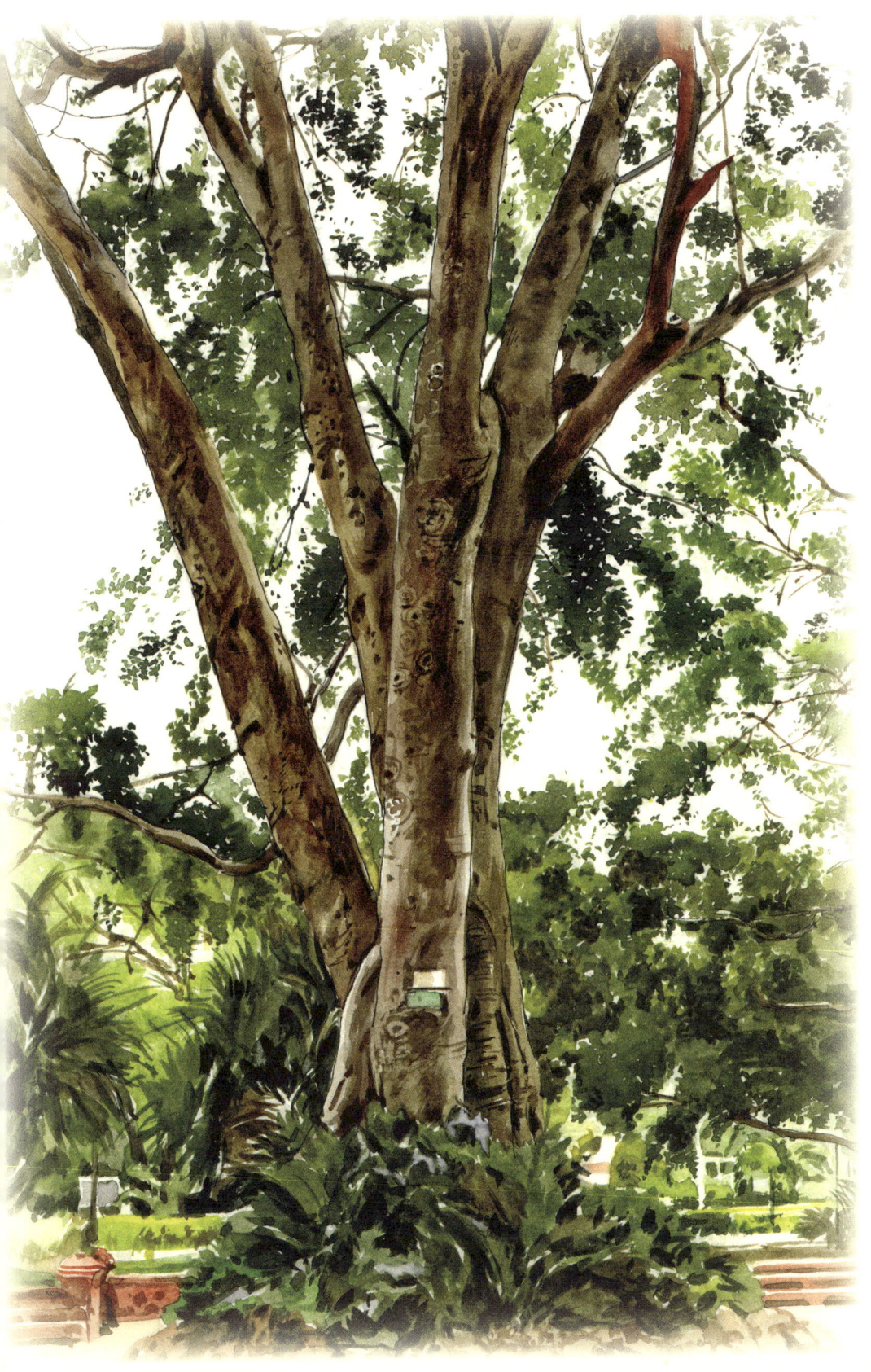

47

The Rain Tree of Taj West End

COMMON NAMES: **Rain Tree (English); Shirisha (Sanskrit); Bāgey Mara (Kannada); Gulabi Siris, Vilayti Siris (Hindi)**

SCIENTIFIC NAME: *Samanea saman* (Jacq.) Merr.

FAMILY: **Legume (Fabaceae)**

WHERE TO SEE: **Taj West End, 41, Race Course Road, High Grounds, Bengaluru, Karnataka**

LATITUDE: **12.984° N**; LONGITUDE: **77.585° E**; ALTITUDE: **Approximately 874 metres**

Rain trees can be seen on numerous avenues of India's IT hub of Bengaluru – their vast leafy crowns coalescing into green tunnels above the busy traffic; they are in parks and public places, their strong branches soaring into the sky and ending in umbrella-shaped canopies. You just cannot imagine Bengaluru without its rain trees. Truth is, it is a migrant like many of the city's people, travelling a long distance from its native home in Central and South America and obviously in love with Bengaluru's climate.

In a city full of grand-looking rain trees, it is almost impossible to single out an outstanding one. Even so, a tree growing in Taj West End on Race Course Road, reputed to be the city's oldest hotel, is noteworthy for several reasons. Hotel sources say that the tree was planted around the year 1848 – about four decades before the birth of the hotel in 1887[174] – but it is not clear who planted it or from where it was sourced. It was around the early 1800s that the British chose to build military barracks in the then-sleepy town of Bengaluru and many enterprising people opened small exclusive establishments for their elite English clientele. One such modest establishment with 10 beds named Bronson's Boarding House was set up in 1887 on Race Course Road by Mrs Bronson, an affable British lady. As the rain tree on the site continued to grow, so did the regular clientele of the inn. Soon, to accommodate all of them, a second wing and then a third was added. The rain tree was just 64 years old when the boarding house changed hands in 1912. The Spencers who purchased it for a princely sum of ₹4,000 rechristened it West End. They also invested in catering to the needs of the growing population of English sahibs and memsahibs who yearned for the milieu of 'life back home'. They added land and laid out gardens and pathways all around to give the surroundings the feel of a tropical forest. New dining halls and billiard rooms came up as the Indian cooks regularly churned out brand new recipes that added flavours to the otherwise bland English fare. This Anglo-

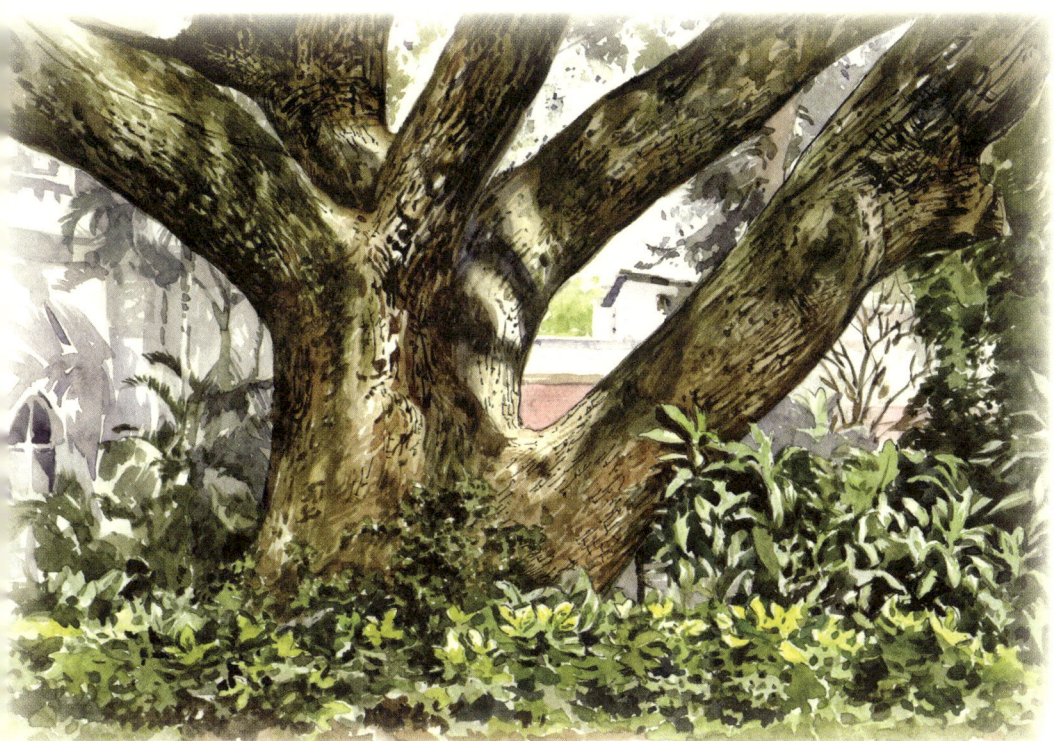

Indian cuisine soon became one of the major attractions. In short, the hotel developed the traits of a true country club.

During all this time the tree kept growing taller, adding a noticeable girth to its trunk, sprouting strong new branches that reached out in all directions. By the time the Taj management acquired the hotel in 1984 the tree, by now a venerable centurion of 136 years, had itself become a star. 'Breakfast under the rain tree' became quite fashionable. Over the years, the rain tree has seen virtually everyone: royalty, the well-heeled, international celebrities and corporate czars have rubbed shoulders under its ample canopy. Sir Winston Churchill came here as a war correspondent. Sir Ronald Ross wrote the cure for malaria here and went on to win the Nobel Prize for Medicine in 1902. David Lean used this place to film *A Passage to India*. Prince Charles was a guest here. For Bollywood superstar Amitabh Bachchan, it was a healing retreat after he was injured on the sets of the film *Coolie*.

The tree giant is close to 176 years now and continues to be a major attraction of Bengaluru's most popular country club. One cannot help but feel charmed standing under its shade. Varieties of birds have made the upper reaches of its crown their home. It is noteworthy that this tree holds the record for age[175] for a rain tree. At over seven metres, its trunk, covered with rough grey bark, is massive and ranks fourth for the thickest rain tree.

Its shiny green, twice-divided leaves are sensitive to light and continue to amuse visitors by folding together from dusk to dawn. The 'sleeping' movements

– technically known as nyctinasty – happen as a response to darkness and occur rhythmically according to a 24-hour clock. Typically, leguminous plants exhibit sleeping movements, although it is seen in some other species as well. Alexander the Great, it is said, was the first to observe this phenomenon. Since then, it has intrigued scientists. Considering that plants do not have muscles and tendons to move leaves around, how do they do it? The puzzle has not been completely solved but we do know that legumes have a special organ called pulvinus at the base of each leaflet. Cells in the pulvinus initiate currents of chemical ions that travel up into the leaflet with clockwise precision. This in turn results in uneven changes in the volume of certain cells on the upper and lower surfaces of the leaflet causing visible movement of leaves.[176]

In season, the tree produces thousands of round flower bunches that look like pink powder puffs or feather dusters. The most attractive parts of the flowers are the long stamens, their bottom half being white with the top half pinkish. The flowering time is also an open invitation to bees and the air is throbbing with their humming. Long live the arborescent centurion!

48

The Golden Champaka Tree of Pushpagiri

COMMON NAMES: **Golden Champaka, Joy Perfume Tree (English);**
Sampigé (Kannada); Champa, Champak, Champaka (Hindi)

SCIENTIFIC NAME: *Magnolia champaca* (L.) Baill. **ex** Pierre

FAMILY: **Magnolia (Magnoliaceae)**

WHERE TO SEE: **Adjacent to Shantha Mallikarjuna Swamy Temple,**
Kumarahalli B.O. Post Office, Pushpagiri, Somvarpet taluk,
Kodagu district, Karnataka

LATITUDE: **12.557° N**; LONGITUDE: **75.868° E;**
ALTITUDE: **Approximately 1,004 metres**

Pushpagiri Wildlife Sanctuary in the Western Ghats is spread over approximately 103 km². Endowed with spectacular scenic beauty it also has the second tallest mountain peak after Thadiyandamol in Kodagu (earlier Coorg) district of Karnataka. Also known as Kumara Parvatha, it is a trekker's paradise. Just seven kilometres further along the route is Kumarahalli, a small and quaint village surrounded by magnificent mountains. Its ancient temple of Shantha Mallikarjuna Swamy enveloped in a silent and tranquil ambience beckons the tired trekker to step in and find inner peace.

Standing next to the temple at the back, a magnificently large sampigé (champak) tree can be seen from afar as it soars loftily over the temple to some

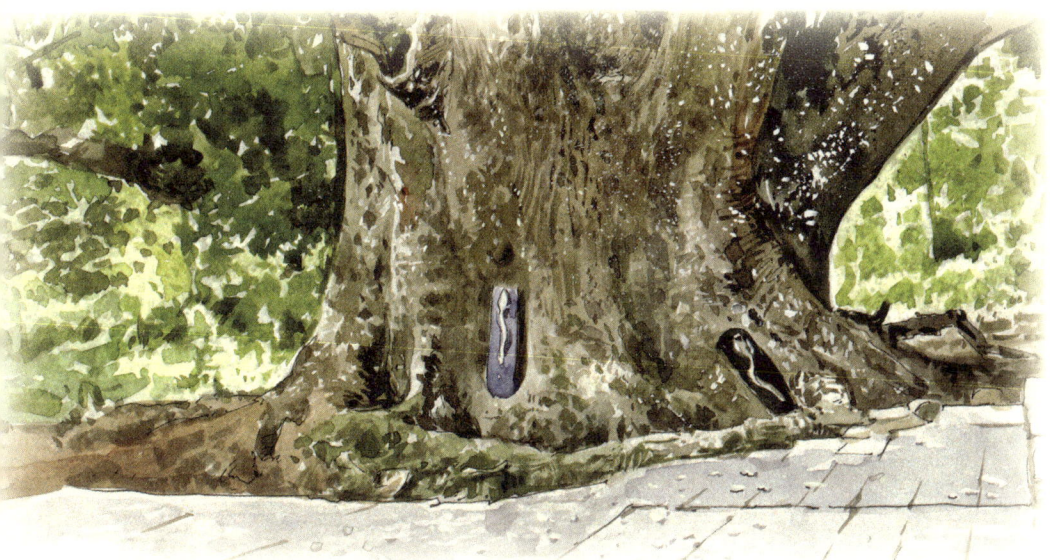

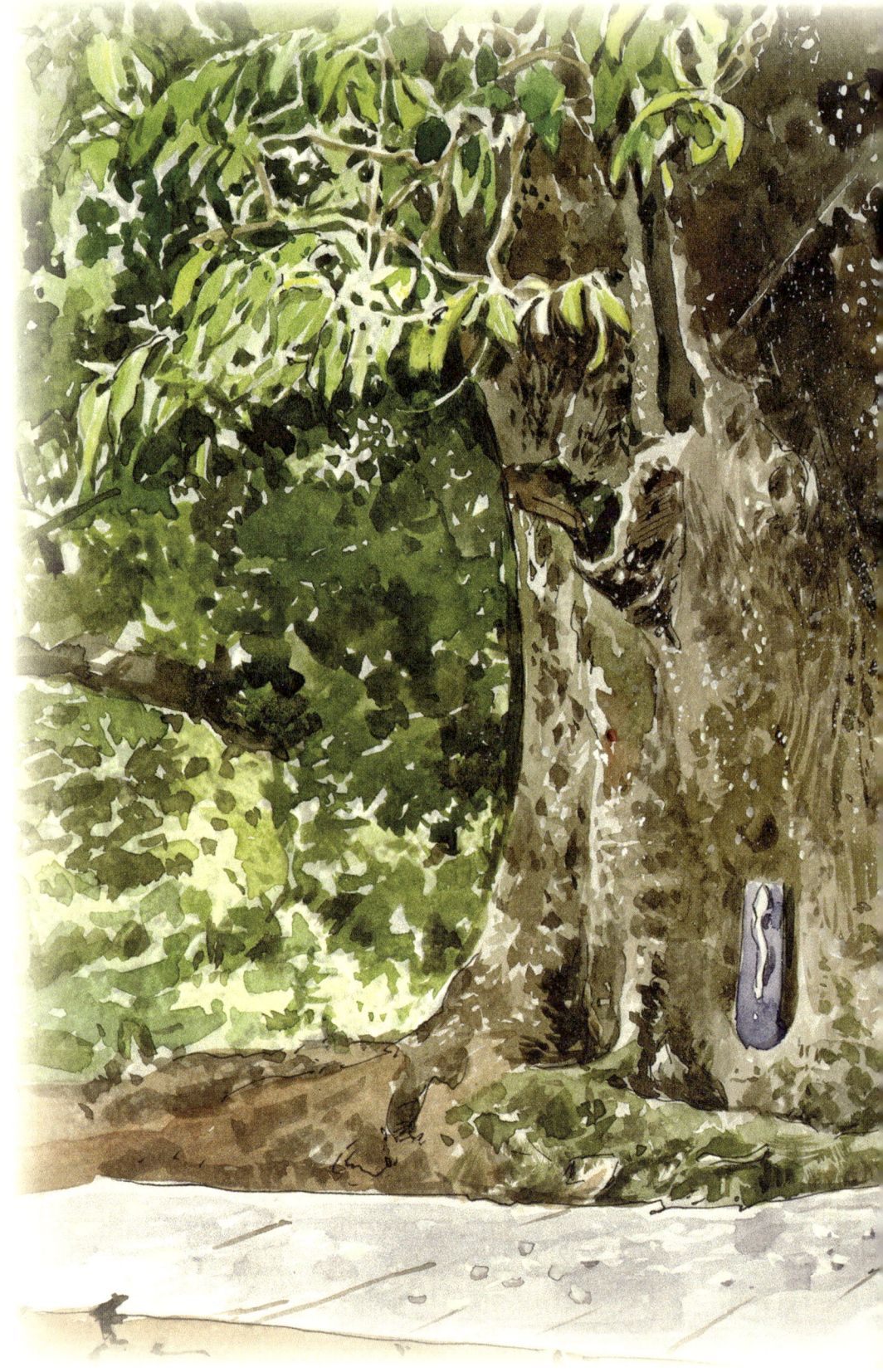

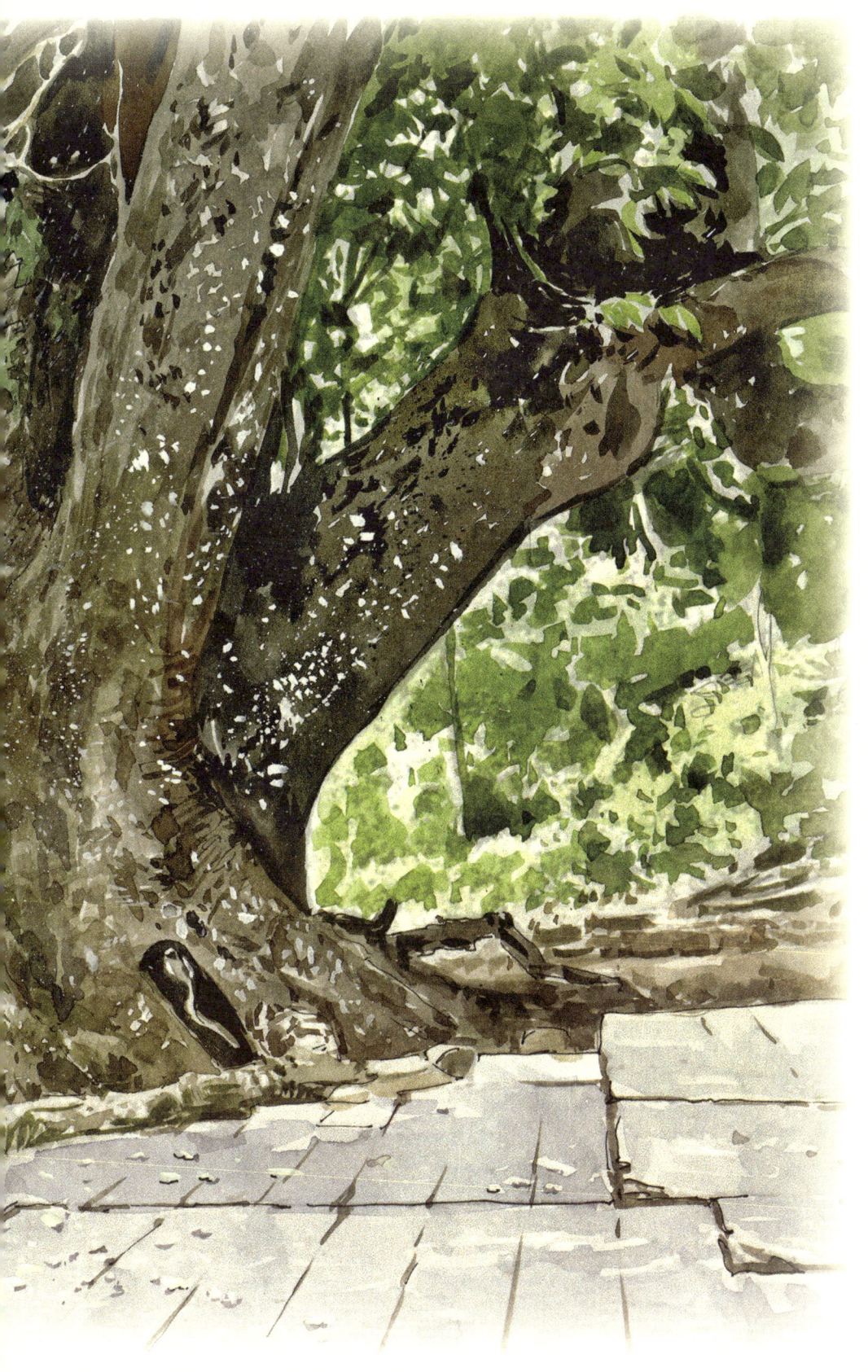

25 metres. Its massive and much-fluted trunk has a circumference of close to 12 metres. Branches arise in all directions bearing a large and spreading crown. Images of the serpent and other gods are placed on a low platform at the base of the trunk. The tree bears characteristically pale green leaves and pale yellow, highly fragrant flowers. During flowering season fallen petals make a thin carpet on the ground below. Its seeds, covered by a bright red aril (a fleshy extra seed coat), are a great attraction to birds.

According to the priest, the temple itself is about 300 years old. The *shivalinga* is said to be an *udbhava murthi* (formed by itself; not made by human hands). A small brick shrine was built at the place and was expanded as more people made contributions. The present edifice is about 17-18 years old. The tree is said to be much older than the temple itself and according to the priest, between 700 and 850 years. Perhaps a more realistic guess would be 450-500 years. A decade ago, its girth was said to be about 11.5 metres.

Noted in these parts for its timber, this handsome evergreen species is even more highly valued as the 'pinnacle of beauty in the poetical fancy of the Coorg bard' because of its beautiful and sweet-scented flowers:

... And among the flowering trees
Is the Sampige the finest;
Thus is Coorg, a string of pearls,
Loveliest among the kingdoms;
Live in it, my friends and prosper![177]

Although rare specimens – 24-30.5 metres tall and six metres thick at the base – have been known,[178] the circumference of the present specimen seems to be exceptional.

49

The Dodda Sampigé of Biligiri Rangan Hills

COMMON NAMES: Golden Champaka, Joy Perfume Tree (English);
Champaka (Sanskrit); Sampigé, Dodda Sampigé Mara (Kannada);
Champa, Champak, Champaka (Hindi)

SCIENTIFIC NAME: *Magnolia champaca* (L.) Baill. ex Pierre

FAMILY: Magnolia (Magnoliaceae)

WHERE TO SEE: Atgulipura, Biligiri Rangan (BR) Hills, Yelandur taluk,
Chamarajanagar district, Karnataka

LATITUDE: 11.947° N; LONGITUDE: 77.184° E;
ALTITUDE: 1,171.3 metres

Dodda sampigé mara, literally 'big champa tree' in Kannada, is one of the huge draws for visitors to Biligiri Rangan Hills, popularly known as BR Hills. *Biligiri* ('white hill' in Kannada) refers to the white rock face of the main hill atop which sits the ancient temple of the presiding deity Lord Rangaswamy, or the white mist that covers it most of the year. Around the temple, an area of approximately 540 km² has been cordoned off as Biligiri Rangaswamy Temple Wildlife Sanctuary. Situated at the intersection of the Western and Eastern Ghats, the sanctuary displays a diversity of vegetation types and supports a rich population of plants and animals. Since 2011, it has been declared a tiger reserve by the Karnataka government.

The hills and associated ranges are home to approximately 20,000 semi-nomadic ethnic people known as Soligas who, according to experts, are perhaps among the earliest settlers in India and, interestingly, exhibit close genetic affinity to two Australian aboriginal populations.[179] They combine animism along with the worship of Hindu deities and consider Lord Biligiri Rangaswamy as their brother-in-law as he is said to have married a Soliga girl. To them, the dodda sampigé with its gnarled and callused trunk is an embodiment of Shiva with his matted locks and can grant all wishes. Twice a year, the Soligas offer ritual worship that includes their traditional gorukana dance around the tree.

The age-wizened tree is situated approximately 12 kilometres from the temple in the evergreen forests of the hills. Local people believe its age to be between 1,000 and 2,000 years, but it is probably about 600 years. The trunk is huge, gnarled and callused, with a girth of between 15 and 18 metres and partially hollowed out. Four main vertical branches arise from the trunk, their tips reaching some 45–50

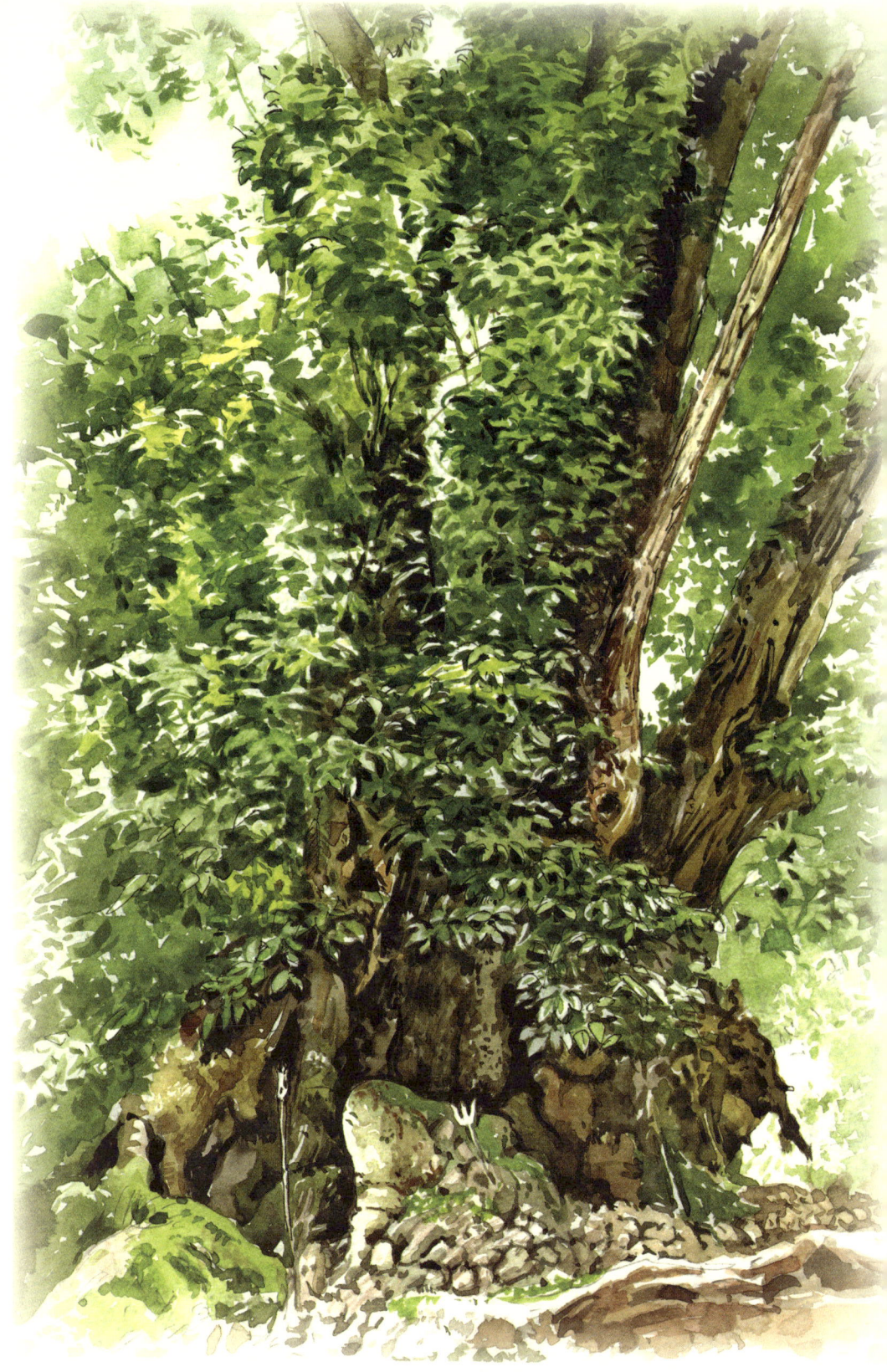

metres into the sky. Branches that fell off in the past have left scars, but new shoots keep sprouting from the callused basal portion of the trunk. The large crown is filled with dark green leaves. The grey and smooth bark is festooned with lichen and green moss. Roots on the levelled ground in front on the tree's eastern side are partially exposed and others can be seen entering Bhargavi River, a tributary of the Cauvery that flows adjacent to the tree's western side. Abundant, highly fragrant, light-yellow flowers appear annually on the tree between March and May. Scores of stone *shivalinga*s smeared with holy ash, along with metal tridents, have been placed by devotees at the foot of the tree on the eastern side.

The Soligas narrate fascinating stories about the tree.[180] Long ago, Sage Parasurama, the sixth incarnation of Vishnu, had to undergo severe penance to rid himself of the sin of beheading his mother at his father Jamadagni's command. Though she was brought back to life, he felt miserable and depressed. He was wandering in the forest and rested awhile under the shade of a tree. A bird's dropping fell on his head and in it were three sampigé seeds. One fell on the ground as he moved away and grew into what is today known as chikka sampigé (literally 'little champa tree').[181] The second fell off at the place where he eventually sat down to undertake penance to Lord Shiva and grew into dodda sampigé. The third seed is still unaccounted for. Meanwhile, Parasurama's mother Renuka could not bear to be separated from her favourite son and took the form of the Bhargavi River to be near him.

According to another Soliga legend, centuries ago a very powerful demon named Shavana was ruling over this region. He grew so powerful that even the gods were reduced to do his bidding. In desperation, they appealed to Lord Shiva and Vishnu and with their active help, managed to slay the demon. Dishevelled and weary, the gods searched in vain for a water body to cleanse themselves of the demon's blood. Sage Parasurama carved a line on the ground with his walking stick and in a trice, it was filled with flowing water to become the Bhargavi River. Goddess Parvati, Shiva's consort, then planted a sapling next to the river which instantly grew into the dodda sampigé. Clean and refreshed, the gods performed the last rites of the demon. Before going their separate ways, each installed a stone *shivalinga* under the sampigé tree to commemorate the event. That is how 101 *shivalinga*s came to be under the dodda sampigé and the Soligas came to worship the tree.

Tree worship and nature conservation is a way of life for the Soligas. Their intimate association with the forest denizens embellished by beautiful legends has ensured that trees such as the dodda sampigé are well protected over centuries.

50

The Speaking Fig Tree
of the Cellular Jail

COMMON NAMES: **Rumph's Fig Tree, Mock Bodhi Tree, Mock Peepal Tree (English); Ashmantaka (Sanskrit); loosely Peepal (Hindi)**

SCIENTIFIC NAME: *Ficus rumphii* **Blume**

FAMILY: **Fig (Moraceae)**

WHERE TO SEE: **Inside Cellular Jail,
Atlanta Point, Port Blair, South Andaman Island**

LATITUDE: **11.6747° N**; LONGITUDE: **92.7479° E**;
ALTITUDE: **3 metres**

If a tree could speak, what would it say? Hard to generalize, but in this instance, we do know exactly what. Standing just inside the entrance to the infamous Cellular Jail in the Andamans almost like a sentry, this grand old Rumph's fig tree has been rooted to the place for well over a century and could tell many a chilling tale. And tell it does, as the anchor of the *son et lumière* (sound and light show) narrating the grim story of this historic jail and its unfortunate inmates. After all, who would know the story better?

The Rumph's fig[182] named after the great German-born botanist and ethnographer Georg Eberhard Rumphius,[183] can easily be mistaken for its more illustrious cousin, the peepal and it often is. In fact, the most popular write-ups on the Cellular Jail identify the tree as a peepal. There are two significant differences though: Firstly, the leaf tip in Rumph's fig is not long and narrow as in the peepal, but short and pointy; secondly, its trunk gives off aerial roots – a feature not very common in its cousin.[184] Sometimes, interspecific hybridization occurs between these two related species.[185]

A native of India and several Southeast Asian countries, Rumph's fig grows commonly in the Andamans. Although there are no records to say when this tree was born, it is known to have been there in 1896, when the construction of the jail began.

After 1857, the British unleashed an orgy of terror and violence sentencing thousands to torture and death. Among the kinds of torture they devised such as tying the 'rebels' to the mouths of cannons and blowing them to bits or hanging them from trees, one was particularly refined: exiling them to a remote island so far away from the mainland as to have no chance of escape. They well knew that transportation beyond the sea was anathema to Indians, especially Hindus. The

British had annexed the Andamans in 1789, but nearly seven decades later no one went there, save the shipwrecked or truly desperate. For nine months of the year, the swampy islands were pounded by monsoons; there were frequent outbreaks of disease and, if that was not enough, there were ferocious tribes – some of them cannibals – in the steaming jungles. A better place to exile the rebels could not be found and that is why the first batch of 200 'grievous political offenders' found themselves on the island on 10 March 1858. From then on for the best part of a century, the Raj deported an estimated 80,000 dissidents and revolutionaries to this remote penal colony[186] which came to be known as *kāla pāni* (black waters in Hindi) and subjected them to hunger, torture, secret medical experiments, forced labour and death.

Trees, they say, are sentient beings. Standing where it does, the Rumph's fig was a silent witness to the fate of the hapless, emaciated convicts from sunrise to sunset: hunger, abuse, cruelty, brutality, floggings and the special punishments – being yoked like a beast to the dreaded coconut oil mill or hung by wrists from the wall. It could not have missed the indomitable spirit and patriotic fervour that refused to flag despite all that the gaolers could throw at the prisoners; or how ultimately, that spirit prevailed and repatriation began in 1937. It silently took in its stride the three years (1942–1945) of Japanese occupation and also Netaji Subhas Chandra Bose's visit in 1943 to the jail premises. Could it have missed the declaration of India's independence in 1947? And how about the event in 1979 when the jail was declared a national memorial? Are we ever likely to know?

What we do know is that this iconic tree has fought its own battles and came out trumps. On 29 June 1998, a cyclone accompanied by torrential rains uprooted it. Aware that the historic fig tree was a major tourist attraction, the Lieutenant Governor quickly put together a team to ensure that it was replanted in its original place and nursed back to health. It had its main branches trimmed off to reduce the crown's weight. The cut ends were smeared with an antiseptic and covered with a muslin cloth to prevent infection. The tree was then hoisted up with the help of a crane and gently lowered into a widened pit filled with fertilizers and root-promoting hormones. To everyone's delight and relief, new shoots started sprouting within thirty days[187] Today, the Rumph's fig tree is again thriving in the jail campus and narrating the story of the Cellular Jail through the *son et lumière*.

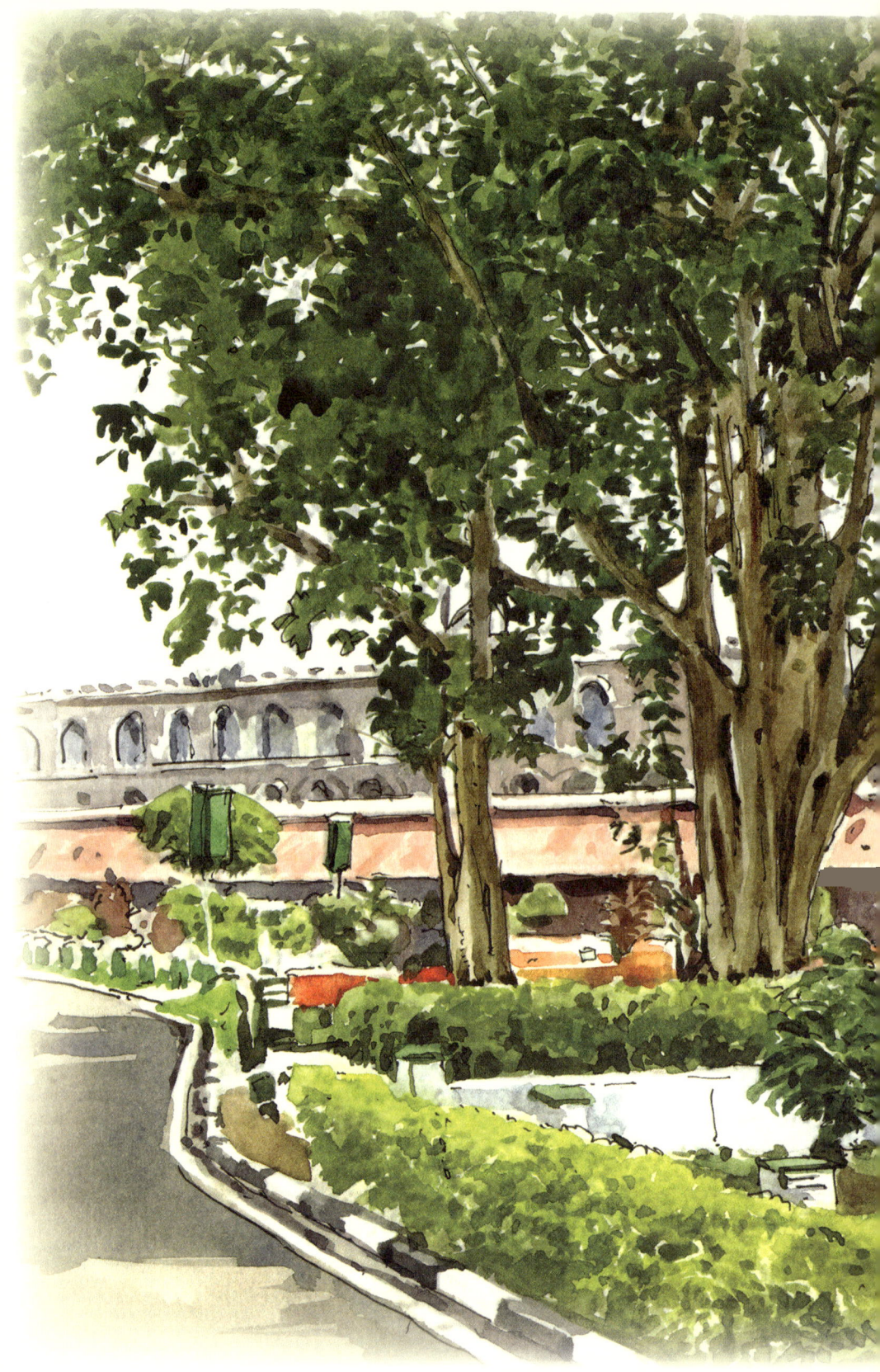

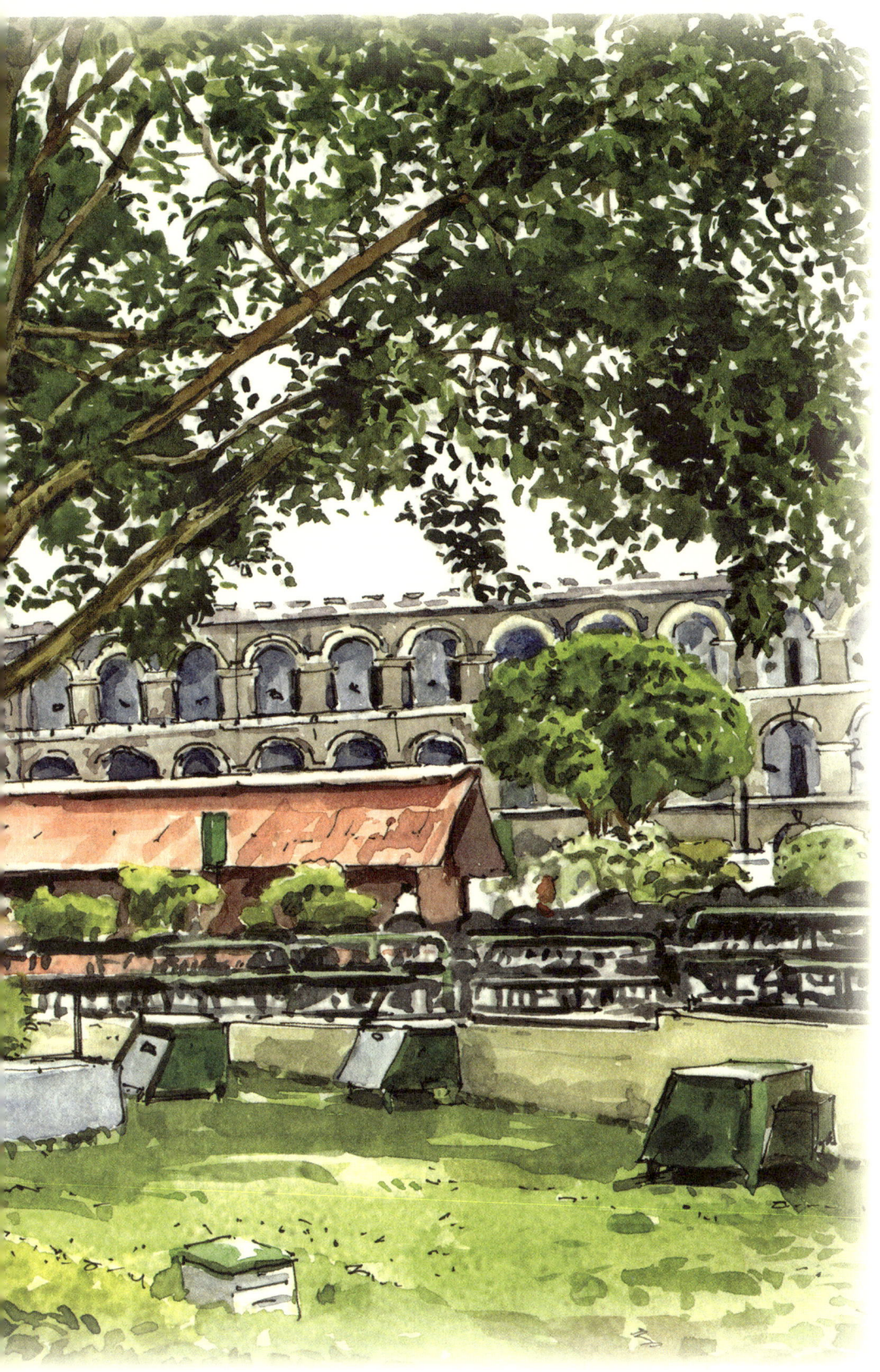

51

The Jamun Tree of Kadugamarathur

COMMON NAMES: **Black Plum, Malabar Plum, Jambolan (English); Jambu (Sanskrit); Naval Maram (Tamil); Jamun (Hindi)**

SCIENTIFIC NAME: *Syzygium cumini* (L.) Skeels

FAMILY: **Guava (Myrtaceae)**

WHERE TO SEE: **On a patta land in Kadugamarathur, Semmanatham panchayat, Yercaud taluk, Salem district, Tamil Nadu**

LATITUDE: **11.8711° N;** LONGITUDE: **78.2264° E;** ALTITUDE: **320 metres**

Perched on the Eastern Ghats at a height of 1,515 metres above sea level, Yercaud is a tourist destination endowed with both pleasant weather and scenic beauty. The mist-shrouded hills support semi-evergreen vegetation dotted with sandalwood and teak. Coffee was introduced here by M.D. Cockburn in 1820, a Scotsman who was then the collector of Salem district and it continues to be the dominant crop. Other cultivated crops include cardamom, black pepper, orange, jackfruit and guava.

There is a famous 32 kilometres-stretch known as loop road that starts and ends at Yercaud Lake and takes one through quiet tribal villages, lush green coffee estates, emerald hill slopes and, if one is early enough, even a gorgeous sunrise. The discerning traveller usually takes this loop road for a joy ride. About 13 kilometres from Yercaud off the loop road near Semmanatham is a small village known as Kadugamarathur with a population of 224, as per the 2011 census. On the edge of a small nameless coffee estate here that has changed hands several times in recent years and just adjacent to a road, stands a strikingly large jamun tree, locally known as 'naval'.

Being native to the Indian subcontinent, the jamun is quite common at this altitude in these hills. It is also widely cultivated in Asia, Africa and South America. The Western world has been aware of the fact that jamun fruits have been consumed by the people in India for long, as noted by Hendrik van Rheede, Dutch colonial administrator and naturalist, who documented its use in his monumental multi-volume work *Hortus Malabaricus* (1678–1793). Probably, the tree was in cultivation much before the fifteenth century in India. *Syzygium,* the genus to which it belongs, is the largest genus in the family Myrtaceae with an estimated 1,200 species across the world and features among the largest genera in terms of the number of species.[188]

This remarkable specimen in Yercaud Hills has an enormous grooved and lobed trunk with a girth of 12.4 metres. Two major vertical branches emerge from

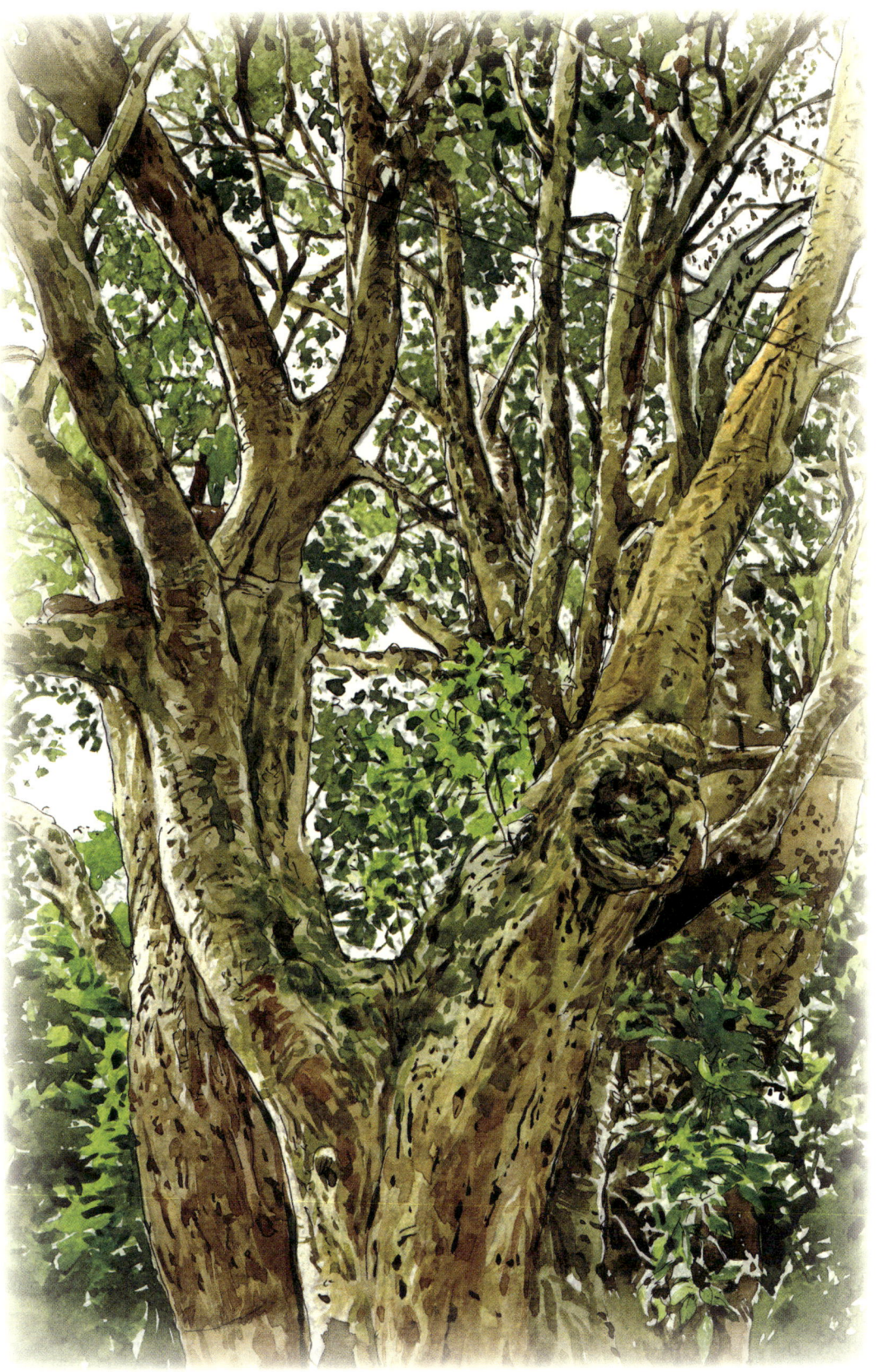

it, each with numerous sub-branches. The bark is yellowish-brown and the foliage is dark green and the lofty crown towers over the rest of the vegetation to a height of 23 metres.

The Tamil Nadu Forest Department has included this champion in its list of heritage trees and reckons its age at 250 years. Some four decades ago, the entire set of branches was pruned as evidenced by several scars and cut branches even now. Remarkably, the tree grew back to its present majestic dimensions. Some of its branches prominently feature honeysuckle mistletoe (*Dendrophthoe falcata*), a hemiparasite that sucks nutrients in the form of water and minerals. Birds and squirrels move on its canopy for shade and doubtless for its fruits in season. According to its current owners, fruiting occurs from July through September. Fruit is contracted out with annual harvests of 300–400 kg. At current rates of ₹200 per kilogram, that adds up to a tidy annual income of between ₹6,000 and ₹8,000 without having incurred any expenditure.

Properly promoted, this champion jamun in the heart of Yercaud Hills will surely be a valuable 'green' jewel in the ecotourism circuit.

52

The Lofty Blue Gum
of Aramby Shola

COMMON NAMES: Blue Gum, Southern Blue Gum, Tasmanian Blue Gum
(English); Thailamaram (Tamil); Safeda (Hindi)

SCIENTIFIC NAME: *Eucalyptus globulus* Labill. | FAMILY: Guava (Myrtaceae)

WHERE TO SEE: Aramby Shola, close to Kandal, Udagamandalam taluk,
Nilgiris district, Tamil Nadu

LATITUDE: 11.4257° N; LONGITUDE: 76.6876° E;
ALTITUDE: Approximately 2,350 metres

About 150 years ago, the Aramby Shola in Udagamandalam was planted with blue gum trees for which the British had imported planting material from Australia. A towering specimen of that bygone age still stands today in the shola, now a protected reserve forest.

Situated in the Western Ghats, the Nilgiri hills are aptly called the 'blue mountains' because once every twelve years the *kurinji* [*Strobilanthes kunthiana* (Nees) T. Anderson ex Benth.] flowers on its slopes, and imparts a blue colour to the hillside. Udagamndalam – Ooty in popular parlance – is nestled at the foothills of the Nilgiris. It was ceded to the East India Company toward the end of the eighteenth century following Tipu Sultan's death after the decisive battle of Mysore. The first English settler was John Sullivan, the then collector of Coimbatore. Soon, a small settlement complete with sprawling bungalows, imposing churches, grand public edifices and lush gardens was established in a bid to produce a facsimile of a 'home' befitting the colonial power. Before long, it became the summer capital of the Madras Presidency.

At the time of British occupation, the hills of Ooty were clothed with natural forests, locally known as *sholas*, comprising patches of dense evergreen trees isolated among grass meadows. Such sholas have been part of the landscape for 35,000 years. Colonial settlers, however, wrongly thought that the grasslands within the mosaics were the result of unsustainable grazing of cattle and deliberate fires started by indigenous tribal communities. Their solution was to set about 'reforesting' the grasslands, primarily through large-scale planting of exotic trees. One of the early problems they faced was the demand for fuelwood. By 1859, an estimated 4,000 fires burned every day on the hills, at least 1,000 of these in the hearth of newly-settled Europeans.[189] To meet the commercial fuelwood requirements of a burgeoning population, they began to steadily extract firewood and timber from the shola forests. Soon, the dwindling supply

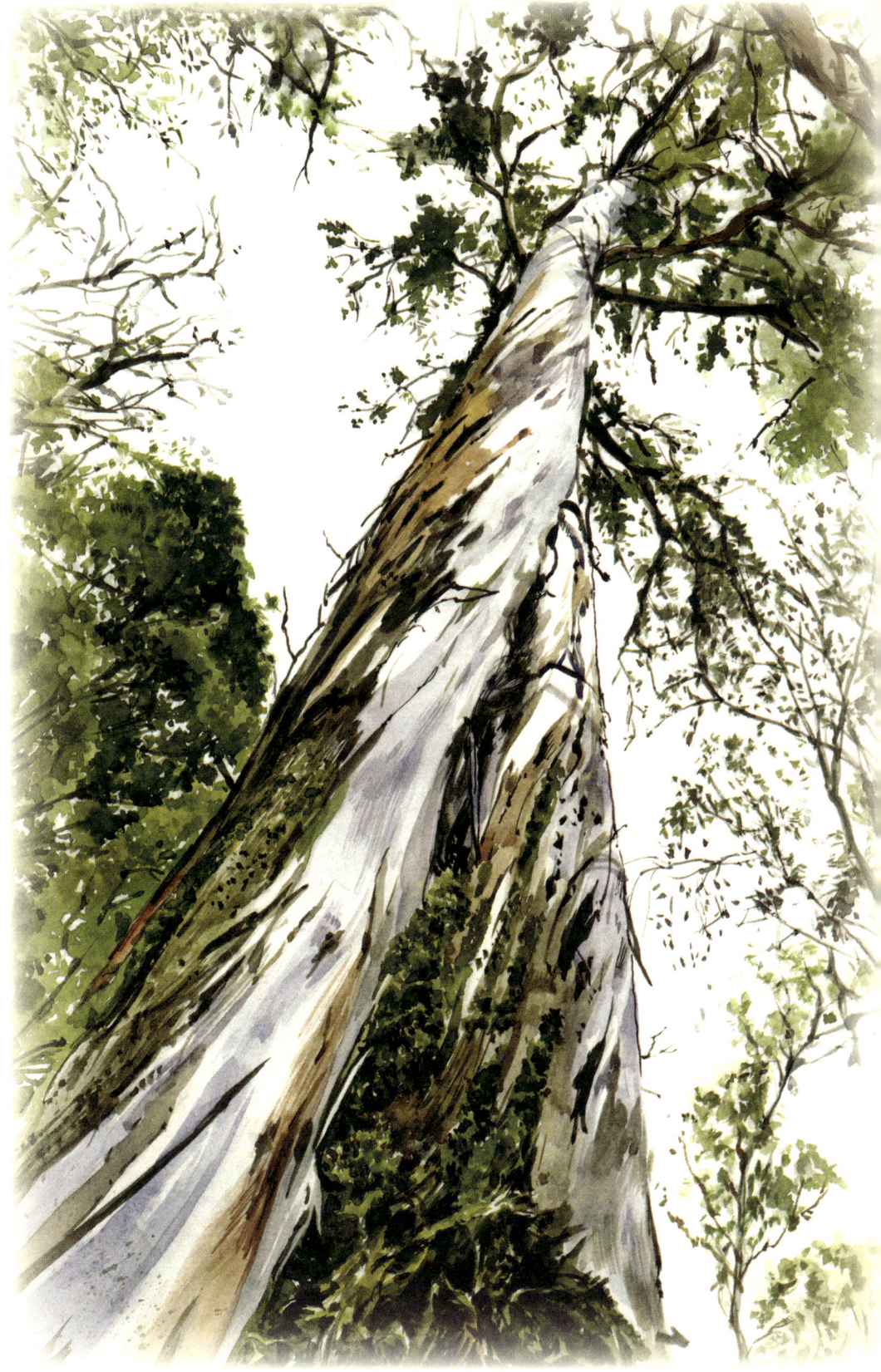

of fuelwood became a matter of concern. Hence, they started cultivating fast-growing exotic species from Australia in a quest for cheap sources of fuel wood. Among the eucalypts, a species of tall, evergreen trees commonly known as the blue gum or southern blue gum (*Eucalyptus globulus*) endemic to south-eastern Australia was the best.[190] It appears to have been a relatively late introduction to the Nilgiris in 1843. However, it is not clear how its seed found its way to India.[191] Some researchers believe that blue gums were introduced around 1832 and the first plantations were raised by Captain H.R. Morgan (then conservator of forests) at his Tudor Hall Estate in 1853.[192]

The Aramby plantation was born only in 1863, the first of the many in the neighbourhood of Ooty 'as a consequence of the foresight of the then governor, Sir William Denison'. Other plantations soon followed and the blue gum could be spotted all over the Nilgiris. Very soon, supplies were so plentiful that not only was the fuel famine a thing of the past, but also fuel wood was available at far cheaper rates than in any other hill station in India.[193] So successful was this exotic that in 1914, just 50 years later, the total acreage under government plantations (pure or mixed with wattle) was 440 ha. By 1951 this had proliferated to 719 hectares with an additional 917 hectares under private ownership.

After Independence, the Forest Department continued this colonial legacy of area expansion of blue gum. It even discovered new uses for the plant: The wood was found to be an excellent raw material for the pulp and paper industry; the leaves, on distillation, could yield a sweet essential oil with a camphor-like aroma useful in making soaps, insecticides and natural medicine. These spawned new industries in the Nilgiris. The state government initiated further large-scale cultivation in 1953 which, by 1970, resulted in 5,600 hectares of blue gum.[194] A decade later, eucalyptus had fallen from grace. It was accused of causing damage to the ecosystem, being a water guzzler and hence depleting soil moisture. Not to mention its inability to support wildlife (as its leaves and fruits are inedible) and producing chemical substances harmful to undergrowth. The last word has not yet been said and the great debate still rages.[195] One immediate outcome of the public outcry that soon followed was that fresh plantations of blue gums were halted after the 1980s. And a recent study has shown that while the plantations have adversely impacted the shola forests, the effect is even more drastic on the grassland in these hills.[196]

The tall blue gum tree in the Aramby Shola – now a respectable 160 years old – is a relic from a bygone colonial era. Its thick trunk with a circumference of 7.5 metres is covered with rough greyish-brown bark like a worn-out stocking at the lower end. Sexually, the champion tree is still active, producing flowers and small fruits in bunches. How and why it escaped the woodman's saw is something we are unlikely to know. Whatever the reasons are, we are glad it did.

53

The Ugamaram Grove of Omāndur

COMMON NAMES: **Toothbrush Tree, Miswak, Mustard Tree, Saltbush (English);**
Pilu, Gudaphala (Sanskrit); Ugamaram, Kunnimaram (Tamil)

SCIENTIFIC NAME: *Salvadora persica* L.

FAMILY: **Miswak (Salvadoracceae)**

WHERE TO SEE: **Annai Kamatchi Amman Kovil, Omandur village,**
Manachanallur block, Tiruchirappalli district, Tamil Nadu

LATITUDE: **11.029° N;** LONGITUDE: **78.668° E;** ALTITUDE: **67 metres**

Barely 30 kilometres from the busy city of Tiruchirappalli is the small and quiet village of Omandur. Prominently sitting in the middle of the village is a *kovil* (temple in Tamil) of Annai Kamatchi Amman or Goddess Kamakshi, the consort of Shiva. In a state known for its temples what sets this one apart is that instead of idols, the deities are represented by the light of oil lamps. More importantly, on nearly 3.24 hectares (8 acres) of open ground surrounding the temple buildings, there is a grove of some 215 old pilu (peelu) trees, locally called ugamaram. They are like the *sthalavriksha*s (temple trees).

The temple is said to be between 300 and 400 years old. The trees have always been there according to the current head priest.[197] Each tree is stately and together they make a fabulous grove. Many of them are at least 200 years old, reaching beyond 12 metre in height. The beautifully gnarled and furrowed trunks are short and often crooked and greyish-white with characteristic rough-and-wrinkled bark. Some older trees here reach girths of 4.5–5 metre. Branches are numerous and drooping bearing a dense crown of oppositely arranged leaves. A few of the trees are singled out for special attention with pieces of coloured cloth tied to the trunk and branches for wish fulfilment. Flowering commences in January and February and small round reddish bead-like fruits appear by March and April, turning black upon ripening.

Common in the arid regions of India, especially on clayey saline soils, this species' global range extends from North and East Africa, Egypt and Middle East right across India to the coastal regions of Sri Lanka. In West Asia and other regions including India, the species has long been valued for its mouth-refreshing and soft fibrous twigs and hence referred to as the toothbrush tree. There are references to its mouth-cleansing property in the *Hadith*. It is reported that the Prophet advised Muslims to use it before prayers, especially Friday prayers, to keep away bad breath.[198] There is a belief that the tree was brought to South Asia by West Asian traders in medieval times as fodder for horses and camels.

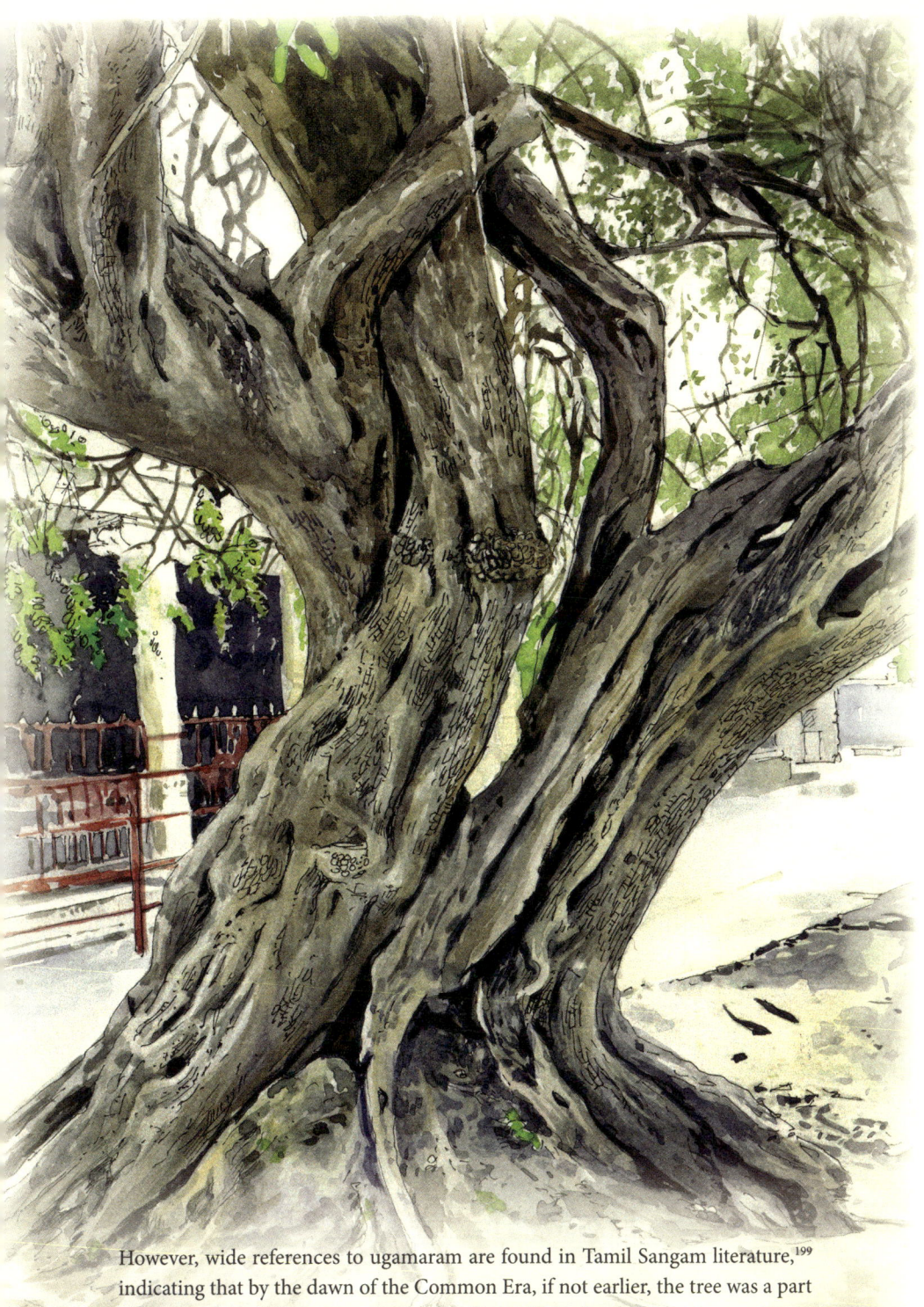

However, wide references to ugamaram are found in Tamil Sangam literature,[199] indicating that by the dawn of the Common Era, if not earlier, the tree was a part of the natural vegetation even in remote forests.[200] Each tree in the grove appears to have an individual 'personality' of its own.

54

The Manjakadambu Tree of Velliangiri

COMMON NAMES: Haldu, Yellow Teak (English); Girikadamba (Sanskrit); Manjakadambu, Mannakatampu, Poonteak (Tamil); Haldu (Hindi)

SCIENTIFIC NAME: *Haldina cordifolia* (Roxb.) Ridsdale

FAMILY: Coffee (Rubiaceae)

WHERE TO SEE: Sri Velliangiri Andavar Temple, Velliangiri, Coimbatore district, Tamil Nadu

LATITUDE: 10.9697⁰ N; LONGITUDE: 76.7244⁰ E; ALTITUDE: Approximately 1,700 metres

A beautiful haldu tree, locally known as manjakadambu, stands in the holy shrine of Andavar and his consort Manonmani Amman amidst the Velliangiri Hills. The hills are part of the Nilgiri Biosphere Reserve on the Western Ghats situated at the junction of the Coimbatore district of Tamil Nadu and the Palakkad district of Kerala. Reputed as the 'Kailas of the South', the seven hills that constitute the range are believed to be the legendary abode of Shiva, whose idol is worshipped as *swayambhu* (self-created) in a cave atop the hills. The hills are a destination not only for thousands of pilgrims but also for adventurous nature enthusiasts who love to trek on its breathtakingly beautiful slopes which are full of verdant forests, natural streams and waterfalls. The manjakadambu tree – known as haldu in north India[201] as well as in trade – is valued for its timber and medicinal uses. It is also worshipped as a symbol of fertility by certain tribal communities of eastern and central India.[202]

Here in the Velliangiri ranges, the temple complex is constructed on a series of landings up the hilly incline with the sanctum at the summit. The charismatic manjakadambu can be seen on the right side as soon as you enter the temple gateway and start climbing the steps. Known to reach heights of up to 45 metres or more, this particular tree is roughly 30 metres tall with massive branching and a large beautiful spreading crown that cannot be missed. As you approach it, it is possible to see the clean and straight trunk, part of which is buried in a brick wall painted alternately in red and white stripes behind the tree, making accurate measurement impossible. The State Forest Department estimates its girth at 4 metres. The leaves are arranged in pairs, a characteristic trait of the coffee family. They are heart-shaped at the base and pointy at the tips. Flowers are tiny and yellow, gathered in globular heads much like those of the kadamb and so are the fruits.

The tree has large buttresses, one of which is visible on the front portion but a concrete platform built on either side of the tree has concealed others that

194

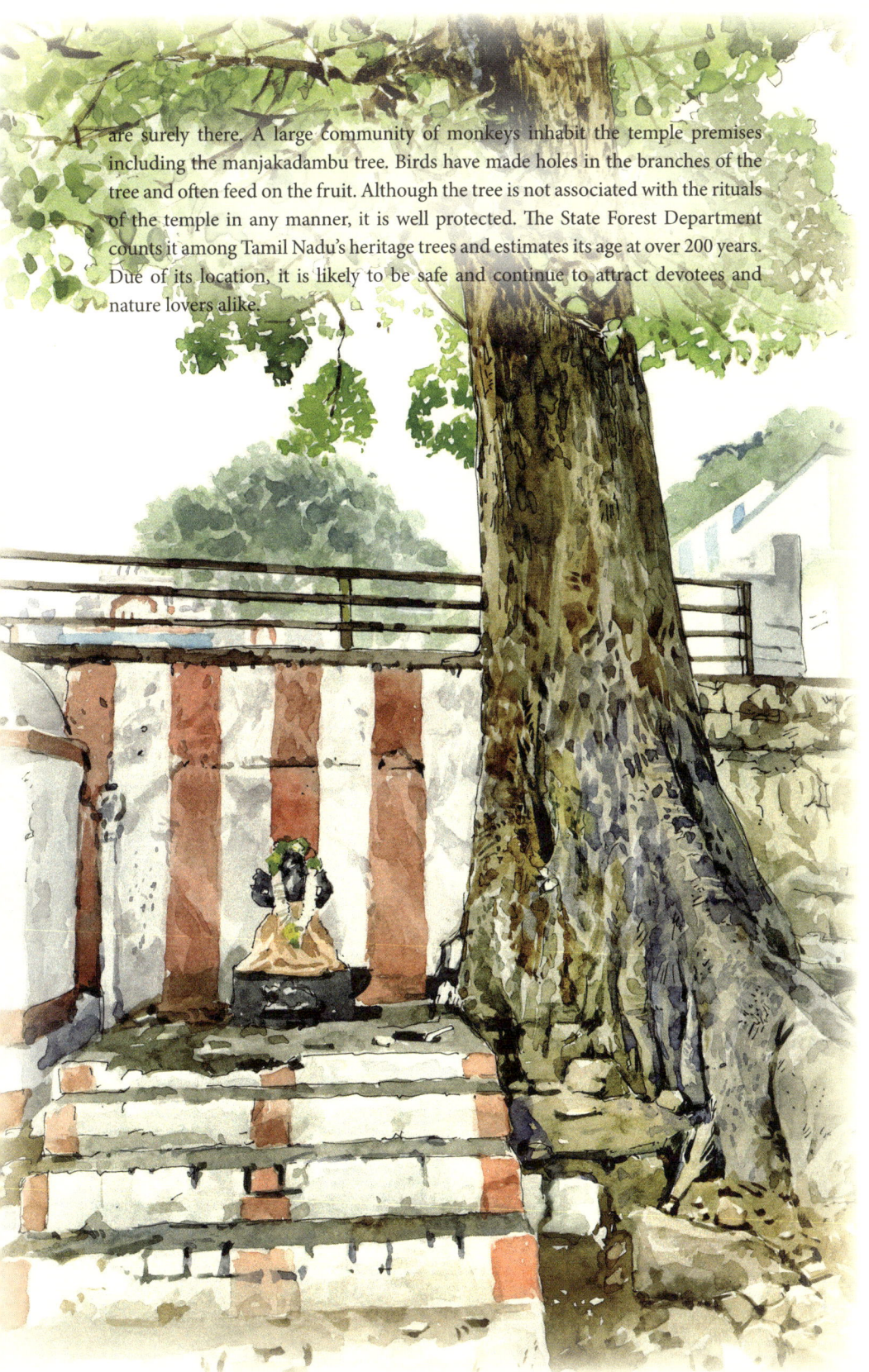

are surely there. A large community of monkeys inhabit the temple premises including the manjakadambu tree. Birds have made holes in the branches of the tree and often feed on the fruit. Although the tree is not associated with the rituals of the temple in any manner, it is well protected. The State Forest Department counts it among Tamil Nadu's heritage trees and estimates its age at over 200 years. Due of its location, it is likely to be safe and continue to attract devotees and nature lovers alike.

55

The Rosewood of Topslip

COMMON NAMES: **Rosewood, Indian Rosewood, East Indian Blackwood, Bombay Blackwood (English); Shimshapa (Sanskrit); Eeti (Tamil)**

SCIENTIFIC NAME: *Dalbergia latifolia* **Roxb.**

FAMILY: **Legume (Fabaceae, Papilionoideae)**

WHERE TO SEE: **Kozhikamuthi Elephant Camp, Topslip, Anamalai Tiger Reserve, Akilandapuram, Pollachi, Coimbatore district, Tamil Nadu**

LATITUDE: **10.4427⁰ N**; LONGITUDE: **76.8511⁰ E**; ALTITUDE: **Approximately 740 metres**

A magnificent rosewood tree inside the Indira Gandhi Wildlife Sanctuary and National Park – better known as Ānamalai Tiger Reserve – is worthy of note for nature lovers. Few people, however, get to see it as it is well inside the reserve.

Entry to the Ānamalai reserve ('elephant hill' in Tamil) is through Topslip – the name reminiscent of an era when timber was slid down the hill slopes.[203] Timber extracted from Thekkadi forests was carted to Topslip through a road of 10.46 km developed in 1850 and skidded down the ghat to the foothills by dry slide. It was then carted to Palghat (now Palakkad), floated down the Bharathapuzha River to Ponnani and then shipped to the Naval Dockyard in Bombay (now Mumbai). The system was in vogue until 1868, when a ghat road from Sethumadai was constructed. There was a time when elephants were engaged in logging here, but since 1994, timber operations are banned and the elephants are mainly used for joy rides or *kumki* operations. These include driving out wild elephants that stray into neighbouring villages, their capture for captivity and subsequent radio-collaring.[204] Only male elephants are used for *kumki* duty. The rosewood tree stands adjacent to their camp.

Although the natural range of rosewood – so named for the smell of its timber when cut – is from the sub-Himalayan tract to the southern tip of India (and the island of Java in Indonesia), it grows best in the Western Ghats of Karnataka, Kerala and Tamil Nadu. Here in the moist deciduous forest near Kozhikamuthi, the tree shares space commonly with teak, bamboos, species of *Terminalia* and dhau (*Anogeissus latifolia*). The first part of its scientific name *Dalbergia* commemorates the eighteenth-century Swedish plant collector Nils Dalberg (and some say, also his brother Carl Dalberg). The specific epithet *latifolia* means broad-leaved in Latin. For very long, rosewood has been sought after as a top-quality timber as it

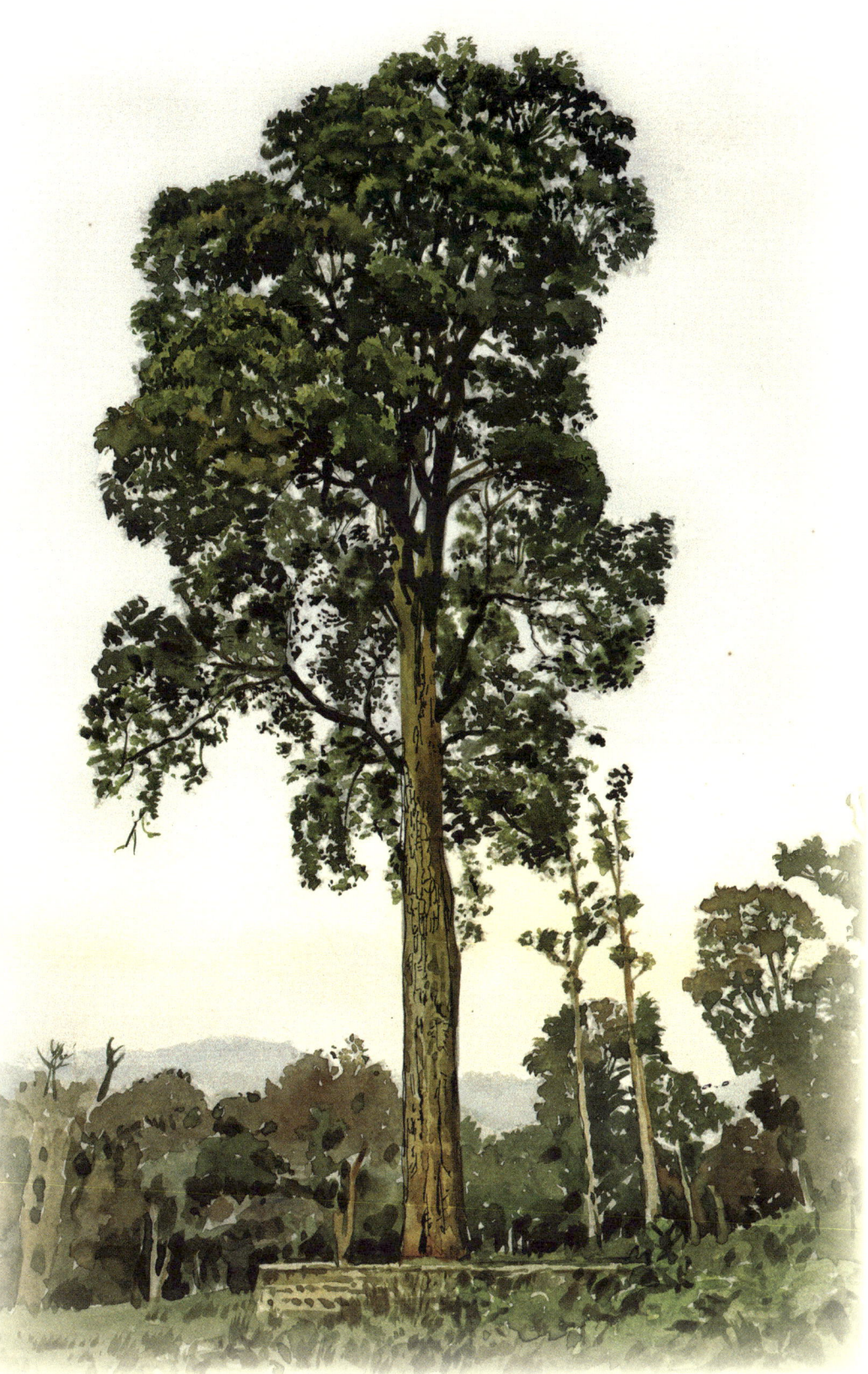

is heavy, close-grained, takes fine polish and used for premium furniture, musical instruments (pianos, clarinets and guitars), and religious objects.

In the Ānamalai, it was called Malabar Blackwood – wherein planks four feet broad were often procurable after all the external white wood had been removed.[205] In their quest for commerce, the British systematically stripped the forests of the Western Ghats along with teak. By the middle of the nineteenth century, not much rosewood was remaining in the Ānamalais. This is why it is not very common to find rosewood trees of this vintage still intact.

Estimated to be around 200 years old, the tall and stately tree can be easily seen on the hill slope from the elephant camp. The forest department has made a circular platform around its base. The clear bole rises majestically upright bearing a dome-shaped crown of lush green foliage. The tree's height at ~25 metres equals that of an eight-story building! The big trunk has a circumference of close to six metres. It takes at least 40-50 years before the timber is good for harvest and going by that yardstick, this tree would be priceless.

Being a timber of high commercial value, rosewood is overexploited. There is a thriving illicit international trade in it[206] and the IUCN has included this species under the 'vulnerable' category. The species is protected under the Indian Forest Act. While domestic needs can be met, the export of logs or sawn timber is banned. Being inside the reserve forest, this tree is well protected.

You are not Atlas carrying the world on your shoulder. It is good to remember that the planet is carrying you.

Vandana Shiva
Author and environmental activist

56

The Ancient Malai Naval
Tree of Kodaikanal

COMMON NAMES: Malai Naval, Kurujaval, Kruthal, Kuruthamaram, Nagay, Pillanjaval (Tamil); Loosely Jamun (Hindi)

SCIENTIFIC NAME: *Syzygium densiflorum* Wall. ex Wight & Arn.

FAMILY: Jamun (Myrtaceae)

WHERE TO SEE: Upper Shola Road, Bombay Shola, Kodaikanal, Dindigul district, Tamil Nadu

LATITUDE: 10.22472⁰ N; LONGITUDE: 77.48306⁰ E; ALTITUDE: 2,124.4 metres

Kodaikanal – the princess of hill stations – is an extremely popular destination for honeymooners. A lakeside resort town nestling amidst the rolling mountains of the Palani Hills in Tamil Nadu at an altitude of 2194 metres, it enjoys a beautiful climate, mist-covered cliffs, meandering rivers and gurgling waterfalls – all of which create an enchanting ambience and flawless setting for a picture-perfect getaway. One meaning of Kodaikanal is 'gift of the forests'.

One example of such a gift is a rare living gem in the form of a gigantic tree of malai naval, a cousin of the jamun tree, located approximately midway between the bus stand and the Kodaikanal Lake in an area known as Bombay Shola. Kodaikanal, being a part of the southern Western Ghats, is also home to a unique mosaic of forests and grasslands. The forests in these parts are referred to as sholas.[207] Bombay Shola[208] is a 25-hectare patch of woods set amidst the increasingly chaotic development of the surrounding municipality. It was so named in 1852 by a Major Partridge of the Bombay Army who camped on its eastern edge near what is now Bryant Park. Designated a reserve forest before Independence, Bombay Shola faces heavy pressure on its resources and space today.[209]

Located on a slope of the Upper Shola Road, just across from the heritage resort Tamara Kodai, the first thing you notice about the malai naval tree is its size. Indeed, it has been described as 'the largest tree in Palani Hills'.[210] At a girth of 15.2 metres and height of 21 metres, this may well be true. Approaching the tree, you can see that its massive trunk is much callused and knobby, with colourful lichens, liverworts, mosses and flowering plants growing in its crevices. A closer examination reveals it to be almost completely hollow with a large gaping, tunnel-

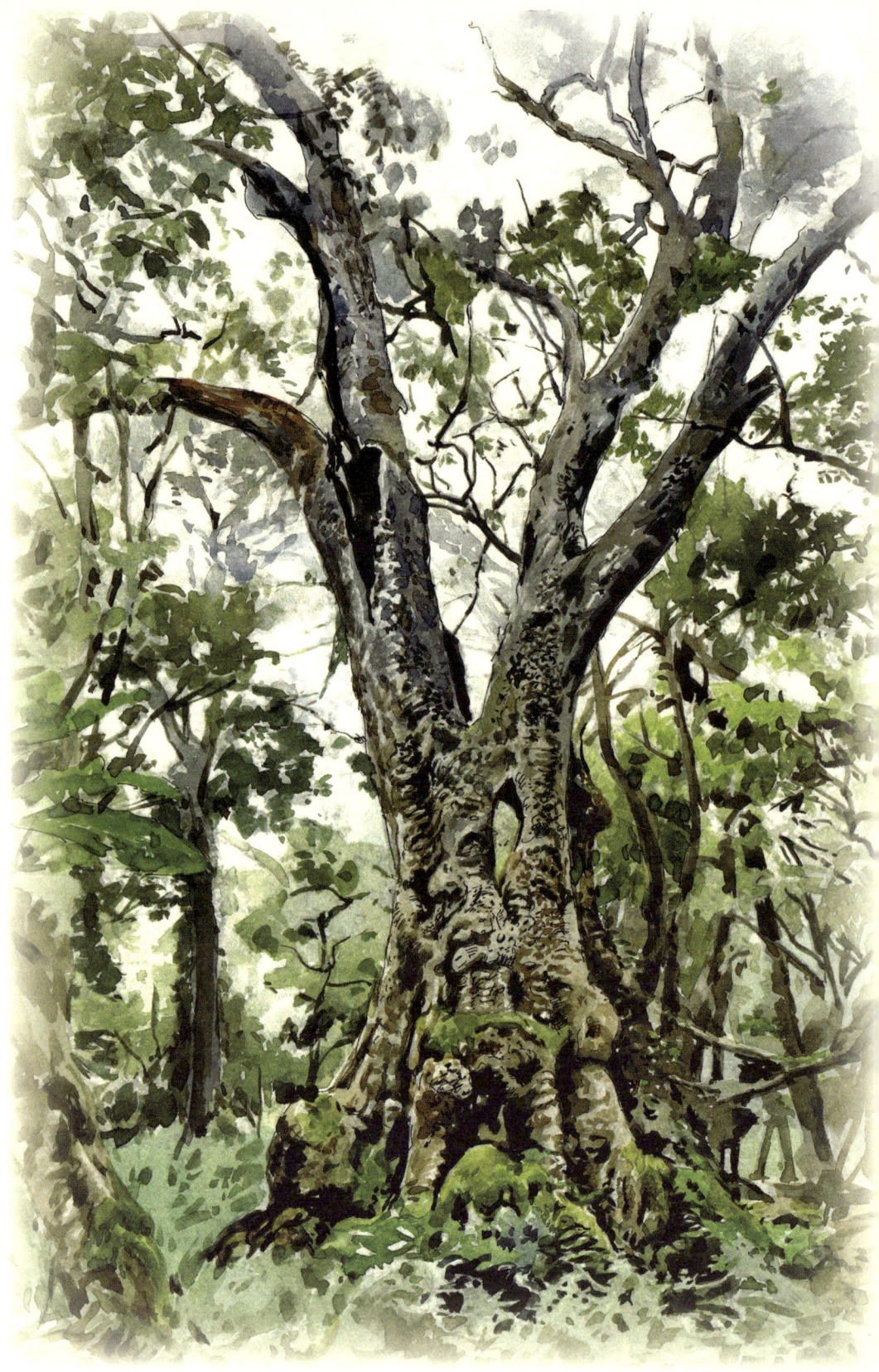

like opening at the base – roomy enough to accommodate five or six persons easily. The hollow of the trunk is created by the rotting away of the wood. Sitting inside the trunk and gazing up, you can see the sky through another aperture above. Two major branches emerge from the trunk at the periphery of this aperture and fork off further into secondary and tertiary branches that bear a large crown. Emerging new shoots and healthy leaves reassure you that the centurion is still thriving. The tree's roots have given rise to new shoots around the trunk through coppicing. Standing under the grand tree champion is a humbling experience. How old might this giant be? It is hard to say without a scientific investigation, but 500 years is perhaps closer to the mark. Old photographs of the tree taken in 1903 and 1994 are available and show the tree to be much the same as it is today.[211]

It is difficult to imagine how this evergreen relic from a bygone era survived so long amid the hustle and bustle of a popular tourist resort. The shola-grassland habitat has been facing intense anthropogenic pressure due to the cultivation of exotic species along with increasing plantation activity. A recent study using satellite imagery and field surveys has revealed that during the past 40 years alone, the Palani Hills lost about 66 per cent area of grasslands and 31 per cent area of sholas respectively to alien plantations and expanding agricultural activities.[212] Although malai naval is native to the high montane Shola forests of Karnataka, Kerala and Tamil Nadu, it is classified as 'vulnerable'.[213] Under these circumstances, the centurion tree would require protection to remain safe and unharmed. Fortunately, there are concerned scientists and voluntary organizations committed to preserving the natural heritage of the area.

57

The Sacred Vannimaram of Kodumudi

COMMON NAMES: Mesquite (English); Shami (Sanskrit); Vannimaram (Tamil);
Jhand, Khejdi (Hindi)

SCIENTIFIC NAME: *Prosopis cineraria* (L) Druce·

FAMILY: Legume (Fabaceae, Mimosoidae)

WHERE TO SEE: Brahma Shrine in the Magudeswarar Temple Complex,
Kodumudi, Kodumudi taluk, Erode district, Tamil Nadu

LATITUDE: 10.07278⁰ N; LONGITUDE: 77.88861⁰ E;

ALTITUDE: 144 metres

An ancient Shami tree – locally known as Vannimaram[214] – in the town of Kodumudi on the banks of the Cauvery River in Erode district of Tamil Nadu is worshipped regularly in the Magudeswarar temple complex as an incarnation of Lord Brahma, the creator and one among the holy trinity of Hindu gods.

The 12-metre-tall tree stands on a square stone podium at a height of 1.5 metres from the ground and has been cordoned off with metal fence above it. The trunk of the tree is much gnarled with several burls at the base and has a girth of approximately 4.6 metres. It bears two main branches, one going eastwards and the other westwards. The stouter eastern branch is relatively vertical but the western branch projects outwards at an angle of 60 degrees and is supported by a metal prop. Leaves are copious on secondary branches, but local people say that the tree does not bear flowers or fruits. Estimations of the tree's age vary, but it is probably about 300 years old. Devotees congregate on the eastern side of the platform to offer votive offerings to a small stone image of Brahma kept at the base of the tree. As is usual in temples of southern India, a priest presides over the worship and devotees are not allowed near the tree.

There is a fascinating story about Brahma in the *purānas*, chiefly the *Shiva Purāna*[215] and the *Kūrma Purāna*.[216] It is said that originally Brahma the creator had five heads (instead of four as he is now depicted). Pleased with the beautiful world he created, he considered himself supreme among the trinity of gods. Once, when an argument broke out between him and Lord Vishnu about who is the ultimate god of the universe, words soon led to a battle and they started showering arrows on one another, engulfing the whole universe in their recklessness. Fearful of the outcome of this unseemly conduct between the two of the trinity, the assembled gods appealed to Lord Shiva to intervene and save the universe. Suddenly, an enormous column of light appeared between the two warring gods.

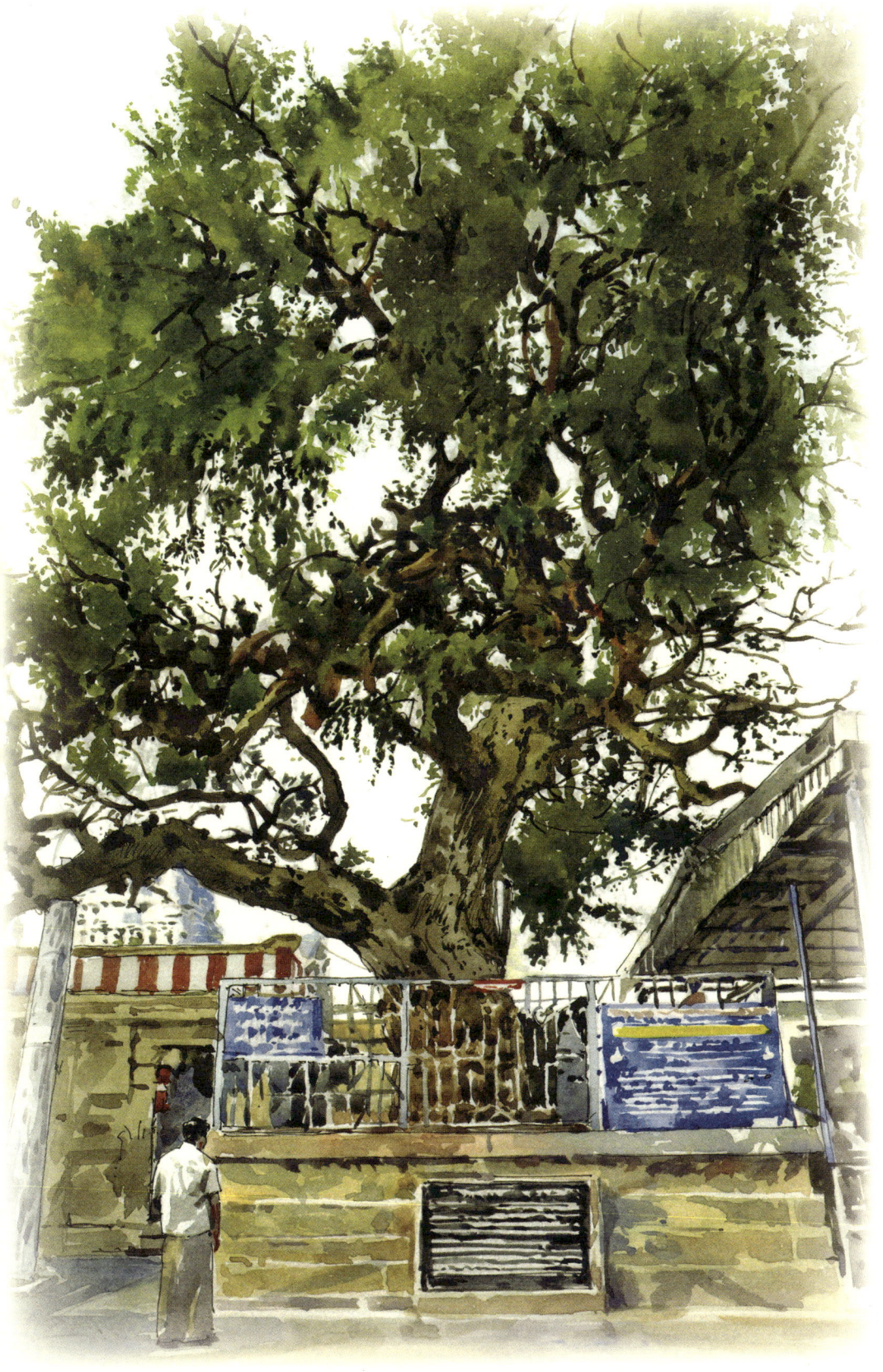

Neither of them could ascertain what it was or how far it extended. Intrigued, they decided to suspend their battle long enough to explore the column, agreeing that the one who could discover either the top or the bottom end of the column would be the undisputed leader. Brahma took the form of a swan and flew away towards the sky to explore the tip of the column. Vishnu assumed the shape of a boar and went off underground into the nether world to find its root.

As Vishnu kept going down without being able to trace the column's end, he realized that it must be a manifestation of Shiva the Supreme. Admitting defeat, he came up and waited for Brahma. Meanwhile, Brahma could not discover the tip of the flame either. But filled with too much ego, he did not wish to lose face and decided to tell a lie. To buttress his claim, he collected a *ketaki* (screw pine, *Pandanus odorifer*) flower along the way, stating that he discovered it at the tip of the column.

No sooner were these false words uttered than the fiery column of light burst open revealing Shiva in his terrifying form. Rebuking Brahma for his greed for glory and his desperation to achieve it even through unfair means, Shiva declared him unworthy of worship henceforth and cut off his fifth head. At once, Brahma fell dead in front of him. Following Vishnu's plea, he was brought back to life albeit with four heads. According to the *sthala purana* (temple legend) of the Kodumudi temple, a contrite Brahma paid his obeisance to Shiva and took the form of a Vannimaram at Kodumudi. And that is the reason devotees worship this ancient tree. Ketaki, the false witness, was banished from Shiva's presence and declared unfit for worshipping him anymore.

Locals also believe that the tree's leaves have water-purifying properties. Pilgrims travelling to Palani to worship Lord Muruga during the Panguni-Uthiram[217] festival carry a pot of holy water with a leaf of the tree from Kodumudi as an offering.

It is not very usual to find old trees of shami such as the one in the temple complex of Kodumudi. The species probably has a much shorter life span, about 150 years. It is the care taken by the temple authorities and the devotion shown by the devotees that have kept this centurion tree alive and healthy for so long.

58

The Chained Tree of Lakkidi

COMMON NAME: **Warty marble tree, Deccan Olive (English); Rudraksham (Sanskrit); Ammakkarām; Bhadraksham, Pilahi (Malayalam)**

SCIENTIFIC NAME: *Elaeocarpus tuberculatus* Roxb.

FAMILY: **Rudraksh (Elaeocarpaceae)**

WHERE TO SEE: **Lakkidi Village, Vythiri taluk, Wayanad district, Kerala**

LATITUDE: **11.5185 ° N;** LONGITUDE: **76.0208° E;**
ALTITUDE: **765.8 metres**

Perched 700 metres above sea level in the picturesque Wayanad district, the village of Lakkidi lies in north-eastern Kerala. On the western side of the road close to the misty ghat pass, as one approaches the village from Kozhikode, stands an unusual tree. A hefty steel chain is wound on one of its branches. Although the tree by itself is neither grandiose nor imposing, the thick steel chain, the small shrine on a cement platform at its base as well as the large black stone lamp in front of the platform are sure to attract everyone's attention. And thereby hangs a powerful and spooky legend.

As late as the early nineteenth century, Wayanad was covered with dense rainforests and inaccessible from the rest of India. Attracted by its natural beauty, the British were eager to build a road through the region but were greatly handicapped as they did not know the topography of the rough mountain landscape. After some failed attempts, a handsome reward was announced by the East India Company to anyone who succeeded in cutting a road across the rugged mountain terrain.

One of the company engineers sought the help of Karinthāndan, a local tribal youth, innately skilled in the ways of the wilds, who pointed out the shortest possible route to the officer. The construction of the road was commenced and at last, accomplished. Not willing to share the reward or the glory with Karinthāndan, the engineer secretly shot him with his pistol and hid the body in the bush. In due season, goods and traffic started moving across the hard-earned mountain road to the delight of the British. However, there were too many accidents along the ghat road. When rumours started spreading that the wandering ghost of Karinthāndan was haunting and waylaying travellers and luring them to death, the terrified locals summoned a priest-sorcerer to find a solution. The sorcerer performed some rituals at the end of which he captured and bound the soul of the dead youth to this tree with an iron chain, still visible after all these years.

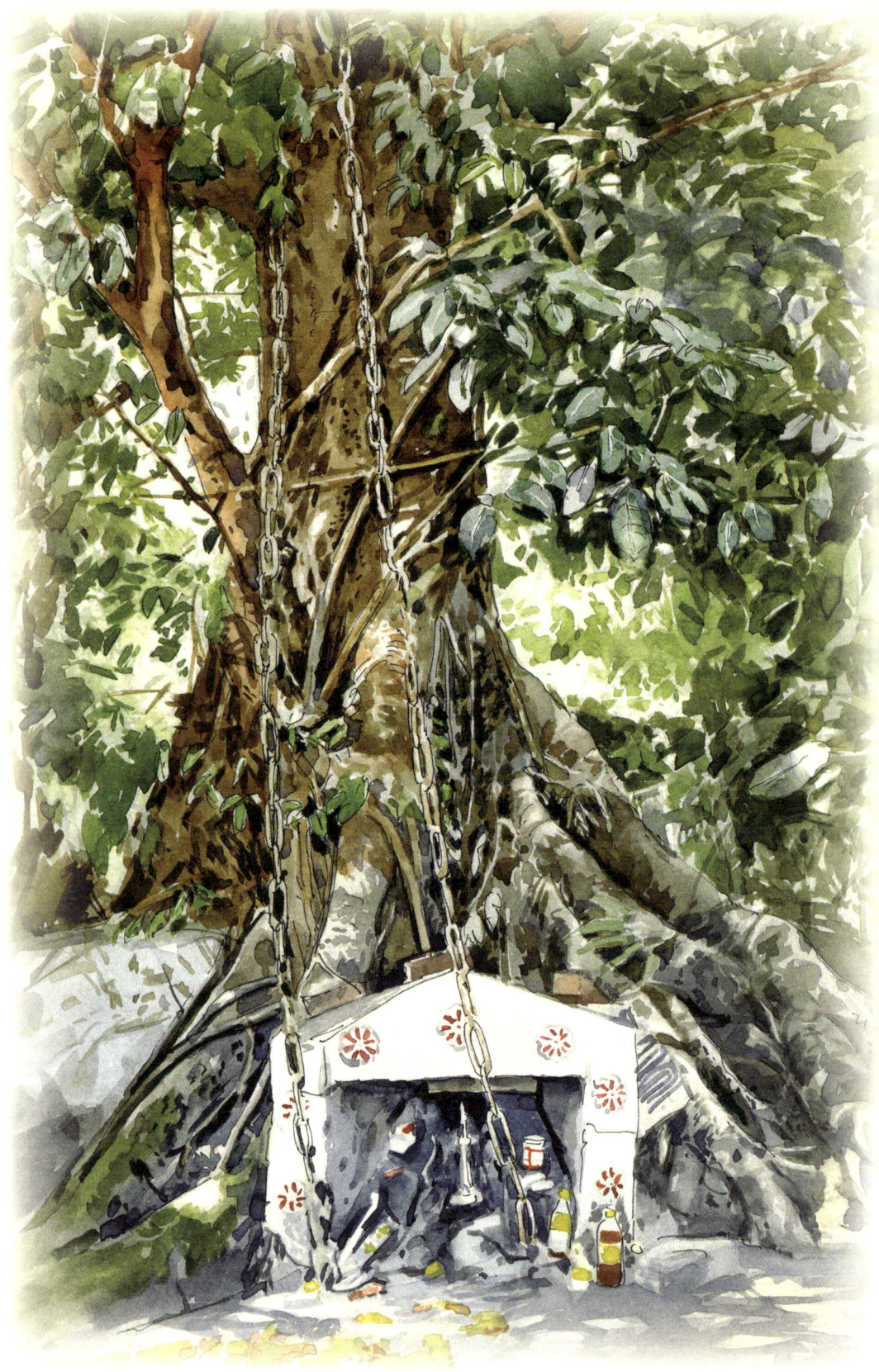

The chained tree has been wrongly identified in the popular media as a fig tree.[218] A white fig plant (*Ficus virens* Aiton) is indeed visible in the foreground, especially at the lower levels but it is obviously perched on the chain tree as an epiphyte. There are also a wild cinnamon (*Cinnamomum sp.*) and a choulmogra tree [*Hydnocarpus pentandrus* (Buch. -Ham.) Oken] growing right next to the chain tree. The real chain tree is in fact what is locally known as Ammakkāram or Bhadrāksham (botanically *Elaeocarpus tuberculatus* Roxb.), a cousin of the Rudraksh tree. This tree is native to the Indo-Malaysian region and occurs in the Western Ghats as well as south and central Sahyadri ranges. It is about 20 metres tall, with prominent buttresses on the trunk. The tree's bark is smooth and mottled with grey and white. The branches are thickly clothed with moss, several species of orchids and two climbing shrubs.[219] The undivided leathery leaves tend to cluster at the end of the twigs. The flowers are white and cup-shaped, with many stamens. The somewhat rounded fruits – technically called 'drupe' – contain a solitary seed that is deeply compressed and warty as in the *rudraksh*. Bhadrāksham seeds are generally used as beads for rosaries, though it is unlikely that anyone collects the seeds from this particular tree. As mentioned before, unknown devotees have built a semi-circular cement platform around the base of the tree. A thick steel chain has been looped to one of its branches with its two free ends firmly driven into the ground in front of the tree. Just behind the iron chain is a small shrine to Karinthāndan.

The haunted lore of Karinthāndan is quite popular in the region. People believe that as the tree grows in height, so does the length of the iron chain. Members of the local Paniya[220] community organize an annual procession on the first Sunday of March in memory of Karinthāndan. There is, however, no proof of the legend. Nobody knows for sure when it actually happened nor recall the name of the British engineer who allegedly killed Karinthāndan. Even so, no one wishes to take a chance with the tree or its ghost bound by the chain. Drivers of passing cars and trucks invariably stop by to light incense sticks and make small votive offerings, praying for a safe passage.

59

Conolly's Teak Plantation

COMMON NAMES: Teak (English); Saka, Sagoon (Sanskrit); Thekku
(Malayalam); Sagun, Sagwan (Hindi)

SCIENTIFIC NAME: *Tectona grandis* L.f.

FAMILY: Tulsi (sacred basil) (Lamiaceae)

WHERE TO SEE: Conolly's Plot, Nilambur, Malappuram district, Kerala

LATITUDE: 11.268⁰ N; LONGITUDE: 76.206⁰ E;

ALTITUDE: 55 metres

A small town in Malappuram, Kerala and close to the Nilgiri range of the Western Ghats, Nilambur's most famous attraction is India's earliest surviving teak plantation, located just about two kilometres away on the Kozhikode-Gudalur Road. Locally known as Conolly's Plot, it is named after Henry Valentine Conolly (1806–1855), the district collector and magistrate of Malabar, who was instrumental in establishing it. In the olden days, you would have been ferried across the Chaliyar River, but nowadays you can just walk through a steel suspension bridge from the main entrance.

The story of how this plantation came to be established is interesting. Soon after Malabar was ceded to the British following the Third Anglo-Mysore War of 1792, the East India Company began tapping all potential sources of profit from its new territory. This was the age of colonial expansion by European powers characterized by naval warfare by rival contenders. The security of the British Empire depended on its naval might which, in turn, relied on a continuous supply of oak timber for building new and better ships as well as keeping them in seaworthy condition for the king's navy. Unfortunately, there was not enough oak in Britain to meet its military demand. The British admiralty, therefore, sent frantic inquiries to its overseas territories requesting a search for a suitable replacement for oak timber and enlisted the assistance of the East India Company. To its delight, an excellent substitute was found in teak.[221]

Teak occurs naturally in the moist dry deciduous forests of peninsular India along the Western Ghats.[222] It is easily worked and has great elasticity of strength – features that make it ideal for use in boat and ship construction. Beginning in 1796, a succession of agencies was established in Malabar to ensure a steady supply of first-class timber for the British Navy. The organized stripping of teak and rosewood (*Dalbergia latifolia* Roxb.) through these agencies, however, led to the ruin of much of the forests, especially in the vicinity of rivers and streams or other means of transport. The advent of the railway added further pressure on the demand for teak for making sleepers.[223] By 1860, Britain had emerged as the world leader in deforestation and not just within its borders. Anxiety over the possibility of the ultimate drying up of the crucial teak supply placed the idea of conservation at the forefront of colonial concerns.

The mission of developing a plantation was entrusted to H.V. Conolly in 1842 and he selected the forests lying to the west of Nilambur, where 3,000 seedlings were grown and 1,000 seedlings picked from the natural forest floor. Among those who assisted Conolly in this task was one Sergeant Graham who, before long, resigned from the appointment although he was offered a higher salary, as he could not stand the 'discomfort of life for himself and his family, nor suffer the indignity of having to take off his trousers before the crowd on the bank every time he had to cross the river'.

In 1844, Chathu Menon, a local man, was appointed as sub-conservator at a monthly salary of ₹50, pending the appointment of an expert arboriculturist. Thus began an association between the two men which lasted until 1855, when Conolly was tragically murdered during the Moplah rebellion. Chathu Menon continued as sole in-charge until 1862.[224] Over two decades, he toiled, raising teak plants and even solving the worrisome problem of seed germination. Between 1844 and 1862, approximately 612.29 hectares (1,512.71 acres) of plantation were raised with an astonishing average of approximately 43.70 hectares (108 acres) each year. Considered in the backdrop of the gruelling and extremely inhospitable conditions prevailing then – the virtual absence of communication with the outside world, constant threat of virulent malaria, inadequate labour force, frequent scarcity of provisions and silvicultural practices at best experimental – this seems to be an amazing achievement! It is not surprising, therefore, that Chathu Menon is regarded as the father of teak plantations' in the region. Today, about 0.93 hectares (2.30 acres) of the original plantation raised by Conolly and Menon in 1846 is preserved at the spot.

Conolly's Plot represents the world's first surviving cultivated area of teakwood and attracts foresters and lay people to pay homage to the two stalwarts. The plantation represents the first step towards systematic forest management in India. It also ushered in a new forestry technique of artificial regeneration of preferred tree species in monoculture plantations. In August 2017, Nilambur teak became the first forest produce to be protected under the Geographical Indications of Goods (Registration and Protection) Act, 1991. Even so, due to a variety of factors – including poor management, slackening supply and growing demand – Kerala has lost its preeminent position as an exporter of premier quality teak and is now a net importer of teak and other timbers (this is true of India as well.) However, this plantation is a beacon that reminds us that with the right mix of dedication, technology, good management and the appropriate policy regime, success is attainable against the odds.

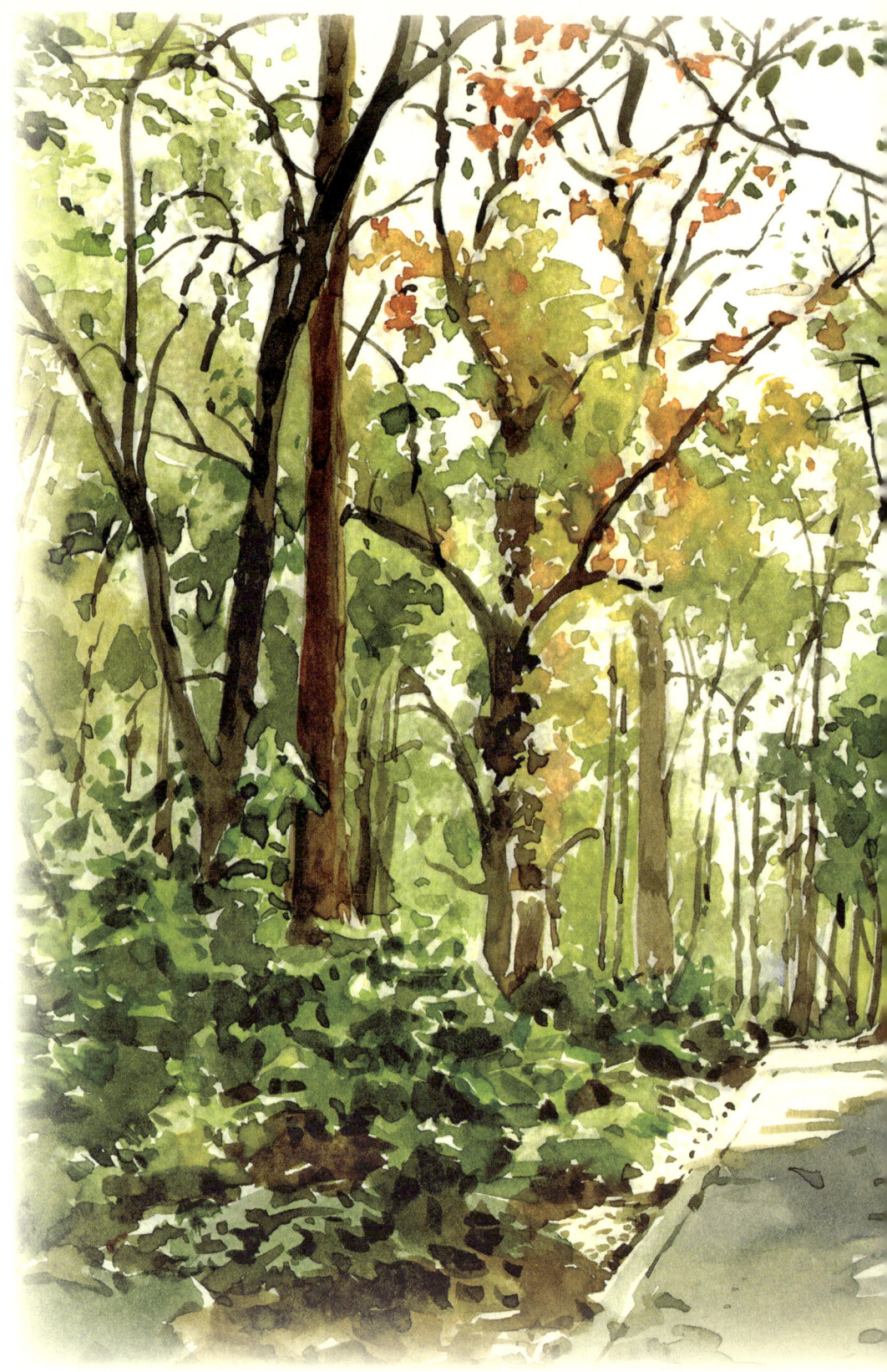

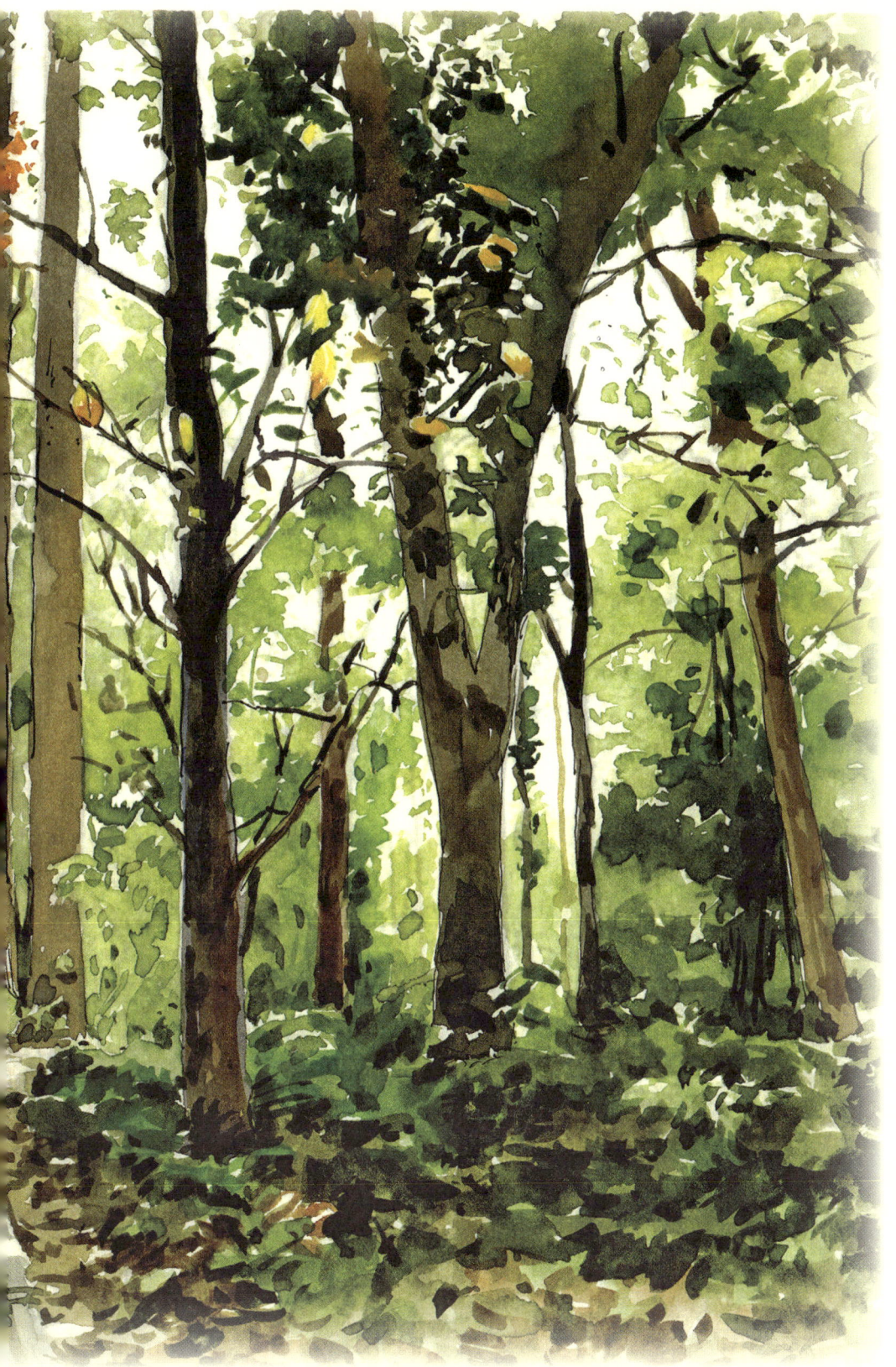

60

The Green Colossus of Conolly's Plot

COMMON NAMES: **Tetrameles, False Hemp Tree (English); Chini/Cheeni, Vella Cheeni, Vellapasa (Malayalam); Jungli Dungi (Hindi)**

SCIENTIFIC NAME: *Tetrameles nudiflora* R.Br.

FAMILY: **Tetrameles (Tetramelaceae[225])**

WHERE TO SEE: **Conolly's Plot, Nilambur, Malappuram district, Kerala**

LATITUDE: **11.270⁰ N**; LONGITUDE: **76.210⁰ E;**
ALTITUDE:**24.8 metres**

Visitors arriving at Conolly's Plot expecting to see ancient teak trees are treated to an unexpected delight. They are greeted by the awesome majesty and arresting beauty of an entirely different tree. Known locally as chini tree, it towers like a green colossus over everything in its vicinity. Its most impressive feature is the immense buttresses flanking the base with a circumference of 33.2 metres at the ground level.[226] The trunk soars some 55.2 metres into the sky and its girth above the buttresses is 4.8 metres.

The tree's bark is smooth, greyish, almost shiny white and conspicuously speckled due to the presence of tiny pores (lenticels) for breathing. Given such impressive credentials, it is a small wonder that the tree is a big draw for visitors. In this species, male and female flowers are borne on different trees and this one is male. This is why although flowers bloom regularly, there is no fruit set. Chini being a native of India,[227] there are several other trees of the species dotting Conolly's Plot but none larger than this one. Realistic estimates of the age of the tree are difficult to make. Unlikely to be as old as the teak trees in its neighbourhood (as chini trees are known to grow rapidly) its seed must have germinated several years after the teaks had established themselves on the plot. Fortunately for us, the tree escaped being cut down over the years possibly because its wood, being soft and non-durable, is not valued as much as timber although it is reportedly resistant to attack by marine boring organisms.

The chini tree has been made famous by the 2001 action-adventure *Lara Croft: Tomb Raider* and symbolizes nature's dominance over man as it straddles the mighty late twelfth-century Buddhist temples of Ta Prohm, Angkor in Cambodia.[228] Locally known as thitpok, its sinuous and serpentine roots are relentlessly twisting and crushing the temple arches and gateways, prying them apart in a mortal embrace over centuries in a grip more powerful than that of stone. Here at Conolly's Plot, however, the tree stands as a picture of grace, poise and tranquillity.

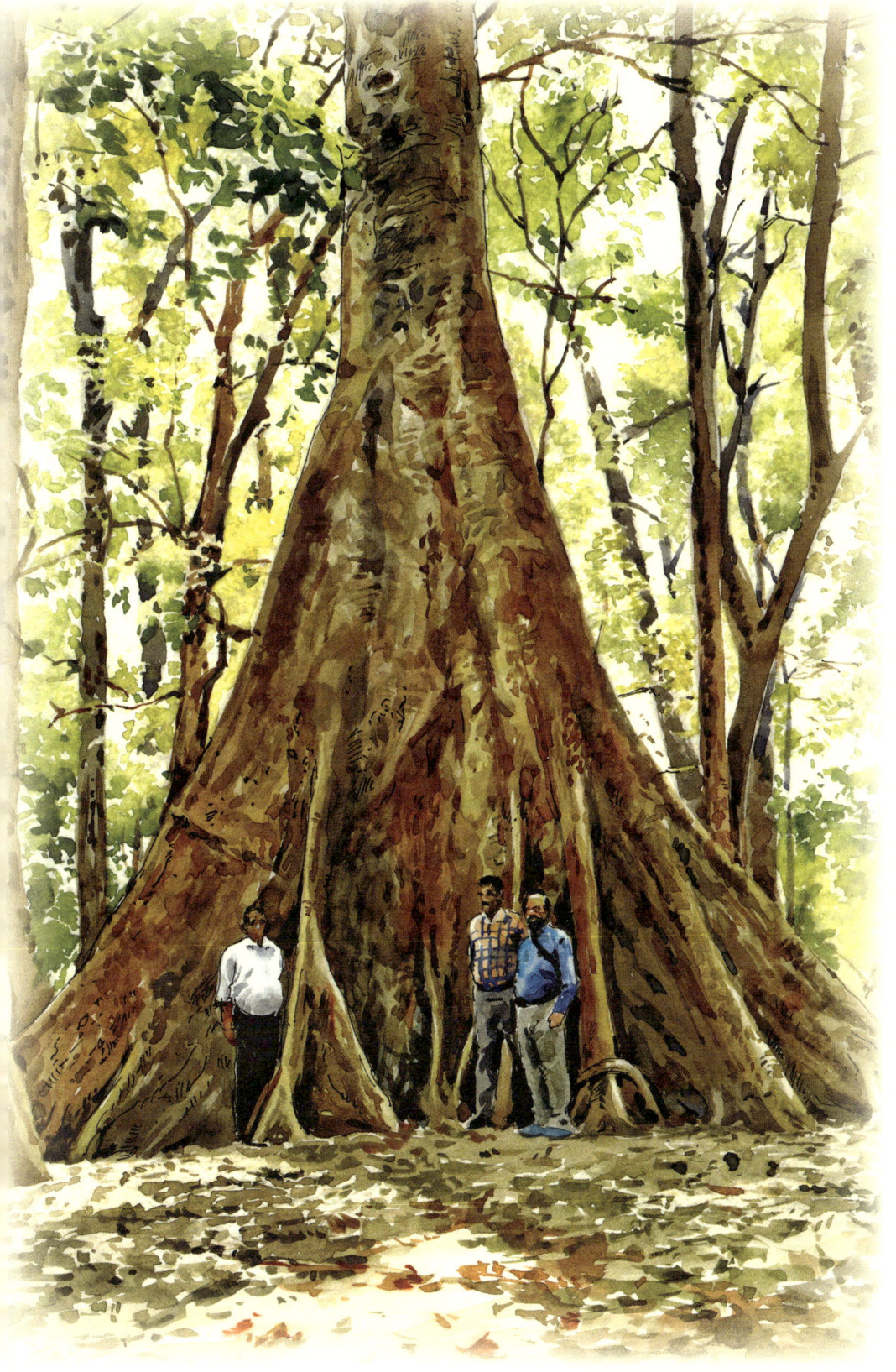

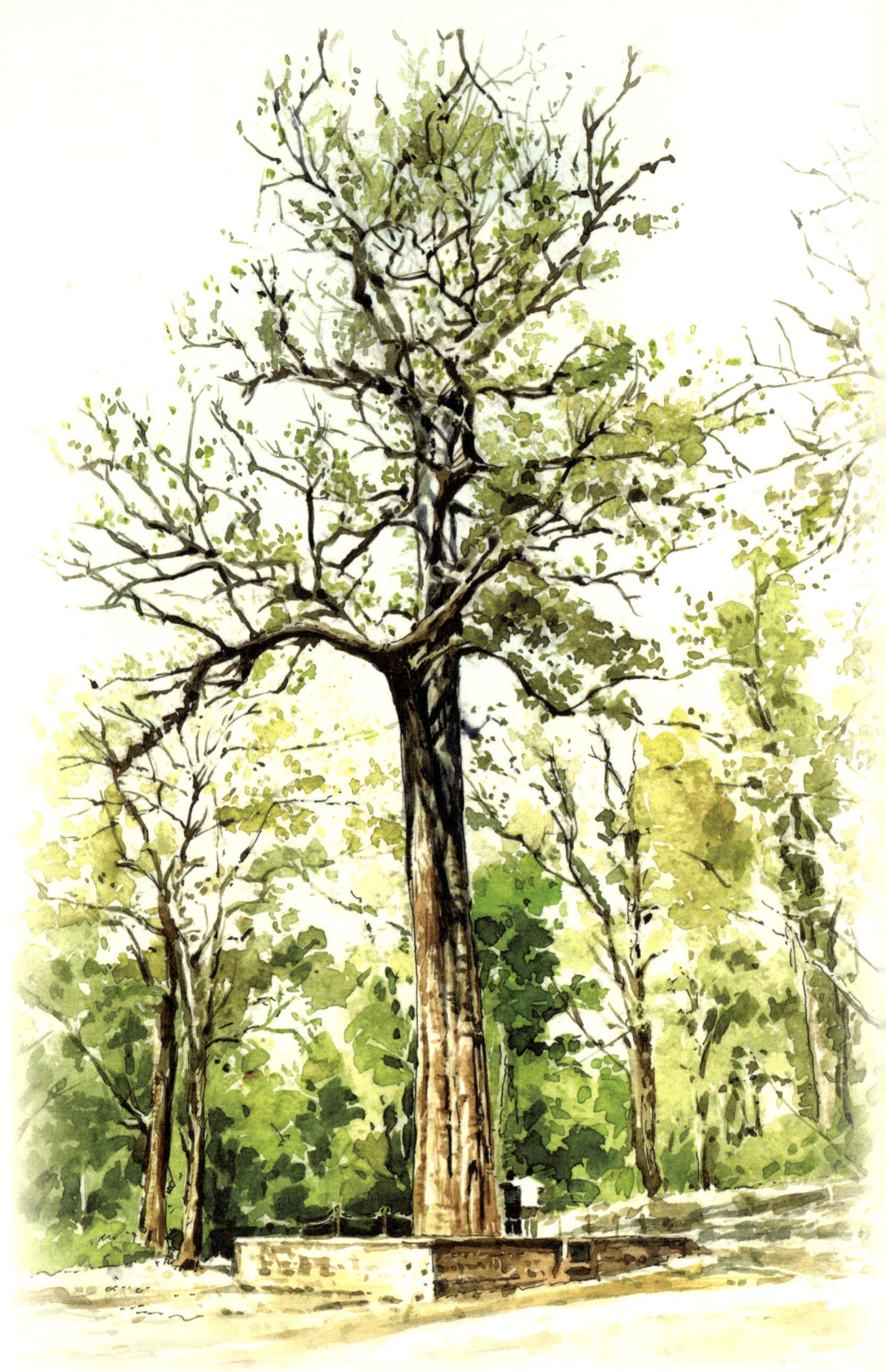

61

Kannimara, the Pride of Parambikulam

COMMON NAMES: Teak (English); Saka, Sagoon (Sanskrit); Thekku
(Malayalam); Sagun, Sagwan (Hindi)

SCIENTIFIC NAME: *Tectona grandis* L.f.

FAMILY: Tulsi (sacred basil) (Lamiaceae)

WHERE TO SEE: Parambikulam Tiger Reserve, Chittur taluk,
Palakkad district, Kerala

LATITUDE: 10. 2300^0 N; LONGITUDE: 764230^0 E; ALTITUDE: 1,438 metres

The star attraction for visitors thronging the Parambikulam Tiger Reserve is the big cats – leopard and tiger. Straddling 285 square kilometres of densely forested Western Ghats to the south of the Palghat Gap, the reserve is sandwiched between the peaks of the Nelliyampathy in the north and Ānamalai in the east. Not all visitors are lucky enough to catch a glimpse of the graceful felines and several go back disappointed, having to make do with the bear, the elephant, the macaque and the flying squirrel. None, however, leave Parambikulam without catching a glimpse of the kannimara teak (in Malayalam, 'kanni' means virgin and 'mara' is tree), aptly the pride of the reserve.

Teak grows here naturally as the warm moist tropical climate is very congenial for growth. Teak has been cultivated for timber production for centuries in India and Myanmar since at least 1840. The kannimara teak is apparently from an older vintage. Considered to be between 450 and 500 years old, it is certainly the oldest wild-grown teak tree in India.

Teak being deciduous, the kannimara is usually leafless from November to February. With the majestic height of a 14-story building at nearly 42 metres, the kannimara has a straight columnar fluted trunk that recalls the Qutb Minar of Delhi. The girth at breast height is 7.24 metres and requires five adults to embrace the tree with outstretched arms. According to forest officials, the tree has grown by 1.85 metres in height and nine centimetres in girth during the last five years. Branches are at the top and are densely clothed with several epiphytic ferns and orchids. Leaves emerge at the end of the dry spell during the monsoon and that is also when the tree starts flowering. The kannimara is healthy and annually produces small white flowers in large terminal panicles. Approximately 15 mm across, its fruits are spongy, enclosed in a persistent calyx and take up to 10-12 weeks to ripen.

Rapacious harvesting during the colonial era for shipbuilding, railway sleepers and other uses resulted in giant teak trees such as the kannimara being regularly and selectively harvested from the forests. Because of this, most natural stands of teak got depleted by the late eighteenth century. How did the kannimara escape the axe during the British Raj? It is perhaps because the local people worship the tree. Parambikulam, even today, is home to four indigenous tribes – Kādar, Malasar, Mala Malasar and Muduvar – who peacefully co-exist with other forest denizens. Historically, this area was part of the Kongu Nadu, the present-day Coimbatore region. The people traditionally worshipped the *saptha kanniayar* or *saptha matrika*s (the seven virgin guardian angels or seven virgin guardian mothers), a practice that is still in vogue today. During the British era, the indigenous people of Parambikulam forests believed that kannimara teak was the abode of the seven guardian angels. According to a popular local legend, when an attempt was made to cut the tree, blood gushed from its trunk.[229] This is how the tree came to be regarded as kannimara. Respecting the popular sentiment, the then-rulers spared the tree. And thus, it has survived to this day.

To the best of our knowledge, kannimara is the oldest surviving teak tree in a natural habitat in India. The tree was decorated with the 'Mahavriksh Award' for 1994 by the Government of India. It is well protected as it is within the tiger reserve and shown off to the visiting public as its prized possession. Popular sentiment is still quite strong for this pride of Parambikulam. In 2008, the forest department nailed a board to the tree announcing that it was measured on 16 May in the presence of the then Kerala minister for forests and went on to say that the tree had a girth of 7.01 metres and a height of 48.5 metres. However, concern that the nail would hurt the champion tree resulted in loud criticism by the tree-loving public, persuading the authorities to take down the board quickly.[230]

62

Dismissed as Extinct and then Rediscovered – After 184 Years!

COMMON NAMES: **Attilippa (Malayalam)**

SCIENTIFIC NAME: *Madhuca diplostemon* (C.B. Clarke) P. Royen

FAMILY: **Chikoo (sapota) (Sapotaceae)**

WHERE TO SEE: **Koonayil Ayiravilli Siva Temple, Paravur, Kollam district, Kerala**

LATITUDE: **8.490° N;** LONGITUDE: **76.4048⁰ E;** ALTITUDE: **Approximately 18 metres**

A pproximately 55 kilometres slightly northwest of Thiruvananthapuram is the town of Paravur in Kollam district. Barely short of a stone's throw from there is the Koonayil Ayiravilli Siva temple tucked away in a *kavu* (sacred grove) dedicated to its presiding deity, Lord Shiva. Kollam district is dotted with *kavu*s – 1,126 to be exact.[231] This particular *kavu* has devotees thronging the Shiva temple. No one, of course, had any reason to spare more than a fleeting glance at the small tree, locally known as attilippa, standing in the temple courtyard. The tree suddenly acquired celebrity status in early October 2020 following media reports that it was a species long-believed extinct and had been rediscovered after a gap of 184 years. A team of botanists from the Jawaharlal Nehru Tropical Botanic Garden and Research Institute (JNTBGRI), Palode, near Thiruvananthapuram identified it as *Madhuca diplostemon*, a threatened species of the Western Ghats that nobody had sighted since it was first collected in 1835.[232]

While the tree's height is just shy of four metres, no one seems to know how old this historic survivor is. Due to some mechanical stress suffered long ago, its deeply-fissured brown trunk is bent at an angle of 60 degrees and has forked into two low branches. Branchlets also arise at an angle holding aloft a spreading crown. Its leaves are smooth and shiny and have a rather leathery feel as is typical for the family, tending to crowd at branch tips. Any injury to the plant quickly produces a thick, milky sap. A low, square cement platform with a tiled surface has been built around the tree. Flowering and fruiting take place from January to March. The tree's flowers are stalked, pale, bell-shaped and held in bunches of three to eight in the axils of the leaves. Each flower has 15-19 stamens arranged in two whorls (hence the specific epithet *diplostemon*). It does produce single-seeded, ellipsoidal berries.

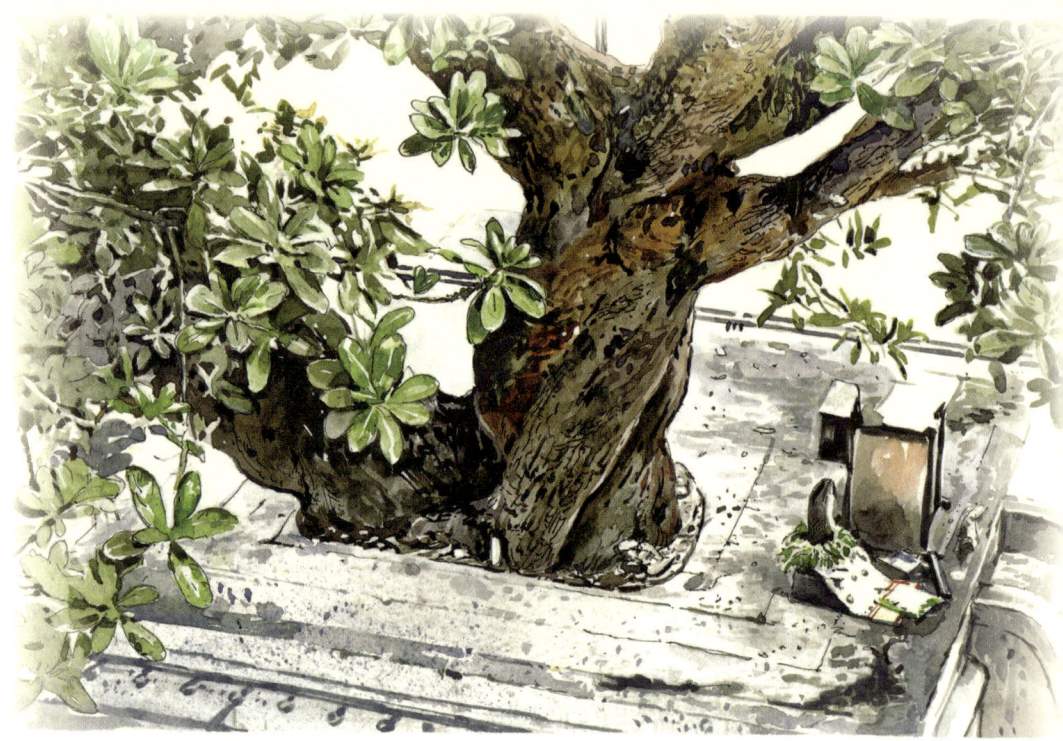

Robert Wight, a Scottish surgeon in the employ of the East India Company was the first to collect sample specimens of the species in 1835. His three specimens had immature flower buds from an unspecified locality in Quilon. He deposited the original dried specimen in the Royal Botanic Gardens (RBG), Kew and the two duplicates in RBG, Edinburgh. He also provided a brief description and a drawing of the plant in his 1848 book *Icones Plantarum Indiae Orientalis*, placing it – wrongly as it turned out – in the ebony family (Ebenaceae) and naming it *Diospyros obovata*.[233]

Forty years later, in Kew, Charles B. Clarke, another British botanist was critically investigating the circumscription and delimitation of Indian Sapotaceae (chikoo or sapota family) for the *Flora of British India*. He concluded that Wight's specimen belonged neither to *Diospyros* nor the ebony family but fitted better under the genus *Isonandra* within the Sapotaceae.[234] He thus changed its Latin name to *Isonandra diplostemon* C.B. Clarke.[235] However, owing to insufficient floral details on account of immature floral buds, he too provided a sketchy description.

The matter was rested for the next hundred years or so until 1960 when Pieter van Royen, a Dutch botanist published a taxonomic revision of the family Sapotaceae for Southeast Asia. He transferred *Isonandra diplostemon* to the genus *Madhuca*, making a new nomenclatural combination *Madhuca diplostemon* (C. B. Clarke) P. Royen. Since then, this has been the accepted name for the species.

Madhuca derives from the Sanskrit root with *madhu* meaning honey, a reference to the remarkably sugar-rich flowers of the group. There are five species in India, commonly known as mahua in the north and iluppai/ippe/ippa/vippa in parts of south India. Pollination and seed dispersal in these species are aided by bats.[236] Their flowers have been used for long in folk medicine and the preparation of local liquor.

Why did *M. diplostemon* remain unseen for 184 years? It turns out that Wight forgot to mention the locality of collection (Quilon/Kollam) on the original specimen submitted to RBG-Kew, mentioning it only on the duplicates housed at RBG-Edinburgh. Usually, botanists specializing in taxonomy consult the original or type specimen on which the description of the species is based. Duplicate specimens are scrutinized mainly in cases where the type specimen is damaged, unavailable or inaccessible. The absence of the crucial entry on location on the type specimen, in this case, would assuredly have put off potential explorers faced with the daunting task of looking for it all over the vast Deccan Peninsula.

Secondly, even the entry Quilon on the duplicate specimens represents a large enough area with luxuriant vegetation those days. Sadly, in the intervening period, development has taken a heavy toll on the area and much of that vegetation is lost. Sacred groves across the district – if sufficiently large – represent perhaps the main, if not the only, refuge for the survival of the remaining vegetation. Until now, there has been no systematic survey of the sacred groves of Kollam. Therefore, it is hardly surprising that a tree of attilippa tucked away in the quiet *kavu* of the Ayiravilli Siva Temple complex went undetected for 184 years. Clearly, more intensive explorations are called for.

Now represented by a solitary individual in just one locality, it has to be treated as 'critically endangered' as per the IUCN Red List criteria. We must indeed be thankful to the owners of the *kavu* for preserving and maintaining this and other species in the sacred grove. Meanwhile, JNTBGRI proposes to initiate experiments to propagate the historic tree. May they succeed!

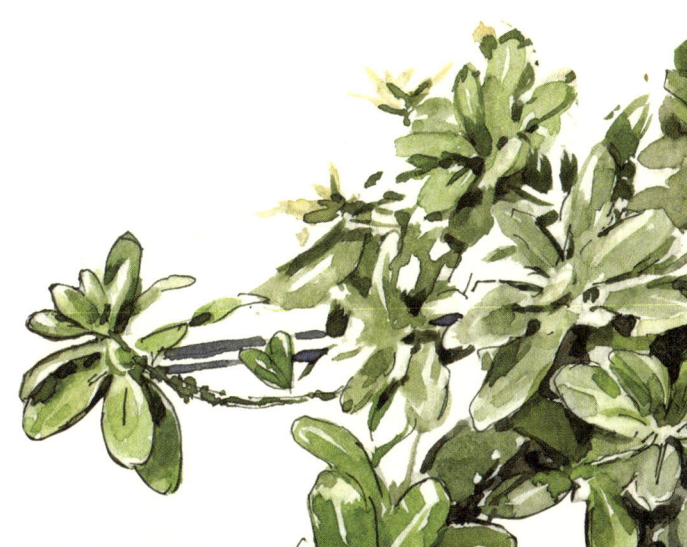

63

The Sacred Rayan Tree of Ranakpur

COMMON NAMES: **Ceylon wood, Ceylon Ironwood (English); Rayan, Khirni (Mewari); Khirni (Hindi)**

SCIENTIFIC NAME: *Manilkara hexandra* (Roxb.) Dubard

FAMILY: **Sapota or Mahua (Sapotaceae)**

WHERE TO SEE: **Jain Temple in Ranakpur village, Pali district, Rajasthan**

LATITUDE: **25.1158⁰ N**; LONGITUDE: **73.4727⁰ E**;

ALTITUDE: **500 metres**

The topmost level of a three-storyed marble edifice is the last place you would expect to find a 500-year-old rayan (khirni) tree. Yet that is exactly where it is, 15 metres above ground in the famous Jain temple of Ranakpur. Built in the fifteenth century amid a garden, the temple is one of India's major Jain pilgrimage sites. The magnificent 31-metre-tall edifice, occupying approximately 3716 square metres, is shaped in the *chaumukha* (four-faced) style comprising 29 halls and 420 pillars of which no two are alike. Undoubtedly, this architectural marvel is the finest example of its kind. An inscription on a pillar close to the entrance of the main shrine chronicles that Deepaka, an architect, constructed the building in 1439 CE on the order of Dharānka, a Jain devotee.[237]

Legend has it that Dharānka Shah Porwal, a minister in Rana Kumbha's court, dreamt one night of an exquisite celestial airplane. He woke up with a burning desire to build a temple to Adinatha in the shape of his dream vehicle. The rana acceded to his minister's plea with the caveat that the complex be named after the monarch. It took 50 years to complete the temple.

The evergreen rayan tree's trunk is massive and entirely hollow and open. The bark is rough and grey. Two major branches emerge from the trunk, each having several secondary branches. The conical crown soars some 10 metres into the sky. Most writings that sing paeans to the architectural marvels of the temple neglect to even mention the tree. And yet, barring this one tree, the ambience is entirely of marble stone. How did this tree come to be so far above the ground? Clearly, its root system must be penetrating quite deep into the soil to draw sufficient water and nutrients for survival under these arid conditions.

It is said that Adinatha delivered his first sermon after attaining enlightenment under a rayan tree and hence they are sacred to Jains. Believed to be contemporaneous with the building of the Ranakpur temple, the tree's age is said to be around 550 years. A scientific investigation of the ancient tree is needed to validate this claim.

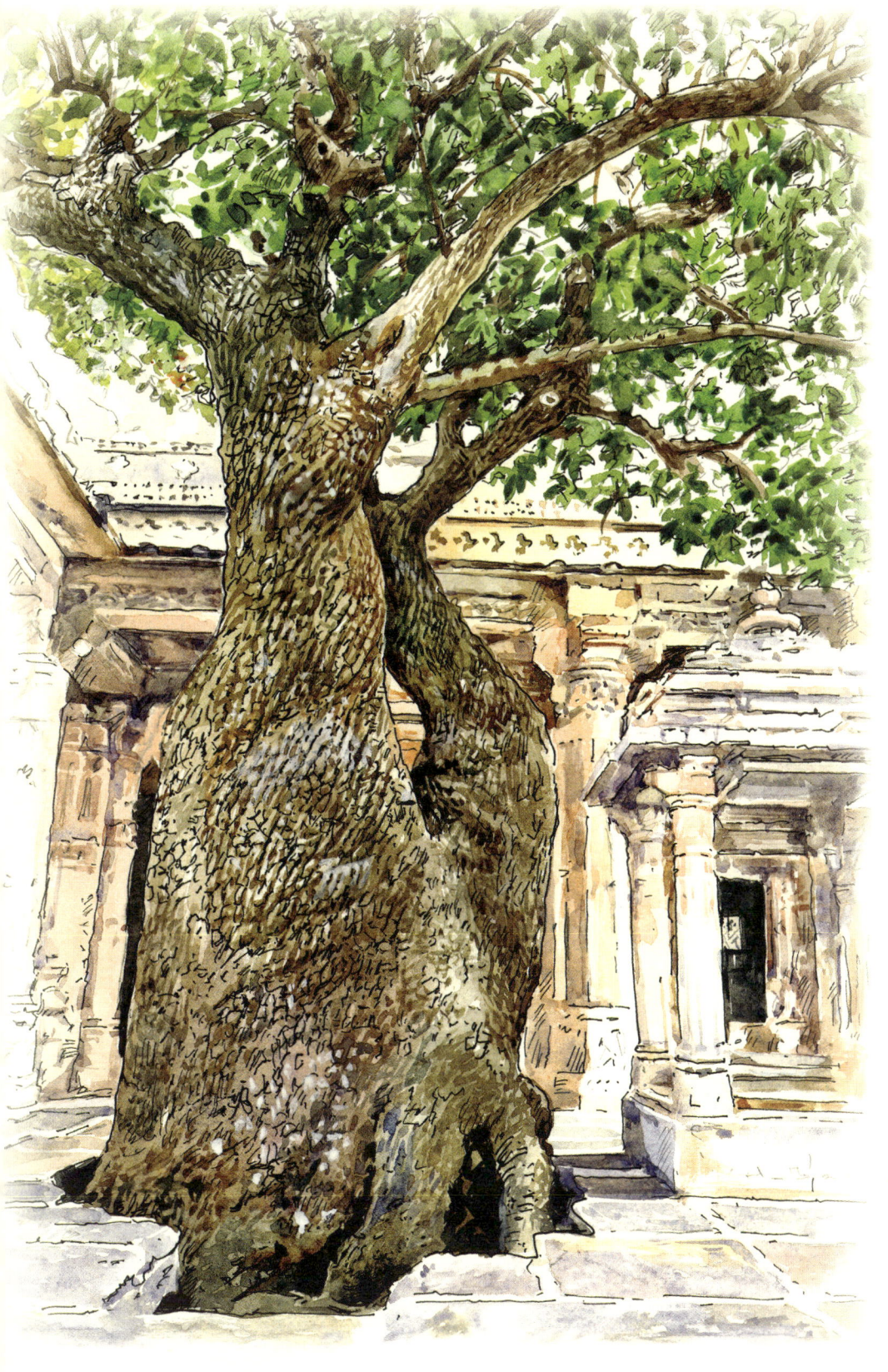

64

The Ancient Limdo Tree of Jaska

COMMON NAMES: **Neem, Margosa (English); Nimbaka (Sanskrit); Limdo (Gujarati); Neem (Hindi)**

SCIENTIFIC NAME: *Azadirachta indica* A. Juss.

FAMILY: **Neem or Mahogany (Meliaceae)**

WHERE TO SEE: **Davalshah Pir Dargah, Jaska village, Vadnagar tehsil, Mehsana district, Gujarat**

LATITUDE: **23.822⁰ N**; LONGITUDE: **72.534⁰ E**; ALTITUDE: **156 metres**

In a small, unremarkable village known as Jaska in the Mehsana district of Gujarat, there is a small dargah dedicated to Davalshah Pir. Both Hindus and Muslims worship Pir Baba, who has a reputation for granting their wishes. Legend has it that a dead man was once miraculously brought back to life here due to the power of the Pir. A mela (fair) is organized during the annual *urs* in July.

Just next to the small shrine on a cemented platform is a large age-wizened neem tree, called limdo in Gujarati. According to the *khādim* of the dargah, the tree is between 400 and 500 years old, but a more pragmatic estimate would be 250 years.[238] The girth of its trunk measures 5.70 metres at breast height and the crown towers over the asbestos roof of the dargah, rising to almost 25 metres. The trunk and the lower ends of branches have burls. A few of the branches are bent low and require the support of wooden stakes driven into the ground.

According to a tree census carried out by the state forest department in 2013, Gujarat has approximately 45 million neem trees with Mehsana, Gandhinagar, parts of Patan, Banaskantha and Sabarkantha districts having the highest concentration in the country. Within Mehsana, one out of every four trees is neem.[239] In 1994, a neem tree in Lunawa village in Mehsana district was selected for the Mahavriksha Puraskar, based on the measurement of girth (5.18 metres). Its exact age was unknown but believed to be approximately 200 years. This tree, considered the largest neem tree in India, stands in Lunawa with a signboard announcing that it was the recipient of the Mahavriksha Puraskar for 1993–94.

Our neem tree at the dargah of Davalshah Pir was discovered later. Found to be larger and older than the Mahavriksha neem, it has been recognized as a heritage tree of Gujarat. A later survey by the State Department of Forests resulted in the discovery of an even larger tree in Bhechada village of Surendranagar district. This 250-year-old tree is reported to be 24 m tall with trunk girth of 6.9 m[240] and seems to be the current record holder neem for girth.

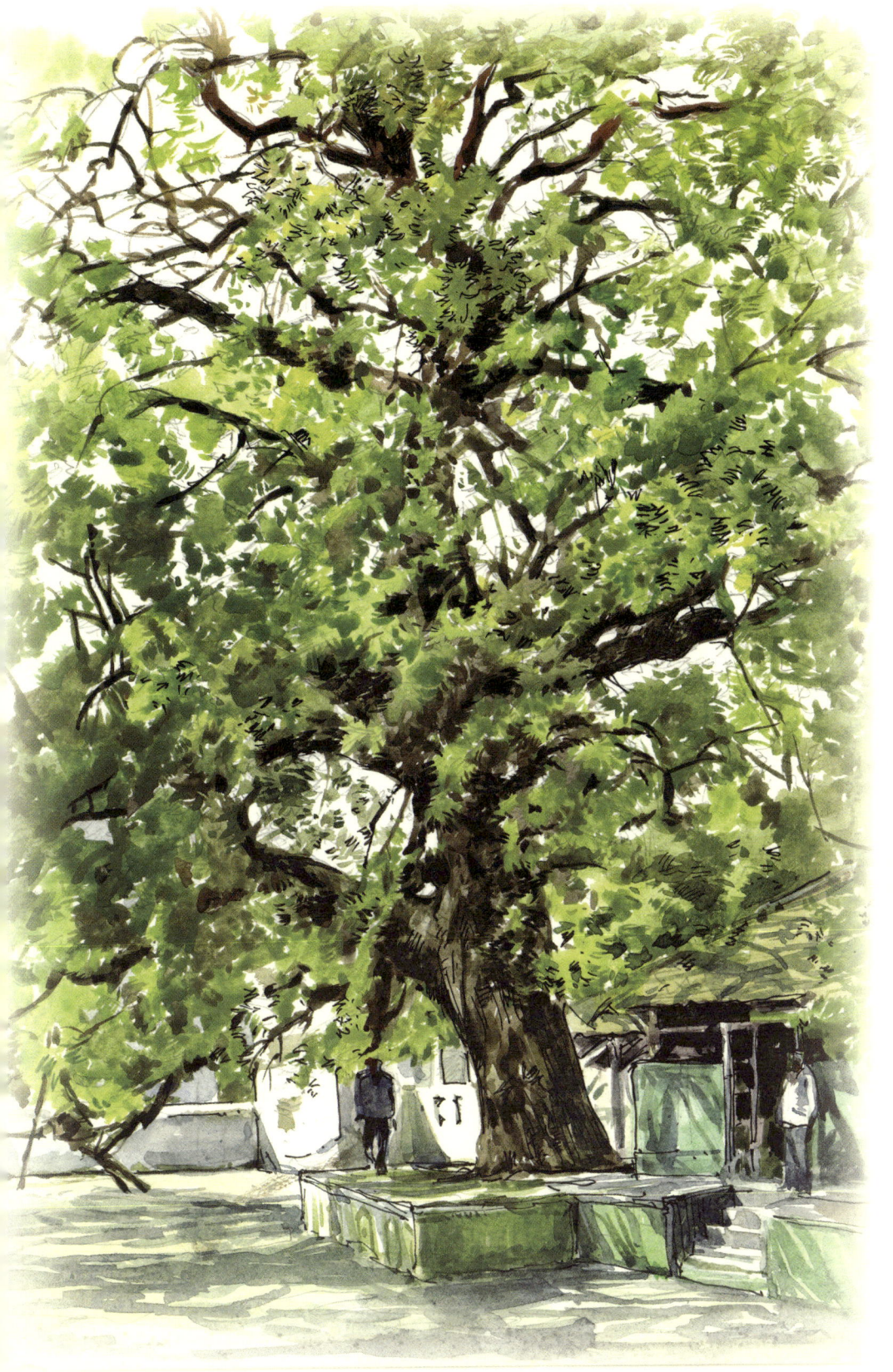

65

The Towering Simado
of Juna Vādiya

COMMON NAMES: **Red Silk Cotton Tree (English); Shalmali, Kantakadruma
(Sanskrit); Simado, Shimalo (Gujarati); Semal (Hindi)**

SCIENTIFIC NAME: *Bombax ceiba* L.

FAMILY: **Cotton (Malvaceae, Bombacoidae)**

WHERE TO SEE: **Juna Vādiya village, Limkheda taluk, Dahod district, Gujarat**

LATITUDE: **22.90778⁰ N**; LONGITUDE: **74.0425⁰ E**; ALTITUDE: **205 metres**

Juna Vādiya is not the kind of place that will feature on your tourist map. A sleepy little village in Dahod district in Gujarat and counted among India's 250 most backward districts, the majority of its residents are Bhil (Bheel) tribal people subsisting on maize, millet, pulses and vegetable farming. However, if you are a tree lover, there is more than ample reason to visit this village as it has the most spectacular red silk cotton tree on display. Locally known as simado, this is the tallest and stoutest of silk cotton trees in Gujarat.[241]

In India, the silk cotton favours the banks of rivers and streams but is equally at home in mixed deciduous, moist deciduous and mixed evergreen forests. Here, in Dahod, there are mainly dry teak forests, moist deciduous forests and scrubs and the simado is a familiar sight, dotting the countryside along with the flame of the forest, neem and acacias.

The simado tree in Juna Vādiya stands at one end of the village, next to the roadside on private property. With its towering height of over 35 metres, it resembles a 14-story building and is easy to spot from a distance. The huge trunk, with a girth of 11.2 metres at breast height, rises straight up without any branches to a height of 18 metres. Spectacularly large buttresses anchor the tree firmly to the ground. It would take more than five people with outstretched hands to encircle the trunk. Recognizing it as a heritage tree, the Gujarat government has helped erect a large rectangular platform paved with marble around the base of the tree. Flowering occurs profusely during February and March when bright scarlet flowers are conspicuous on the leafless tree. The fleshy flowers contain copious quantities of nectar and are visited by a host of birds and insects. Interestingly, the tree here is not associated with any deity or any folktales either. The owners of the land are happy to leave the tree alone as it stands at the edge of their farm close to the road. The local people are also aware that the state government attaches significance to the tree.

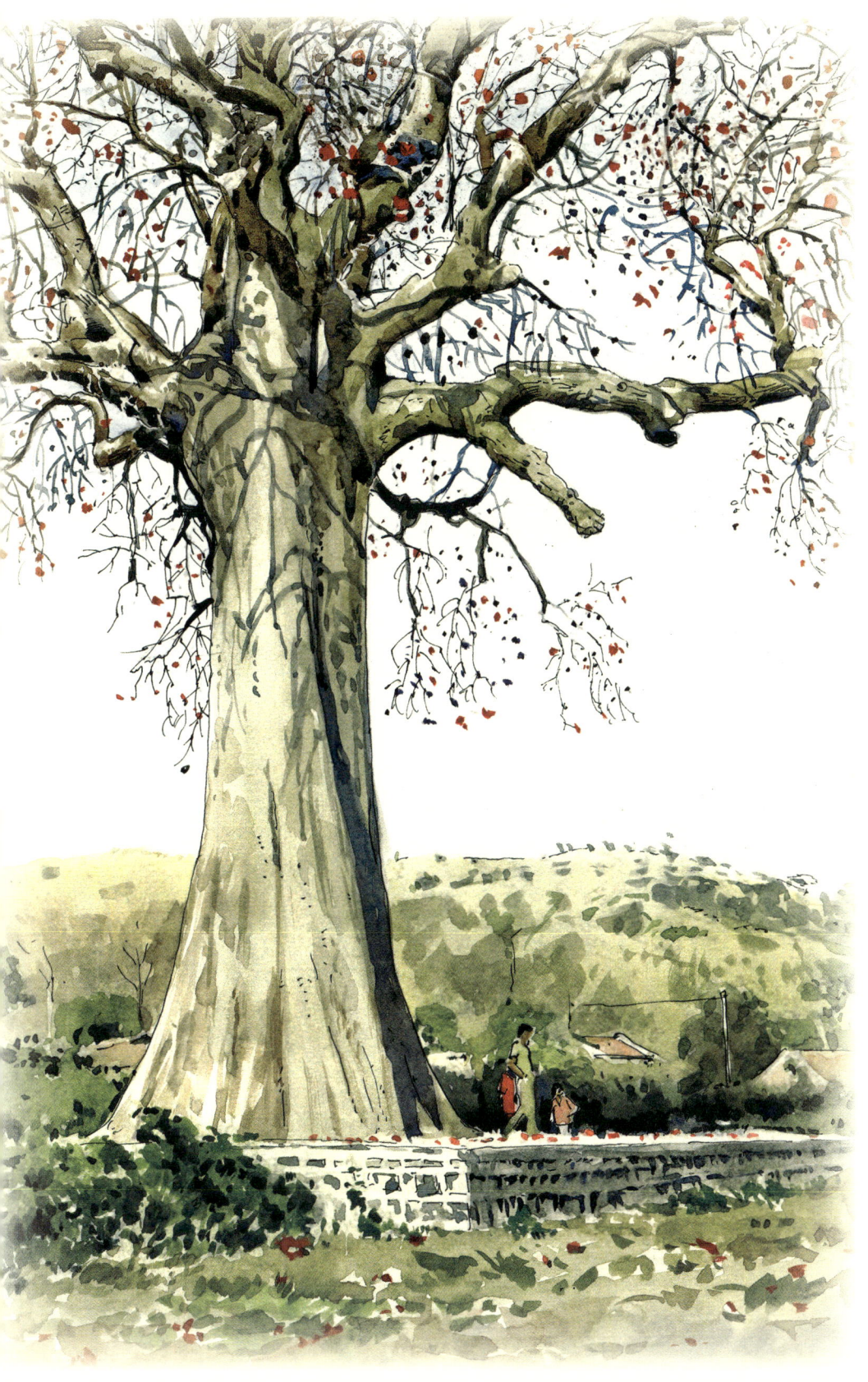

66

Vallabh Vad of Ras Village

COMMON NAMES: **Banyan (English); Nyagrodha (down-growing)
and Bahupada (many-footed) (Sanskrit); Vallabh Vad (Gujarati);
Vat Vriksh, Bargad (Hindi)**

SCIENTIFIC NAME: *Ficus benghalensis* L.

FAMILY: **Fig (Moraceae)**

WHERE TO SEE: **Ras High School, Ras village, Borsad taluk,
Anand district, Gujarat**

LATITUDE: **22.3447° N; LONGITUDE: 72.8297° E;**
ALTITUDE: **29 metres**

Mention 'Ras' and anyone familiar with India's freedom struggle will recognize it as the place from where Sardar Vallabhbhai Patel was arrested during the salt satyagraha in 1930. The banyan tree under which his arrest took place is known as Vallabh vad. Sadly, both Ras and the banyan tree gradually receded from public memory, much like Sardar Patel himself.[242] This is beginning to change in recent times with much more public acknowledgement of Sardar Patel's legacy, though as yet rarely of this key place in his long public life. Vallabh vad still survives in Ras.

Sardar Patel was Mahatma Gandhi's right-hand man and had meticulously chalked out the destination, route and the entire logistics for the Dandi march, even handpicking the satyagrahis who would walk the 388 kilometres from Sabarmati to Dandi with Gandhi. Well before the event, Patel decided he would tour the entire route to drum up support for the civil disobedience movement. Everywhere, his speeches cast a spell on the people. All over Gujarat, there was a feeling of revulsion against British oppression. The government realized it was necessary to get him out of the way quickly. On 7 March, five days before the satyagraha was to commence, Sardar Patel arrived in Ras village and said, 'Sisters and brothers, are you ready for the satyagraha?' to the thousands of enthusiastic villagers who had gathered under the shade of a large banyan tree outside Ras. A loud 'yes' rented the air. This did not amount to a speech and yet he was arrested by the first-class magistrate of Borsad,[243] and sentenced to three month's imprisonment.

The news of Patel's arrest shook the entire population of Gujarat and the country in general. Gandhiji used Patel's arrest to great advantage during the march against the British. His arrival in Ras on 19 March was celebrated with mass resignations of village officials and other forms of protest.

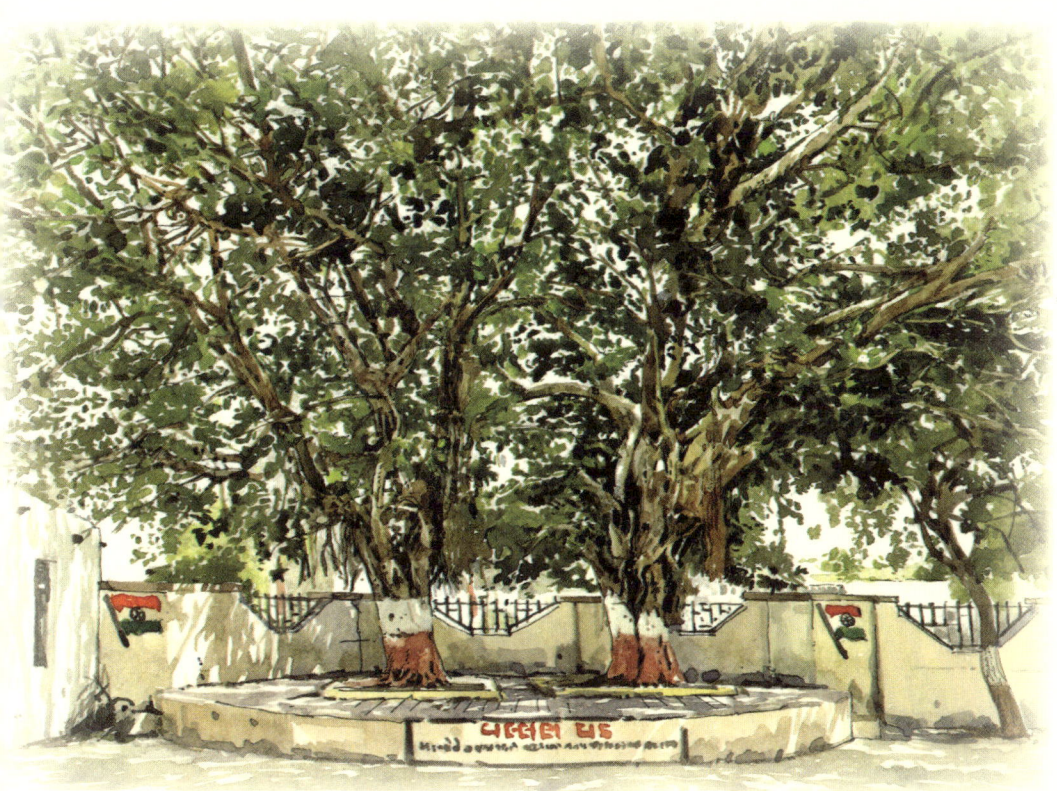

Ras' border has grown beyond the historic banyan tree today. A high school came up in 1952, enclosing Vallabh vad within its compound. Ninety years ago, the tree must have been large enough for many people to gather under, but a storm uprooted it. Two branches of the tree were planted side by side in the original place and a circular platform paved with interlocking tiles was built around them. Details of when the original tree fell and who decided to replant two branches of that tree are not available. Painted in red letters on the platform, there is a simple announcement in Gujarati: 'Vallabh Vad. Sardar was arrested first time on 7 March 1930 from this Vallabh Vad in Ras'. Much like the man they seek to commemorate, the trees have not attracted much attention or publicity and stand there in silence.

67

Kabir's Gigantic Banyan

COMMON NAMES: **Banyan (English); Nyagrodha (down-growing), Bahupada (Sanskrit); Kabir Vad (Gujarati); Vat Vriksh, Bargad (Hindi)**

SCIENTIFIC NAME: *Ficus benghalensis* L.

FFAMILY: **Fig (Moraceae)**

WHERE TO SEE: **Kabirvad Island, Bharuch district, Gujarat**

LATITUDE: **21.762⁰ N;** LONGITUDE: **73.142⁰ E;**
ALTITUDE:**13.3 metres**

Just about 18 kilometres east of the ancient city of Bharuch in Gujarat, the Narmada River bifurcates to accommodate a small island of silt. On this island grows a gigantic banyan tree that is among the largest trees on earth. To reach the tree, one must catch a medium-sized motorboat on Narmada's western bank and sail across to a makeshift jetty on the island. Even a seasoned tree gazer would feel humbled by its complicated maze of thick pillar-like roots and the vast canopy. Known as Kabir vad, the tree is associated with the fifteenth-century mystic-saint, poet and reformer Kabir Das, whose teachings influenced Hindus, Muslims and Sikhs alike. Legend says that Kabir visited here in the late fifteenth century.

There are two legends associated with the origin of the banyan. According to the first, the tree sprouted from a discarded plant twig used by Kabir as a toothbrush while on the island. According to the second, two brothers were desperately searching for a spiritual guru to mentor them. After a 12 year unsuccessful wait, Kabir visited the island and the brothers traditionally welcomed him by washing his feet. They threw the water on a dried-up stump of a banyan, which miraculously sprouted buds and leaves and became Kabir Vad. The brothers had found their guru.

The British artist, writer and explorer James Forbes, who was collector in both Bharuch and Dabhoi, spent 'many delightful days with large parties' during 1772 under the 'Cubbeer Burr' on the island of the Narmada River. He described the giant banyan as 'forming a canopy of verdant foliage impenetrable to a tropic sun, extending over a circumference of two thousand feet...' Forbes also produced a sketch of the tree.[244]

In 1794, Thomas Maurice, a British historian, observed more than 350 false trunks, each one 'thicker than an English oak tree', in addition to 3,000 smaller stems. He noted that locals said the tree was 3,000 years old, suggesting it existed long before Kabir himself. This raises the possibility that it was the same banyan

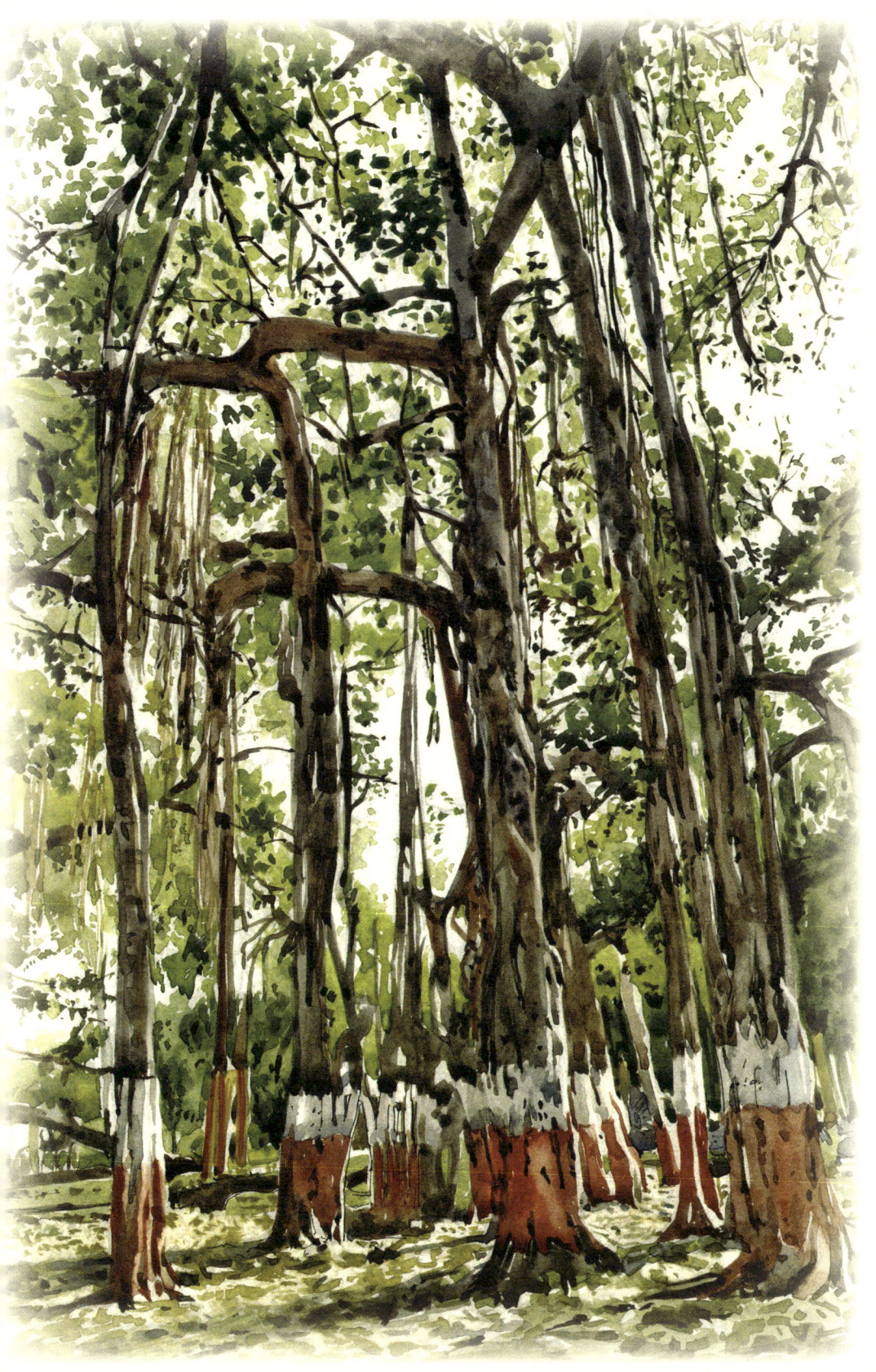

that Nearchus – one of Alexander the Great's generals – and his army encountered on the Narmada in 326 BCE.[245] Under its vast canopy, 7,000 troops could camp. While this is hard to confirm, it is probably the oldest banyan in India. A scientific determination of the tree's age will clear up several questions.

As mentioned earlier, Kabir's banyan is also one of the largest banyans on earth in terms of net canopy coverage – i.e., after subtracting visible clearings caused by either natural phenomena or human activity – and ranks second at 17,520 m^2 after Thimmamma marrimanu of Andhra Pradesh at 19,107 m^2. However, based on the gross area, it is the largest (20,985 m^2) followed by Thimmamma marrimanu (20,190 m^2). The main trunk has been lost over the years and now several independent trees exist with conspicuous gaps in the canopy allowing other tree species to grow.

In its long life this centurion tree has been battered by floods and soil erosion and each time it seems to have fought back recovering lost ground. Today, however, there are increasing signs of human intrusion and damage. Several buildings have come up under its canopy – a new two-story shrine to Kabir built in 1989 adjacent to the existing one, a multi-story guesthouse added to accommodate devotees during the annual temple festival and a residence for the temple priest and his family. Much of the soil in the clearing has been paved with tiles. The annual religious fair around *Kartik Shuddh Purnima* (July–August) is said to attract around 40,000 pilgrims. Even if only half that number visit, it is greatly stressful for the centurion tree. Whether and for how long, it will be able to absorb the onslaught of new anthropogenic stresses is a question that begs an answer.

68

The Banyan Tree of Dandi

COMMON NAMES: **Banyan (English); Nyagrodha (down-growing) and Bahupada (many-footed) (Sanskrit); Gandhi Vad (Gujarati); Vat Vriksh, Bargad (Hindi)**

SCIENTIFIC NAME: *Ficus benghalensis* L. | FAMILY: **Fig (Moraceae)**

WHERE TO SEE: **At Dandi, near the National Salt Satyagraha Memorial, Jalalpore taluk, Navsari district, Gujarat**

LATITUDE: **20.8844° N**; LONGITUDE: **72.8033° E**; ALTITUDE: **12 metres**

Dandi rose to national and international fame in 1930 when Mahatma Gandhi chose this coastal hamlet in western Gujarat as the destination for his salt satyagraha. It was here on the morning of 6 April that the Mahatma famously lifted a small fistful of salt from the sea to defy the oppressive British Salt Act. Close to this site, where now stands the National Salt Satyagraha Memorial, is another historic landmark – a banyan tree under which the Mahatma held meetings and delivered speeches. The tree served as the venue for prayer meetings throughout Gandhiji's stay in Dandi.

On the eve of the historic day, Gandhiji addressed a large gathering before breaking the salt law. Seated under the tree on a rickety chair that evening he delivered a powerful message challenging the unfair British taxation system and setting the tone for the long struggle ahead.[246] The next morning, everyone was up early and the prayer meeting under the tree was markedly solemn. Speaking on the occasion, Gandhiji announced that were he to be arrested, the satyagraha would continue to be led first by Abbas Tyabji and later by Sarojini Naidu.[247] The group then proceeded to the sea and there, amid chants of '*Gandhiji ki jai*' and '*Vande Mataram*', the sage of Sabarmati picked up a fistful of saline mud and famously announced, 'With this salt, I am shaking the foundations of the empire'. This was the signal for the commencement of civil disobedience all over the country.

Over three decades following the Dandi March, the tree continued its existence in relative obscurity. In 1961, an enclosure with a low wall and a decorative gateway was built to honour the unity of religions. A small statue of the Mahatma seated on a pedestal along with quotations from his speeches (also carved in stone) was installed under the tree. Prime Minister Jawaharlal Nehru inaugurated the memorial and dedicated it to the nation in a ceremony that year.

Towards the end of 1982, a severe storm pounded the coastal districts of Gujarat, leaving a trail of destruction in its wake. The toll included hundreds of human lives and lakhs of heads of livestock. One of the casualties was Gandhiji's

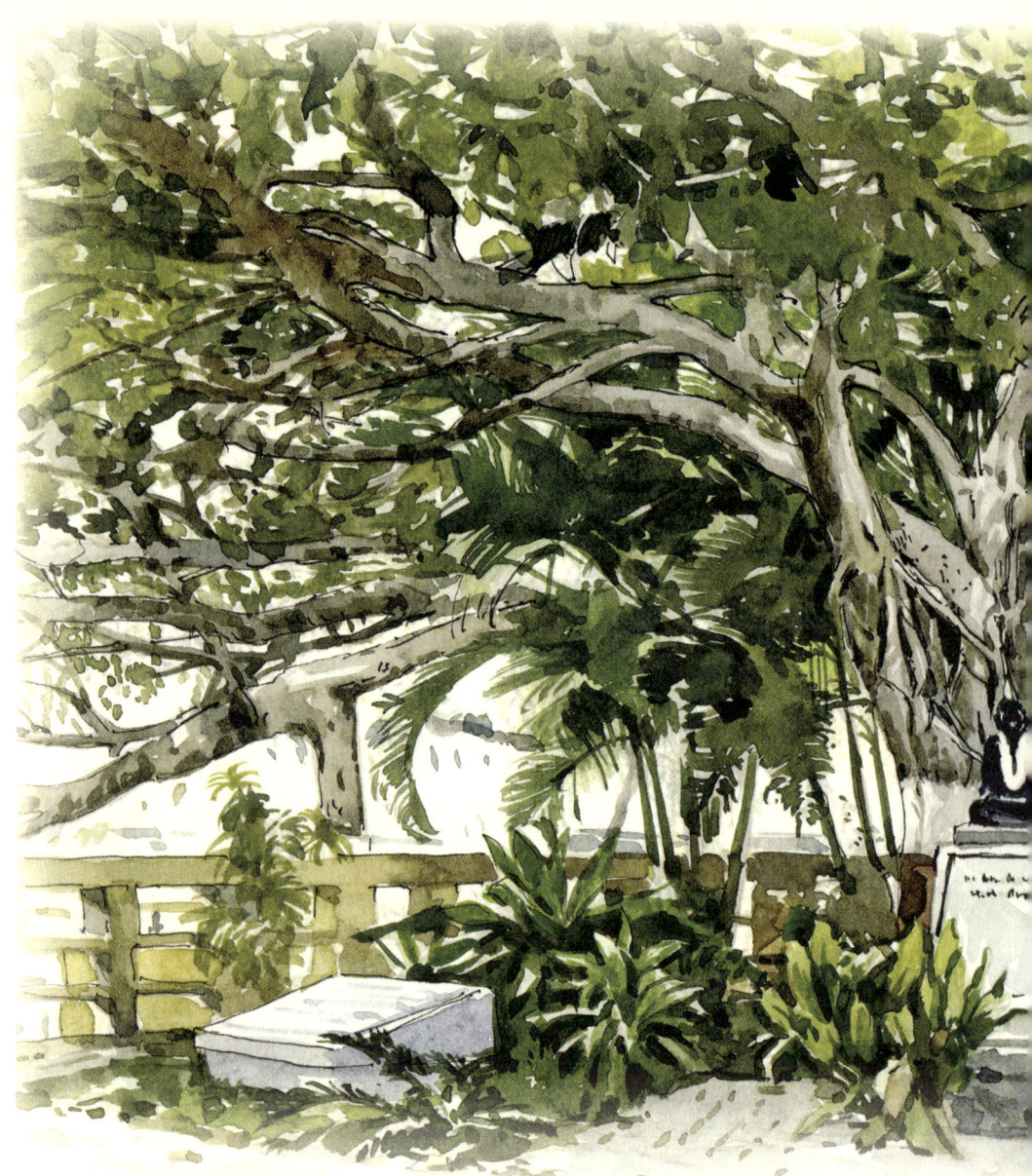

banyan, which collapsed at the site. For the young Gandhians in Dandi, the tree was not only historic but also sacred. Since all government agencies were preoccupied with the devastation of the cyclone and no one had time for the tree, Dandi's Yuva Mandal decided to take matters into its own hands. First, a few cuttings from the original tree were readied for planting at the site. The memorial complex was spruced up. Shri Morarji Desai, former prime minister and a staunch Gandhian, was invited to replant the tree. At that time Desai was out of power and retired from active politics. A function was organized on 6 November 1982. After performing a brief puja, Mr Desai replanted one of the rescued branches at the site where the original tree had stood. Dhirubhai,[248] a Gandhian of Dandi who was

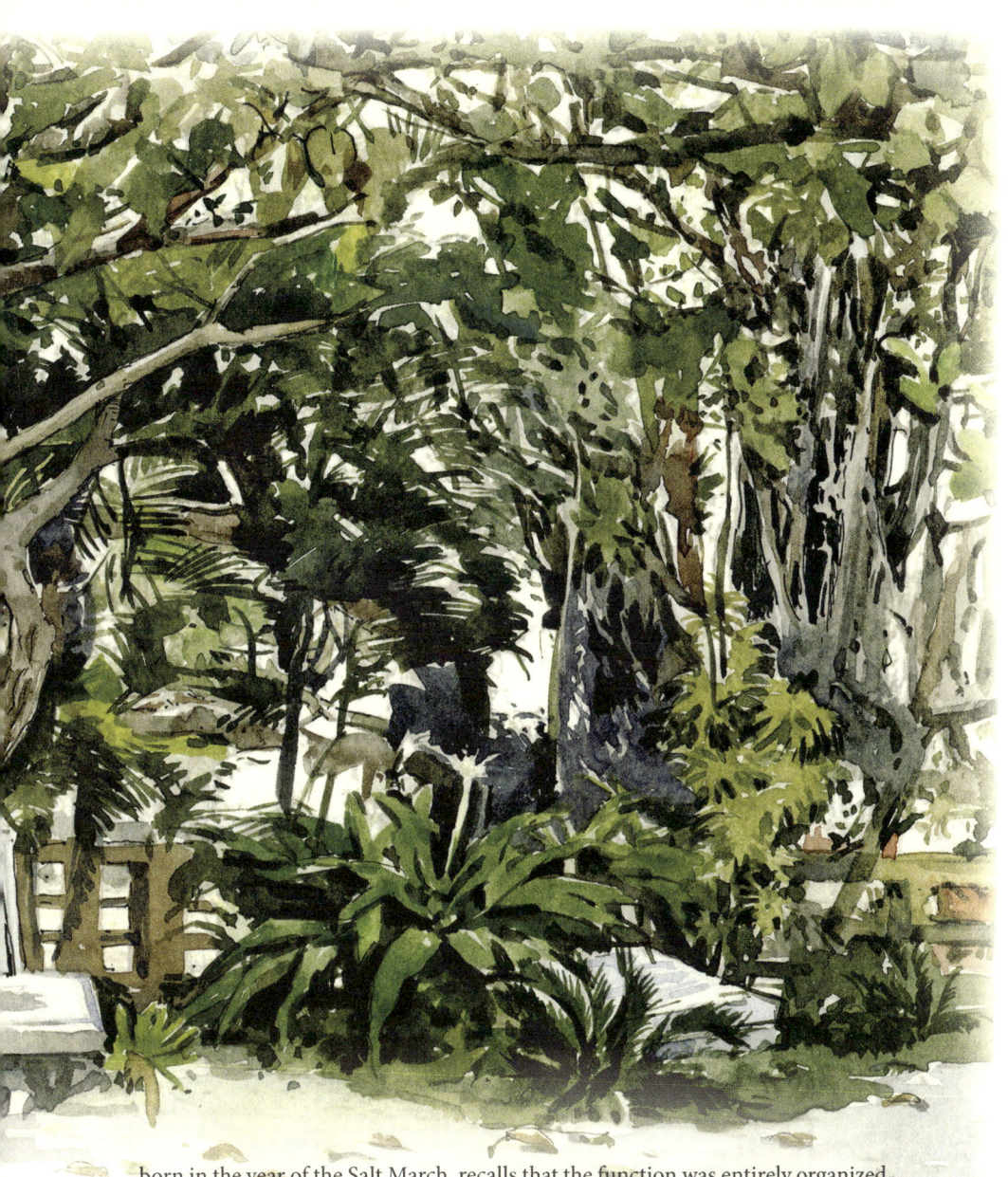

born in the year of the Salt March, recalls that the function was entirely organized by the people of Dandi. It is the replanted tree that stands at the site today. When Professor Thomas Weber, an Australian scholar, writer and Gandhian visited Dandi while retracing the entire route of the Salt March during March-April 1983, the old tree was missing.[249] For over nine decades, Gandhi vad has continued to inspire generations of people from across the world who keep coming to Dandi. For some strange reason, the sprawling National Salt Satyagraha Memorial, inaugurated by Prime Minister Modi on 30 January 2019 does not include this historic tree.[250] Yet, it is the only living, still throbbing visceral connection in Dandi to the great man.

69

Chalto Ambo – The 'Walking' Mango Tree of Sanjan

COMMON NAMES: **Mango (English); Aamra (Sanskrit);**
Ambo (Gujarati); Aam (Hindi)

SCIENTIFIC NAME: *Mangifera indica* L.

FAMILY: **Mango or Cashew (Anacardiaceae)**

WHERE TO SEE: **Sanjan, Umbergaon taluk, Valsad district, Gujarat**

LATITUDE: **20.2021⁰ N; LONGITUDE: 72.8034⁰ E;**
ALTITUDE:**11.6 metres**

For connoisseurs of history, Sanjan – a small town on the banks of Varoli River in Gujarat – is familiar as one of the earliest Parsi settlements in India. According to historians, a section of the Zoroastrian community fled from Persia (present-day Iran) in the seventh century (or even up to the tenth century, according to other experts) to escape persecution by Muslim invaders. The descendants of these migrants known as Parsis (or Parsees) landed in Sanjan in today's Gujarat and sought refuge. The benevolent local ruler Jadi Rana is said to have given them asylum and permission to practice their religion. The grateful Parsis assimilated themselves into the local population 'like sugar in milk'. The Sanjan memorial column, popularly known as *Sanjan Stambh*, was built by the community in 1917 to commemorate their historic arrival and was opened to the public three years later.

Not far from the stambh is an ancient mango tree that is far older, though much less heralded and has its own saga to narrate. According to local legend, it was planted by the early Parsi settlers who brought the fruit from their homeland.

Unlike any other known mango tree, this tree likes to 'walk'. Hence the sobriquet *chalto ambo* in Gujarati. It is believed that it has 'walked' several kilometres from the seashore to its present location. Unfortunately, there are no records to establish where exactly the tree was originally planted. For decades now the tree is on farmland owned by the family of the late Valli Ahmed Achchu. The 'walking' happens thus: a branch of the tree droops down and touches the ground, growing parallel to it. Over a period of time, roots emerge at the point of contact and anchor the branch firmly in the soil. New shoots and leaves sprout from the grounded branch. Meanwhile, older portions of the tree wither away and die. This phenomenon keeps recurring at periodic intervals as if the tree is

234

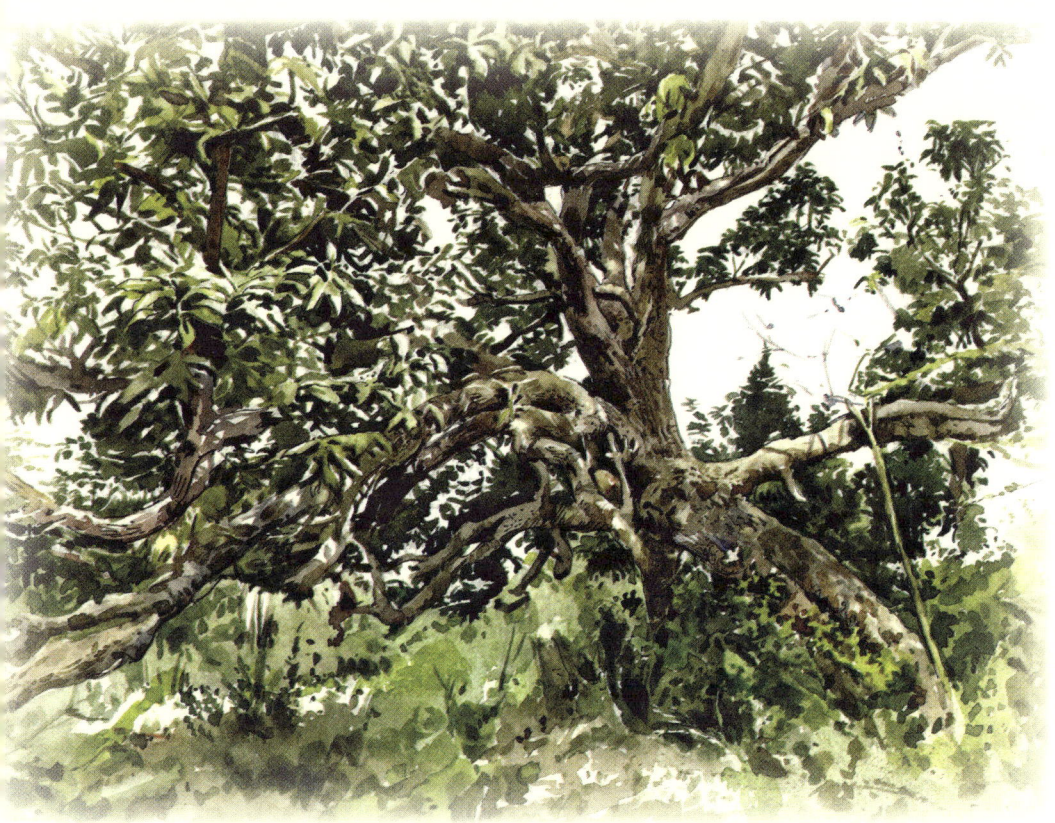

'walking into India'.[251] Villagers say that during the last 250 years, it has moved in stages by nearly 200 metres. Interestingly, the tree's movement is unidirectional, towards the east. Due to its legendary association with early Parsi migrants, its age is generally considered to be 1,200 to 1,300 years.

As is common for mango, this tree also exhibits alternate bearing, i.e., more fruits are produced every other year.[252] However, the fruits are smaller with a shelf life of just two to three days. The flesh is flaming red when ripe and tastes different from the local varieties. Seeds germinate but seedlings do not survive for long. Attempts to produce grafts by the state forestry department have not been successful.

Although there is no recorded historical link between the Parsis and the tree, the walking mango tree may have been planted by Zoroastrians as a pious *vaqf* ('bequest') or *yād-bud* ('memorial') for trees are sacred in Zoroastrian tradition as symbols of immortality and were often planted around fire temples.[253] This one has decided to 'walk' some distance at its own pace. One wonders though about where exactly it is headed.

70

The Twin Sen'trees'of Matunga

COMMON NAMES: **Malay Padouk, Narra (English)**

SCIENTIFIC NAME: *Pterocarpus indicus* **Willd.**

FAMILY: **Pea (Fabaceae)**

WHERE TO SEE: **Pavement adjacent to Siddhachal Apartment 2, Plot No. 447, Dr Babasaheb Ambedkar Road, King's Circle, Matunga, Mumbai**

LATITUDE: **19.029⁰ N;** LONGITUDE: **72.857⁰ E;**
ALTITUDE: **Approximately 3 metres**

ommuters speeding over Matunga on the busy King's Circle flyover (officially known as Nanalal D. Mehta flyover) on Dr Babasaheb Ambedkar Road towards Sion-Dharavi will probably not be able to spare a glance towards Siddhachal Apartment building on their left. If they did, they will not fail to see two majestic Malay padouk trees standing like a pair of sentries at attention on the pavement adjoining that building and towering over the flyover. Introduced into India from its natural home in South/Southeast Asia, it is incidentally the national tree of the Philippines.

The two extraordinary trees of Matunga have earned the moniker of 'emperors' and are featured among the sen'trees' of Mumbai.[254] Although both trees are probably of the same vintage – they could be 150 years old or older – the one nearer to King's Circle is conspicuously taller (>30 metres) and stouter (approximately 6.45 metres) than its companion (approximately 25 metres and approximately four metres respectively).

The crowns of the two trees touch and together create a giant canopy providing welcome shade for people moving on the pavement below. Their feather-like leaves are dark-green and shiny. When in bloom, the trees make a stunning display with numerous fragrant golden-yellow flowers on slender pedicels. Compared to the abundance of flowers, the fruits are noticeably fewer under Mumbai conditions. They are flat, one-seeded orbicular pods with membranous wings that help in dispersal. Indeed, the generic name *Pterocarpus* signifies 'winged fruit' (Greek – *pteran* meaning wing and *karpos* meaning fruit). The fruits do not dehisce though.

In their natural home in South and Southeast Asia, Malay padouks are highly valued for their timber. The specific epithet *indicus* refers to East Indies by which name the present Malay Archipelago was known earlier. Excessive exploitation of timber and use in folk medicine have resulted in the species being currently placed in the endangered category of the IUCN Red List of threatened species. In India,

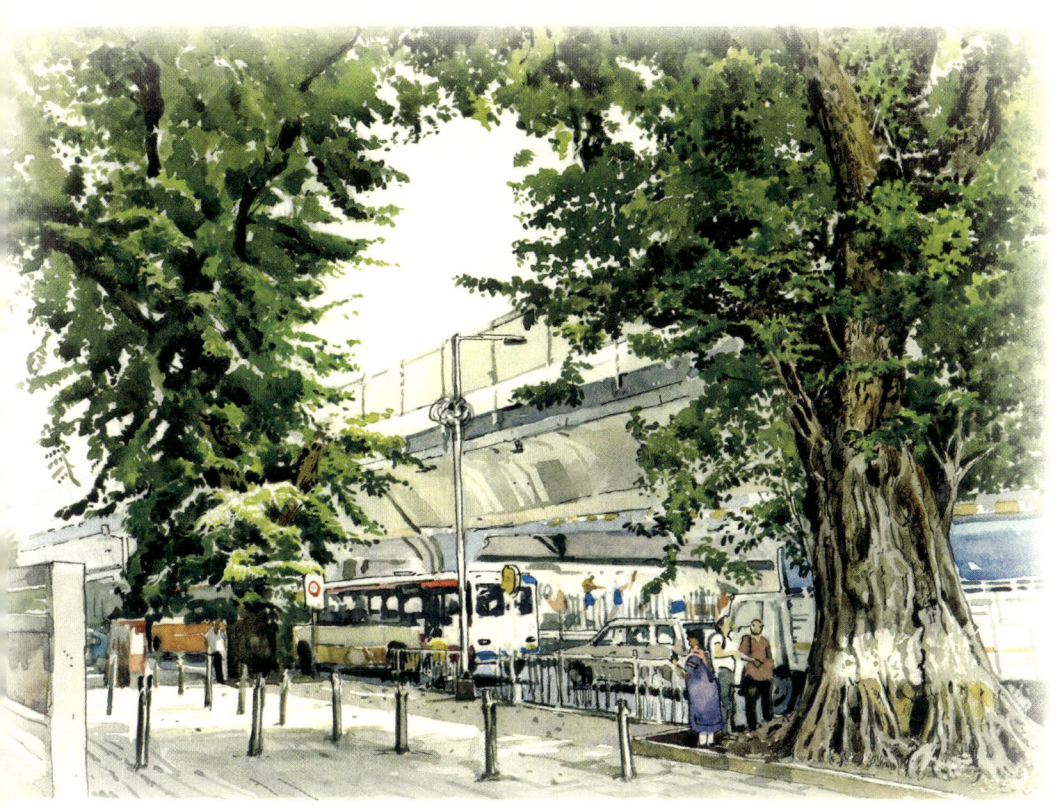

though, it is mostly cultivated.[255] Though three other specimens of this species can be seen in the city at Jeejamata Udyan, Cooperage Garden and BSNL building near Churchgate, the twin giants of Matunga are by far the finest examples. Long-lived and slow-growing, these centurions have witnessed the metamorphosis of young Bombay of the bullock-cart era into today's burgeoning Mumbai of flyovers and metro rail. Though hectic residential and commercial activities pose a threat to their survival, some measures ought to be taken soon by the state authorities to preserve these historic trees.

71

Livingstone's Legacy

COMMON NAMES: **Mahogany, True Mahogany, West Indian or Cuban Mahogany (English); Mahogany (Hindi/Marathi)**

SCIENTIFIC NAME: *Swietenia mahagoni* (L.) Jacq.

FAMILY: **Neem/Mahogany (Meliaceae)**

WHERE TO SEE: **Both trees stand on K. Dubash Marg, Kala Ghoda, Fort, Mumbai**

LATITUDE: **18.927⁰ N and 18.928⁰ N**; LONGITUDE: **72.832⁰ E**; ALTITUDE: **Approximately 3 metres**

ort' refers to the part of the original Bombay – now Mumbai – that is the hub of art and commerce. The legacy of the colonial era is evident everywhere in this precinct in its old and famous landmarks. The foundations of some of the great buildings were laid when Sir Bartle Frere was the governor of the Presidency (1862–67). The only living evidence of that period today are a few ancient trees, still gamely battling for survival amidst all the 'development' around them. Of these, two ancient mahogany trees in Kala Ghoda, the southern tip of the Fort area, are believed to have been planted by the famous Scottish explorer and missionary David Livingstone during his visit to Bombay in 1865.

The trees stand on the footpath on the busy Kaikhushru Dubash Marg (formerly Rampart Row) close to a row of buildings on one side. One tree is closer to Dr Salim Ali Chowk, behind St Andrew's Church and the other is a little further away, closer to the statue of Kala Ghoda. In their native environment in the Americas (from Mexico to Brazil), mahogany trees grow to an immense size – as tall as 45 metres and between three and four metres in girth – but here they are much stunted. Haphazard and mindless disfigurement of their beautiful crown and limb over the years – obviously to prevent damage to the buildings nearby and accommodate the teeming traffic on this road – have added to their sorry look. The bark is characteristically grey-brown, fairly smooth in younger parts but appears rough and furrowed elsewhere.

Did the famous African explorer really plant these trees? Historically, we do know that Livingstone transited through Bombay, entering the harbour in June 1865[256] in his little steamer, the *Lady Nyassa* concluding a voyage of 2,500 miles (approximately 4,000 kilometres) lasting 45 days from Zambesi, East Africa. He was the governor's guest. However, it was with great difficulty that he could persuade the reticent Livingstone to accept his hospitality. During his stay, he is believed to have planted the mahogany plants at the request of the governor.[257]

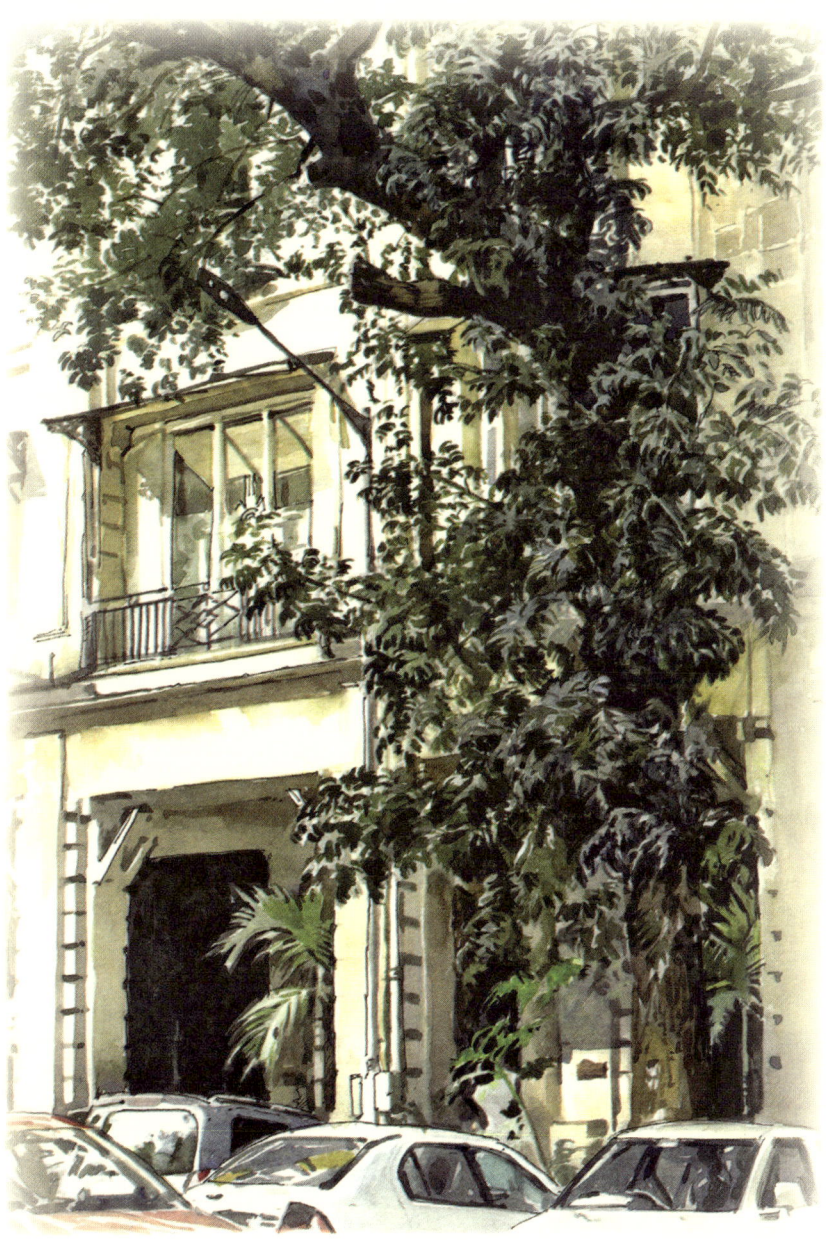

This is entirely possible, given his closeness to him and the fact that he spent more than a month in the city and visited Matheran and Nashik in between. Even so, evidence that he did plant the trees is hard to come by. If Livingstone did plant them, they ought to be close to 150 years now, but it is difficult to guess their age. In its native home, mahogany is known as the 'King of the Forest'. In many parts of India, the species is popular as an avenue tree (other than being the source of great timber). Mumbai city would, in any case, do well to preserve and protect these historic centurions and generate some public awareness around them.

72

Ajanvruksh of Alandi

COMMON NAMES: **Charmavriksh (Sanskrit); Ajānvruksh, Datranga, Tamboli (Marathi); Chamrod (Hindi)**

SCIENTIFIC NAME: *Ehretia aspera* **Willd. (=** *Ehretia laevis* **Roxb.)**

FAMILY: **Borage (Boraginaceae)**

WHERE TO SEE: **Dnyaneshwar Samadhi, Alandi Devachi, Pune district, Maharashtra**

LATITUDE: **18.67695⁰ N;** LONGITUDE: **73.89658⁰ E; ALTITUDE: 5,68.4 metres**

Alandi, a small town located on the banks of the Indrāyani River on the northern edge of Pune is famous for the samadhi (tomb) of Sant Dnyaneshwar, the thirteenth-century Marathi saint, poet, philosopher and yogi of the Nath-Vaishnava tradition. His philosophical masterpiece *Bhāvārthdeepika*, which came to be known as *Dnyāneshwari*, is at the forefront of the Bhakti movement in Maharashtra.

Marking the holy samadhi is a tree that is referred to as ajānvruksh. Although a native species of Asia and found throughout India, it is usually not cultivated or worshipped, except at other temples associated with Nath saints within Maharashtra. Here, this tree is considered sacred by its association with Dnyaneshwar's samadhi. A medium-sized deciduous tree with several pale trunks simultaneously emerging from the ground, as is common for this species, some of which are bent and emerge at various angles from the ground. The temple authorities have cordoned off the trunks within a stone and metal enclosure, presumably to keep people from disfiguring them. Together, the trunks bear a large canopy of shiny leaves providing dense shade below.

The age of the tree is a matter of conjecture, but it is undoubtedly old. According to local legend, Dnyaneshwar elected to go into the state of *sajeevan/sanjeevan* samadhi (samadhi while still alive) in Alandi at the age of 21 in December 1296. At that time the ajan tree was probably not there. Nearly 300 years later, Eknath (1533–99) a poet, mystic and social reformer was visiting Alandi (then known as Alankāpuri) with his followers. Legend says they were resting during the night on the banks of the Indrāyani when Dnyaneshwar appeared to him in a dream complaining that the root of an ajan tree was throttling his neck and suffocating him. He ordered Eknath to dig into the samadhi and cast out the root.[258] This was easier said than done. In those days, the area was densely wooded and filled with lush undergrowth. To Eknath's dismay, several samadhis were dotting the area.

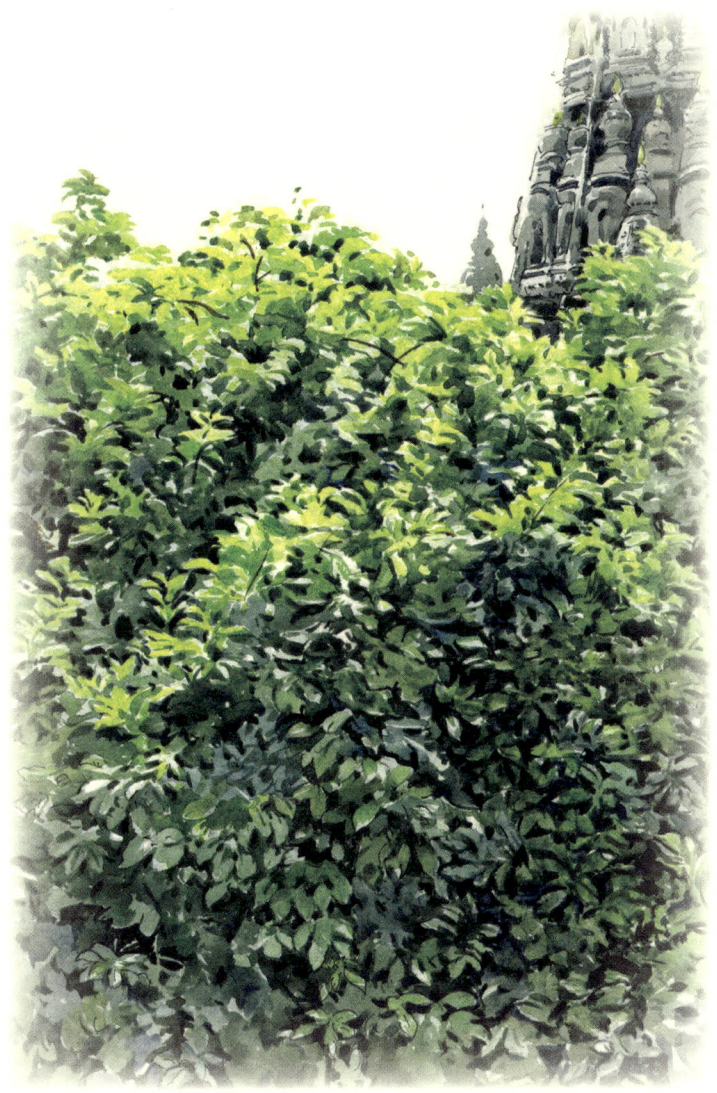

Which one was Dnyaneshwar's? Eknath had to perform certain austerities before the correct one was revealed to him. He prostrated himself before Dnyanadeva's samadhi and got rid of the offending root. Apparently, Dnyaneshwar also directed Eknath to edit *Dnyāneshwari* to eliminate 'impurities' that had crept in through the centuries. Although replicas of *Dnyāneshwari* did exist, the copies were full of mistakes in grammar and meaning, thus making the work virtually unknown to common people. Eknath meticulously corrected these errors and restored the original meaning of the text.[259] It is entirely possible that the reference to the root of ajānvruksh at his throat was a metaphor signifying Dnyaneshwar's pain and anguish at the pollution of his magnum opus. Be that as it may, today, the ajānvruksh is still there and devotees of the temple-samadhi can be seen fervently reciting *Dnyāneshwari* under its shade.

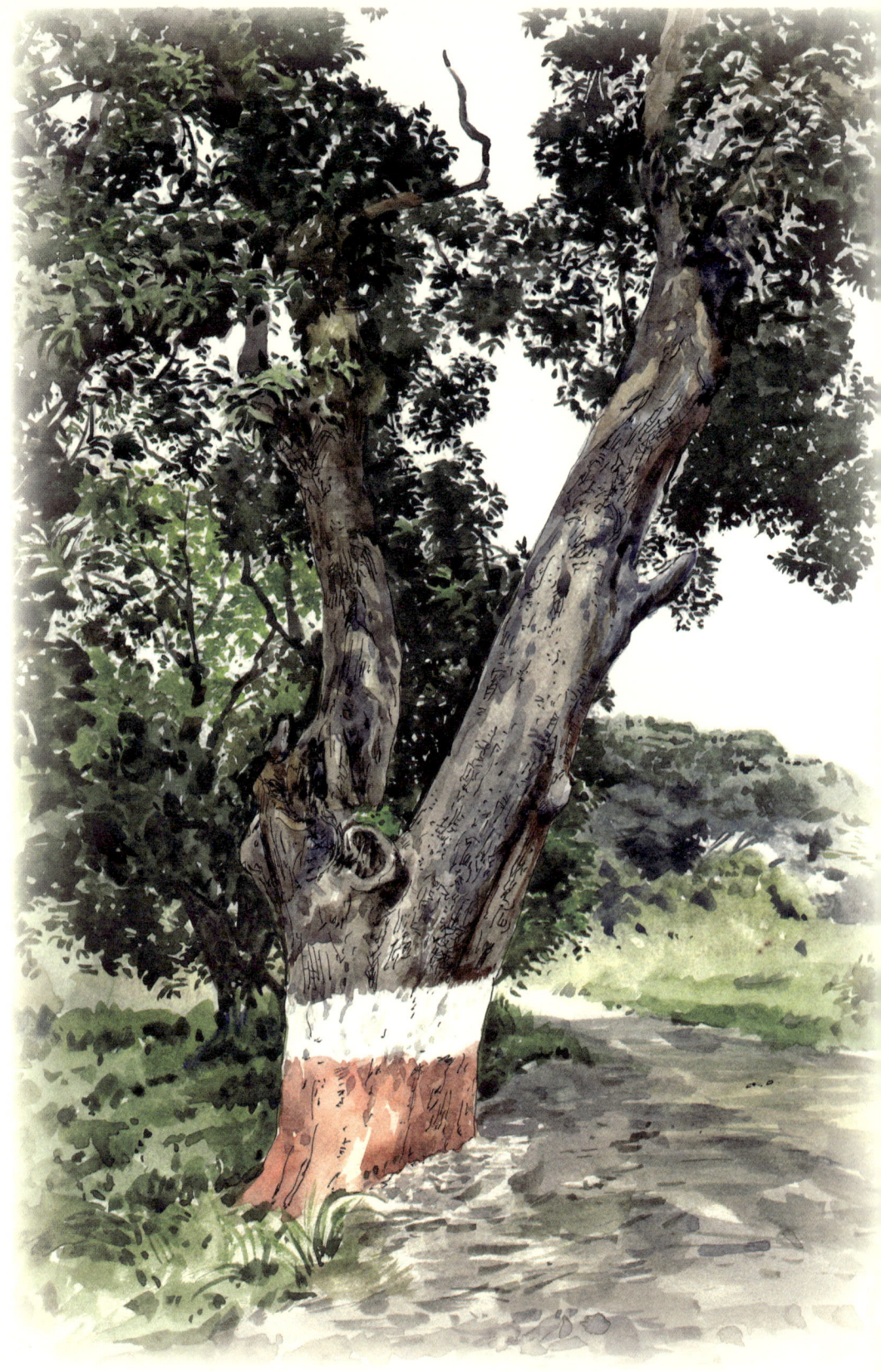

73

Peshwa Bajirao's Mango Tree

COMMON NAMES: Mango (English); Aamra (Sanskrit);
Amba (Marathi); Aam (Hindi)

SCIENTIFIC NAME: *Mangifera indica* L. | FAMILY: Mango (Anacardiaceae)

WHERE TO SEE: **Mahatma Phule Krishi Vidyapeeth Zonal Agriculture
Research Station and Fruit Research Station, Ganeshkhind,
Aundh-Khadki Road, Pune, Maharashtra**

LATITUDE: **18.567⁰ N**; LONGITUDE: **73.825⁰ E**;
ALTITUDE: **Approximately 560 metres**

Shrimant Bajirao II, the thirteenth and the last Peshwa (r. 1796–1818) of the
Maratha Empire is remembered mostly for all the wrong reasons today.
However, he was very fond of mangoes and probably had green thumbs.
That explains why he ordered the planting of a million mango trees in and around
Poona (now Pune),[260] his capital. Visiting Pune on 31 January 1818, Lt Colonel
Frederick Fitzclarence, a British Army officer, recorded that he had a most
extensive view of the city and its mango groves from the top of the Shaniwar
Wāda palace, adding that, 'From the back of the position to the city extends a
grove of mango trees in rows and in every part round the city are similar orchards.
They were planted by the order and at the expense of the Peishwah and their
number, we were told exceeded 300,000'.[261] It is said that the Peshwa had long
caravans employed to bring mango seeds from Goa. Of the several plantations
that Bajirao II created, an old mango *tope* (grove) on the banks of the Mula River
near Ganeshkhind is remarkable. A large tree from this ancient plantation still
survives there.

The British East India Company converted this *tope* into a botanical garden
under the aegis of the Royal Horticultural Society of India, Calcutta (now
Kolkata) in 1853. Later, the garden was taken over by the government of the
Bombay Presidency and became a premier centre for the study of taxonomy,
horticulture and cultivation of medicinal plants when George Marshall Woodrow
(1846–1911), the noted botanist, took charge in 1873. Indeed, William Mudd,
curator of the Cambridge Botanical Garden and a member of the Prince of Wales's
entourage in 1876, gushingly voted it as 'the prettiest I saw in India'.[262]

Woodrow had already served as a gardener at the Royal Botanic Gardens
in Kew, England, for seven years when he travelled to India to take charge of
the experimental garden. He set out to establish the best practices for mango
cultivation. For example, good quality fruit was produced in loamy soil three feet

deep comprising 5 to 10 per cent of lime and enough peroxide of iron to give the soil a reddish tinge. He was quick to note the poor quality of the average mango fruit raised from seed. During Peshwa Bajirao's time, all the mango trees were raised from seed and the progeny did not breed true to type. Grafting of a desirable variety to a vigorous stock was the best answer. In his book titled *Mango: Its Culture and Varieties*, Woodrow recorded 78 famous prevailing mango varieties, 22 of those from Pune. One among the latter – a 'first-class sort' variety named 'Pyrie' with soft creamy pulp and delicious delicate flavour – was from the Ganeshkhind gardens.[263]

Today, the lone representative of the Peshwa-era mango tree is probably 200 years old and stands in a corner of the Ganeshkhind garden, now the Zonal Agricultural Research Station of the Mahatma Phule Krishi Vidyapeeth. The trunk is slightly tilted towards the unmetalled road and two main branches reach out – one ascending upright and the other at an angle. Three other branches seem to have been lost to the vagaries of nature leaving behind hollowed-out crevices where they were once attached. The large trunk is close to five metres in circumference with dark grey bark broken into trapezium-like pieces. The tree's fruit is referred to by the garden staff simply as 'Peshwa mango', whose pulp is slightly fibrous. However, there is no evidence to back the claims in the popular press of the mangoes being exported to Queen Victoria.

Sadly, no records are available on the garden's lone surviving tree from the heydays of the Peshwa period to inform us of the role it served in those days. In today's Pune, the champion tree goes mostly unnoticed. In a recent heartening development, the Maharashtra government has approved a proposal to declare 58.6 hectares of the Ganeshkhind garden as one of the biodiversity heritage sites of the state, citing the ancient mango tree among the major reasons for doing so.[264] Hopefully, with this development, the garden will be upgraded with better outreach programmes, so that the historical tree will finally get its due.

74

Umaji Naik's Peepal Tree

COMMON NAMES: **Sacred Fig Tree (English); Ashwattha (Sanskrit);**
Peepal Tree (Hindi)

SCIENTIFIC NAME: *Ficus religiosa* L. | FAMILY: **Fig (Moraceae)**

WHERE TO SEE: **Pune District Tehsil office, Khadakmal Ali Road,**
Shukrawar Peth, Pune, Maharashtra

LATITUDE: **18.507° N; LONGITUDE: 73.858° E;**
ALTITUDE: **Approximately 560 metres**

Amidst the clang of rickety typewriters and arguments on wills, revenue and taxes, few people have the time or inclination to notice a peepal tree in a quiet corner of the District Tehsil office in Pune's busy Shukrawar Peth. Judging by its size and spread, the tree seems to be standing there for over two centuries, towering over the surrounding single-story buildings that must have been built at various times in the past. A small rectangular boundary surrounds the base of the tree with a signboard announcing in Marathi: 'Umaji Naik was hanged by the British Government on 3 February 1832 on these premises. His dead body was suspended from this tree for three days to strike terror in the minds of people'.

Umaji Naik was among the earliest freedom fighters to rebel against the British government, well over a quarter century before the 1857 revolt. He was born in the Berad or Ramoshi community who, during the Maratha reign were entrusted with the security of their forts. However, following the British conquest of the Maratha Empire in 1818, the administration of the Bombay Presidency found no use for their services and took away even their hereditary lands, driving many among them into a life of crime.[265] The colonial powers were quick to retaliate, declaring the entire community a criminal tribe and imposing severe restrictions on their movement. The simmering discontent soon led to open rebellion against the government. Umaji joined the rebellious group and soon became its leader.

Between 1825 and 1830, Umaji and his comrades carried out a series of daring attacks on government personnel and properties, looting treasuries and killing officials. Declaring himself a king, Umaji issued a manifesto calling for people to execute every European they came across. The British placed a bounty of ₹10,000 on his head but still failed to capture him. In the end, he was betrayed by one of his colleagues, a fellow Ramoshi,[266] and was captured and brought to Pune and hanged on 3 February 1832. His dead body is said to have been suspended

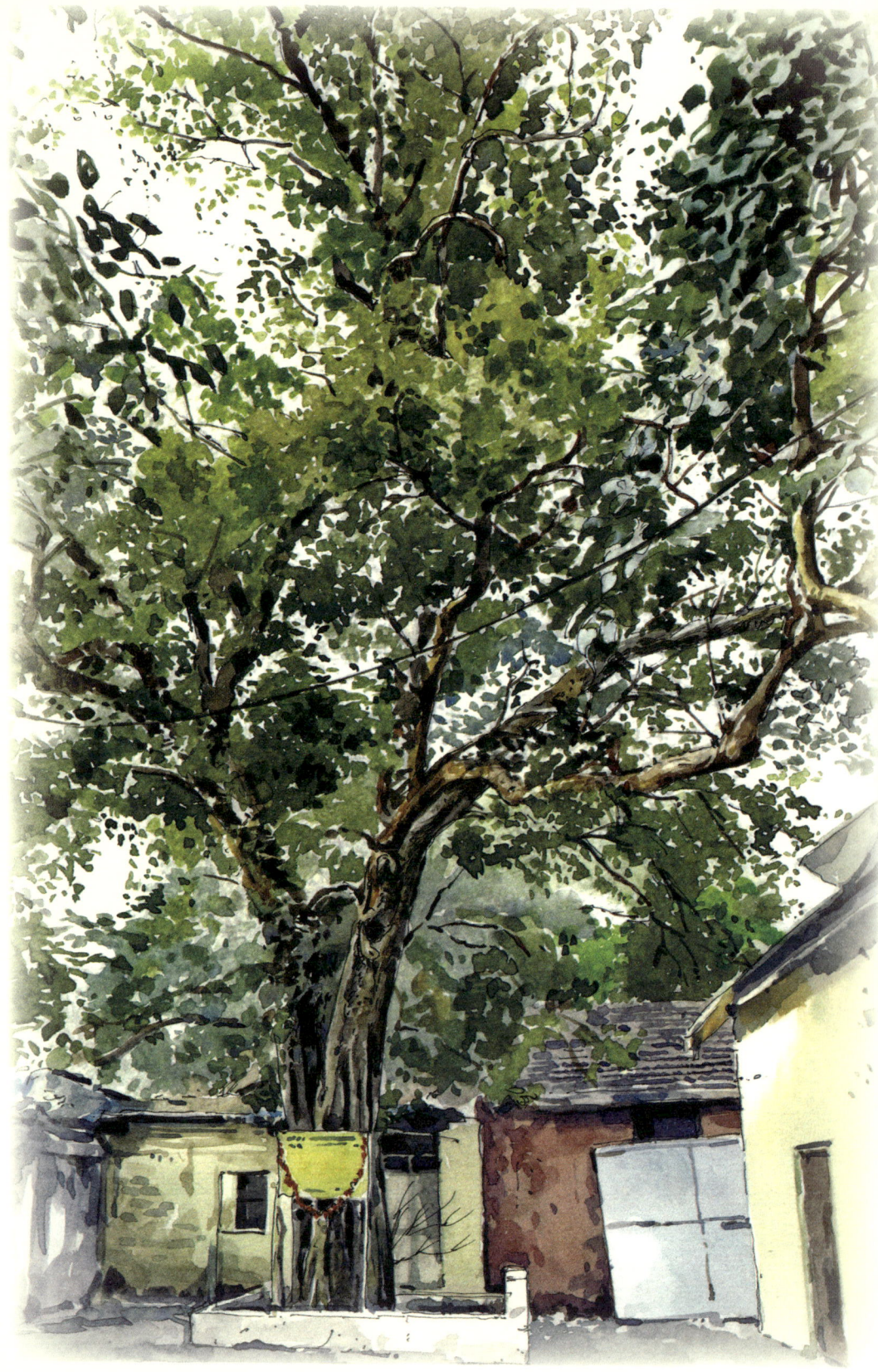

from the peepal tree just outside the jail. Contemporary British accounts do not mention this[267] but it fits into a pattern of the racial arrogance and callous brutality of the British Empire during the nineteenth century. Colonial executions were calculated to strike terror in the native mind.

A police station still functions on the premises, although the adjoining jail that held Umaji Naik is no longer in use. There is also a two-story building that functioned as the colonial court where Naik was sentenced to death. A monument for the freedom fighter has been built next to it.

The Adya Krantiveer Umaji Naik Kshatriya Ramvanshi Sanghatana is striving to keep alive his story.[268] The group is also lobbying to convert the entire premises, including the peepal tree and prison area, into a monument-cum-museum dedicated to Naik and the history of the Ramoshis.

Live close to nature and you'll never feel lonely. Don't drive those sparrows out of your veranda; they won't hack into your computer.

Ruskin Bond
Author

The Plumeria of Parvati Hill

COMMON NAMES: **Frangipani, Pagoda Tree, Temple Tree (English); Chapha,
Dev Chapha (Marathi); Champa (Hindi)**

SCIENTIFIC NAME: *Plumeria rubra* L.

FAMILY: **Oleander (Apocynaceae)**

WHERE TO SEE: **Peshwe W$\bar{a}$da, Parvati Hill, Pune, Maharashtra**

LATITUDE: **18.4966⁰ N**; LONGITUDE: **73.8413⁰ E**;
ALTITUDE: **640 metres**

For many Punekars, climbing Parvati Hill is part of a daily exercise regimen. Rising to 640 metres above mean sea level, the hill represents the second-highest peak after Vetal Tekdi (approximately 792.5 metres). Once you climb the 100-odd stone steps to its summit, the panoramic view of the Pune city takes your breath away. If you are in time, there is the additional bonus of a beautiful sunrise or sunset thrown in. The Peshwas of Pune constructed a temple complex here, the chief of which is dedicated to Shiva or Devadeveshvar, with others including Bhawani/Parvati, Vishnu Narayan and Kartikeya.[269] They also built a small and unpretentious *wāda* (palace), part of which is now a museum. The *wāda* has a small courtyard whose three sides are enclosed by buildings and the fourth by a wall. In the centre of this courtyard stands a plumeria/frangipani tree with a circular platform around it made of rubble. Locally known as chāpha, it is claimed to be Pune's oldest tree.[270]

Old it certainly is, with a gnarled and wrinkled short trunk and two main branches. Even the secondary and tertiary branches give off the impression of age, clothed as they are with creased brownish bark. Its long dark green leaves – broad at the tip and tapering at the base, with prominent veins – tend to crowd at the ends of branches. Break or injure one and a trickle of milky juice (latex) appears at the cut ends. Flowers are borne in terminal bunches and have white petals with the inner sides of their bases tinged yellow. Though flowering is profuse, fruit set does not occur here.

There are no records to say when the tree was planted. However, a black-and-white photograph of Parvati's top portion taken in the year 1890 is available in the *wāda*, which depicts the upper portion of the frangipani tree along with parts of the *wāda*, the Devadeveshvar temple and the spire of the Vishnu Narayan temple. It is obvious that the photograph has been taken from atop the walls of the *wāda*. The crown of the chāpha tree is seen to be obscuring all but the top three tiers of Vishnu Narayan temple. If you stand at the same spot in the *wāda* today and view

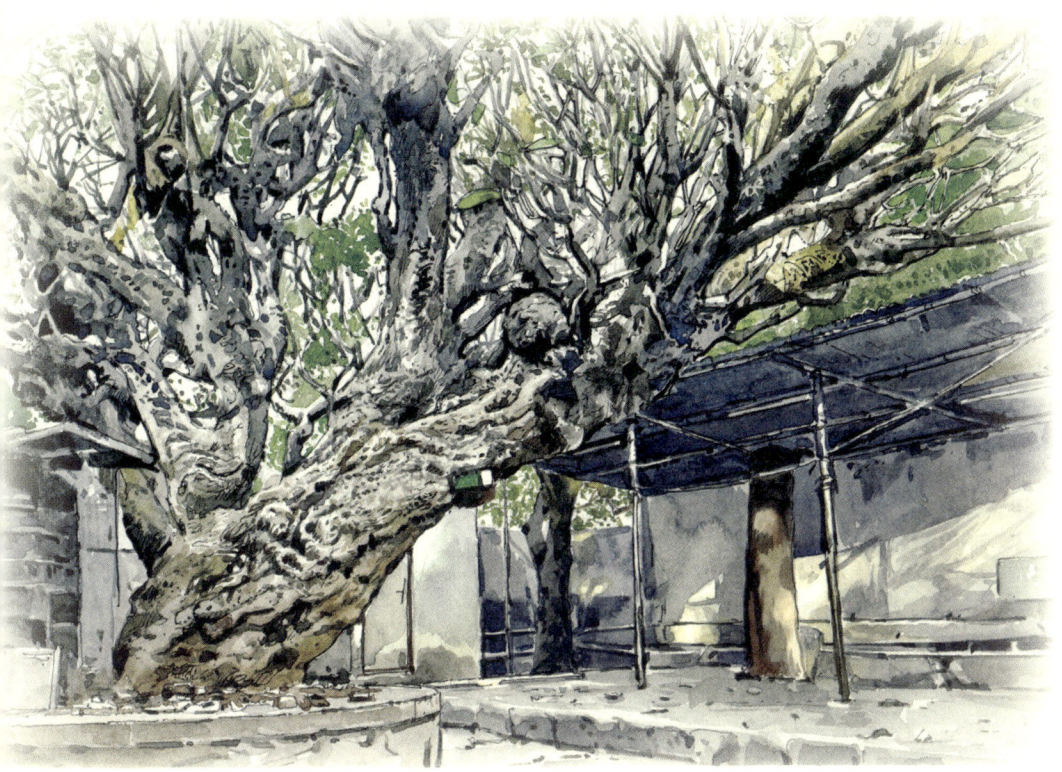

the tree from the same angle with the Vishnu Narayan temple as background, you can see it has gained a height of a little over a metre now.

It seems that Balaji Bajirao, the third Peshwa, built the main temple in 1749 and consecrated the idols there to fulfil his mother Kashibai's wish. He also constructed the private *wāda* for change and recreation in 1755.[271] If the tree was planted sometime during or after that period, it could be between 200 and 250 years old. Originally from Central and South America, the species has been widely cultivated in the tropics. It must have been introduced into India very early, possibly by French or Portuguese Catholic missionaries. Interestingly, Hendrik van Rheede, the colonial administrator of the Dutch East India Company – well-known for his monumental treatise *Hortus Malabaricus* (published between 1678 and 1703) – did not mention *Plumeria*. Even the celebrated naturalist Carolus Linnaeus did not mention it as growing in India in his famous work *Species Plantarum* (1753) (although he did include other plants from the country). It is possible that it did not exist here before 1750. By 1787, however, it was found growing abundantly and its bark was used for treating intermittent fever, just as that of cinchona.[272]

Its presence within Parvati's *wāda* suggests that it was planted for its year-round abundance of fragrant white flowers for worshipping the family deities of the Peshwas. In that regard, the tree continues to be useful even now.

Acknowledgements

I would like to claim that the entire credit for writing this book belongs to me. But that would be far from the truth. A book such as this is always the outcome of several people's invaluable help, unwavering support, and indispensable assistance. Writing it has been a blessing and a cathartic experience for me. It presented me with an unprecedented opportunity to see different parts of this great country – both mainstream and off the beaten track, meet and interact with ordinary and extraordinary people from different walks of life, and compelled me to read on subjects that, otherwise, I might not have done.

At a time when I was hesitant and unsure about undertaking a project of this magnitude for more than one reason, I was getting on in years, the job involved a great deal of travel crisscrossing the country, and there was no institutional support – Dr Kamal Bawa, Distinguished Professor, University of Massachusetts, Boston, and President Emeritus, Ashoka Trust for Research in Ecology and the Environment, provided the initial nudge of encouragement, with a voluntary and unconditional offer of a small travel grant from his funds at ATREE. I gratefully accepted the offer. As the project got going, my self-doubts began to clear. I realized that my needs were frugal and that the project was doable. This strengthened my resolve to continue and finance the project myself. I returned the travel grant with a note of appreciation to Dr Bawa so that he could help those who most needed it. However, I will always cherish and remember his encouragement and continued support that gave me the momentum. I owe him my first thanks and deep gratitude for his belief in me.

Professor Mahesh Rangarajan (Ashoka University), Professor Harini Nagendra (Azim Premji University), and Dr M. Sanjappa (former Director of the Botanical Survey of India) not only found time in their busy schedules to read through an earlier version of this book but also provided their thoughtful suggestions and comments. Their contributions have been invaluable, and I am deeply grateful for their support. The assistance and expertise on dendrochronology and radiocarbon dating of Indian trees from Drs C.M. Nautiyal, S.K. Basumatary, K.G. Misra and Santosh Shah, all from the Birbal Sahni Institute of Palaeosciences, and Dr Hemant Borgaonkar, at the Indian Institute of Tropical Meteorology have helped me greatly. Nautiyal and Basumatary's generosity in allowing me access to their unpublished material on the Wishing Tree of Assam is particularly appreciated.

When this work started, I had not factored in the generosity of many friends, acquaintances, and even total strangers. Their courtesy and kindness have been a truly humbling experience. Collectively, they assisted with their knowledge, expertise, local hospitality, or in other ways. It is impossible to name all of them, but I will mention a few below in alphabetical order:

Andaman & Nicobar – Rasheeda Ismael; **Andhra Pradesh** – Jagannath Raju; **Australia** – Thomas Weber (La Trobe University, Melbourne); **Bihar** – Sanjukta Datta; **Delhi** – Md Aslam, Mantu Das, Anamika Gambhir, F.A. Khudsar, T. Madhan Mohan, Dennys Leighton and Sudeshna Mazumdar-Leighton, S.M. Qasim, Radheyshyam and Vandana Sharma;

Gujarat – Anuj Ambalal, Jitendra Gavali; R.D. Kambhoj, Dhirubhai H. Patel, Vinay Raole, Rajkumar Yadav; **Himachal Pradesh** –Ramesh Chandramouli; **Jammu & Kashmir** – Y.S. Bedi, Mehraj Din Bhat, Ahmed Bilal, Khurshid Ganei, A.K. Koul, O.P. Sharma "Viidyarthi", Ram Vishwakarma, B.A. Wafai; **Karnataka** – K.M. Bopanna, Shankar Dandin, Jyotsna Dhawan, K.N. Ganeshaiah, M. Jagadeesh, S.S. Kulkarni, C. Madegowda, S. Narayanaswamy, the late K.P. Pandiyan, Syed Shah Sibghatulla Quadri, Krishnam Raju, Keshava Reddy, Girish Rao, K. Shankara Rao, Siddappa Shetty, B.S. Shylaja, Vemagal Somashekhar, M. Umashaanker; **Kerala** – Nirmal Babu, U.M. Chandrashekara, Jose Kallarackal, the late Harilal Kochugovindan, R. Prakashkumar, P.N. Ravindran, A.B. Remashree, E.S. Santhoshkumar, C. Satheesh, S. Shailajakumari; **Maharashtra** – Sakina Gadiwala, P.K. Ghanekar, Phiroza Godrej, Sunil Jadhav, A.S. Kothari, P. Lakshminarasimhan, Maneck Mistry, Behnaz Patel, M.S. Pendse, Sachin Punekar, Pankaj Sekhsaria, Rajendra Shinde, V.S. Supe, Sashirekha Sureshkumar; **Manipur** – Bunny, H. Kipgen, Y.S. Rajashekar, Dinabandhu Sahoo, Biseshwari Thongam; **Meghalaya** – A. Bhattacharjee, Nripemo Odyuo, B.B.T. Thom; **Nagaland** – Zawei Hiese, Nesatalu Hiese, Doneipa Khale; **Rajasthan** – Vinod Maina, K Ravikiran; **Sikkim** – Usha Lachungpa, and D. Manjunatha; **Tamil Nadu**: K. Anbalagan, J.S. Britto, Madhumita Dasgupta, K. Gurumurthi, S.S. Hamid, S. Rajan, S. Uma, R. Yasodha; **Telangana** – M. Ahmedullah; **Uttarakhand** – Rakesh Ruskin Bond, Ruskin Bond, LMS Palni; **Uttar Pradesh** – S.K. Barik, Arti Garg, A. Shukla, Pramod Tandon; **West Bengal** – A. Chattopadhyay, A.A. Mao, and M.U. Sharief. Thank you, all.

Professor Emeritus C.R. Babu (Delhi), Professors K. Dharmalingam (Madurai), Raghavendra Gadagkar (Bengaluru), D. Narasimhan (Chennai), B. Ravindran (Bhubaneswar), Hema D. Sane (Pune), Dr Chandrima Shaha (Delhi) and the late Dr J.D. Mehta IFS (Almora), have constantly stood behind me, cheering and encouraging. I am truly grateful for their unstinted support.

My thanks and appreciation are due to Sagar Bhowmick for converting my photographs into marvelously nuanced paintings. He has brought out the essence of the trees themselves as well as their ambiance while uncomplainingly accommodating my constant demands.

At Roli Books, Priya Kapoor, the Director has always been highly upbeat, supportive, and committed to this project from the word go. It has been a great pleasure knowing her and working with her. Ushnav Shroff and Sneha Pamneja deserve a big shout-out. Ushnav employed his editing skills meticulously, efficiently, and without fuss, always with a patient and sensitive ear. With her design expertise, Sneha ensured that the book arrived in very pleasing and attractive attire.

Finally, my daughters Seetha and Meghana, and my son-in-law Tanmoy Goswami have been my first readers, unsparing critics, and admirers by turns. I have greatly benefitted from their incisive comments. Even my six-year-old grandson Advaith has been exhorting me to finish the book quickly so that we could do 'more interesting' things. My wife Geetha has accompanied me on several tours, sharing the joys of tracking and mapping the iconic trees. She has also shown great forbearance and understanding during the writing of this book. I could not have completed this without their direct and indirect help and support.

251

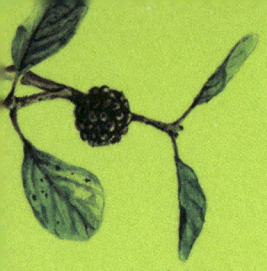

Index

Abdulla Nursery 89
Accelerator Mass Spectro-
    metre (AMS) 27, 136
Acharya Jagadish Chandra
    Bose Indian Botanic
    Garden, Kolkata 116, 119
African baobab tree
    22–23, 29, 34, 36, 86, 95,
    136–37, 144, 154
African bushland 144
Ajānvruksh 36, 240–41
Alerce Milenario tree 18
Alerce tree 14, 18
Al Muqarrar tree 89, 92
Āmla (gooseberry) tree 66
Ammāchi plāvu tree 15
Āmra (mango) trees 36,
    66, 71, 89, 92–94, 234,
    235, 243–44
Ashmantaka tree 182
Ashvattha (peepal) tree
    35, 62–63, 66, 72, 74, 76,
    83, 100–101, 105–06, 112,
    131, 182, 245, 247
Aspen trees 17
Attilippa tree 35, 217, 219
Autumn wood 24

Baad-e-naseem 52
Baba Budha ber tree 33,
    67–68
Bagh-e-Naseem 52
Baheda tree 35, 107
Bahupada tree 82, 119, 138,
    147, 158, 226, 228, 231
Bakhor Bēngana Rakshya
    Samiti 123
Bakhor Bēngana tree
    122–23
Bakula tree 149, 151
Bald cypress tree 19
Bamboo 127, 130
Banyan tree 39, 82–83,
    105, 119–21, 138, 147,
    158, 226, 228, 230–31

Baobab tree 22–23, 29, 34,
    36, 86, 88, 95–96, 136–37,
    144–45, 153–54
Baobab wood 88
Bargad tree 138, 147, 158,
    226, 228, 231
Bedda nut tree 107
Bee tree 33, 60, 158–59
    Guru Nanak 33, 60
Benjamin's fig tree 170
Bengaluru Environment
    Trust 37
Ber Baba Budha tree 67
Ber tree 14, 33, 60–61,
    67, 68
    ancient 67
    oldest 61, 68
Big banyan tree 14, 39
Big champa tree 179
Bijbehara chinar tree 57
Biodiversity 14–15, 17,
    37, 244
Biological Diversity Act
    2002 37
Birbal Sahni Institute of
    Palaeosciences (BSIP),
    Lucknow 22, 26, 28,
    80, 122
Birch tree 30
Black pepper 186
Black plum tree 186
Blue gum tree 142–43,
    156, 189, 191
Southern 142, 189
Tasmanian 142–43, 189
Bodhi tree 14, 19, 36, 62,
    100–101, 182
Boin (chinar) tree 47
Bombay Blackwood 196
Botanical gardens 15,
    119–20, 243
Botanical Survey of India
    96, 121
Botanic Gardens
    Conservation

International 15
Bristlecone pine tree 14,
    18–19, 23
Californian 14
Buddhism 100, 110, 113
Bulletwood tree 149
Būruga tree 34

Cardamom 186
Cedar tree 12, 28
Centre for Social
    Forestry and Eco-
    Rehabilitation 96
Centurion deodar tree 66
Centurion tree 20, 40, 201,
    204, 230
Ceylon ironwood 30, 220
Ceylon Wood 98, 220
Chained tree 205
Chalto ambo tree 36, 234
Champion tree 16–17, 48,
    126, 191, 216, 244
Chamrod tree 240
Charasu marabo kaji
    tree 131
Charmavriksh 240
Chaurasi deodar tree 64
Chikka sampigé tree 181
Chilgoza pine tree 29
Chinar tree 31, 34–35,
    47–48, 52, 54–55, 57, 145
of Bijbehara 55, 145
of Naseem Bagh 52
oldest living 48
saplings of 52
Chinar garden 35, 52
Chinchini (tamarind) tree
    33, 35, 66, 144, 162–63
Chinese weeping cypress
    tree 110, 113–14
Chini tree 212
Chitebu kaji tree 134
Choulmogra tree 207
Citizen tree wardens 43
Climate and Environmental

Sciences Laboratory,
Paris 18
Coastal Redwood 20
Coco de mer tree 116–17
Coffee 186, 194
Coniferous Wood 28
Conservation
    efforts 43
    policies and management
    guidelines 15
    strategies 20
Cooperage Garden,
Bombay 237
Council of Scientific and
    Industrial Research
    (CSIR) 49–50, 86
Cubbeer burr tree 228

Dara Shukoh Garden 55
Dargah 22, 40, 98,
    144, 222
Datranga tree 240
Decay 27–28, 68, 101, 127
measurement of 27
Deccan Olive tree 205
Deforestation 39, 209
Dendrochronology 23,
    28, 30
Dendroclimatology 28
Deodar (Devadaru) tree
    34, 49, 64, 66, 79–80
of Jageshwar 79
oldest 80
Devva hunasé tree 144
Divine jasmine tree 122
Dodda sampigé mara tree
    32, 179, 181
Double-coconut palm
    tree 116
Dukh Bhanjani Ber tree 67

East Indian Blackwood 196
English oak tree 228
Eucalyptus tree 20, 39,
    142, 155, 189, 191

False Hemp Tree 212
Farmer, Jared 16
Fig (peepal) tree 62, 72,

74, 100, 131, 134, 138–39,
    170, 182–83, 207, 245
Fir tree 26
Foxtail pine tree 19
Frangipani tree 248

Gandhiji's tree 72
Gandhi vad tree 231, 233
Gentle giant tree 169
Giant banyan tree 21, 32,
    34, 38, 119, 228
Giant honeybee 158, 166
Giant kapok tree 169
Giant sequoia/coast
    redwood 19, 21, 35,
    49–50
Giant teak tree 215
Giant toon tree 77
Gigantic banyan tree
    138, 228
Girikadamba tree 194
Golconda baobab tree 137
Golden champaka tree
    175, 179
Gold tree 165
Gorakh Imli tree 136,
    144, 153
Gorakshi tree 86, 95,
    144, 153
Government Botanical
    Garden, Karnataka 169
Grassland 189, 191,
    199, 201
Great banyan tree 39,
    119–21
Groves 22, 33, 35, 79–80,
    162–63, 192–93, 217,
    219, 243
    of tamarind trees 33
    sacred 35, 80, 217, 219
Guava 186
Gudaphala tree 192
Guinness Book of World
    Records* 21, 33–35,
    126, 138
Gulabi Siris tree 172
Gum tree 142, 155–56, 191
    hybrid 156
    of Nandi Hills 155

Guru Arjan Dev 68
Guru Nanak 33, 60, 61

Habitation 131, 134
Hadith*, the 192
Haldu tree 194
Harsingar tree 86
Hātiyan ka jhad 23–34, 136
Himalayan cedar/deodar
    tree 29–30, 34, 64, 79
Himalayan hemlock
    tree 30
Himalayan mulberry
    tree 69
Himalayan pencil cedar
    tree 22–23, 25, 28–29
Himalayan white pines
    tree 49
Hinduism 71
Hindus 12, 36, 47, 58, 182,
    222, 228
Hindu, The* 37
Hollow tree 47
Honeysuckle mistletoe
    72, 188
Horticulture 89, 243
Huccha hunasé tree 144

Iconic trees
    conservation and
    management of 15–17,
    20, 37, 41, 43, 126,
    181, 209
    criteria for determining 41
    living 45
    protection 13, 17, 40–41,
    43–44, 74, 151, 201
Imli tree 95, 136, 144,
    153, 162
India
    determining tree age
    in 28
    forest management in 209
    gardens in 52
    iconic trees 22
    landmark trees of 37
    living and breathing
    monuments 44
    living trees in 31

loneliest tree 50
oldest scientifically dated
  living tree in 22–23, 31
remarkable trees 38
Indian flying fox 167
Indian Forest Act 198
Indian Forester, The* 57
Indian freedom
  movement 54
Indian Institute of
  Integrative Medicine
  (IIIM) 49–50
Indian Institute of Science
  (IISc), Bengaluru 166
Indian Institute of Tropical
  Meteorology, Pune 28
Indian Medlar tree 149
Indian National Trust for
  Art and Cultural Heritage
  (INTACH) 37
Indian Rosewood 196
Indian rubber tree 35, 127
International Tree Ring
  Data Bank 20
International Union
  for Conservation of
  Nature (IUCN) 50, 198,
  219, 236
  Red List 219, 236
Irish oak tree 25

Jackfruit tree 15, 186
Jambolan tree 186
Jambu tree 186
Jamun tree 186, 188, 199
Java Atti tree 170
Java fig tree 170
Jeevi tree 170
Jhand tree 202
Joy perfume tree 175, 179
Jujube tree 60, 67
Jungli Dungi tree 212
Juniper tree 28

Kabir's banyan tree 230
Kabra tree 170
Kadamb (kaim) tree 84, 194
Kadiri tree 21, 32, 34, 138

Kàlidàsa 71
  Ritusamhãra* 71
Kalpavriksha 86
Kanjilal, Upendra Nath 122
  Flora of Assam* 122
Kannimara teak tree 14,
  33, 215–16
Kantakadruma tree 224
Kapittha (lemon) tree 66
Kapok tree 167
Kashmiriyat* 47
Khãra tree 153
Khejdi tree 202
Khirni tree 32, 36, 98, 220
Kintoor baobab tree 86, 88
Kittur tree 147–48
Kruthal tree 199
Kũrma Purãna* 202
Kurujaval tree 199
Kuruthamaram tree 199

Laachi ber tree 67
Lalbagh Botanical Garden,
  Bengaluru 167, 170
Landreth, Jenny 16
Late wood 24
Lawrence, Sir Walter 52
Limdo tree 222
Little champa tree 181
Living tree-ring
  chronology 25

Mahavriksha neem
  tree 222
Mahavriksha Award/
  Puraskar 143, 149, 151,
  216, 222
Mahogany tree 58, 77,
  222, 238
Malabar plum tree 186
Malai naval tree 33,
  199, 201
Malay padouk tree 236
Manjakadambu tree
  194–95
Marabo tree 132
Margosa tree 222
Maulsari Mahavriksha
  tree 149

Methuselah tree 14, 18–19
Miswak tree 192
Monkey Bread tree 144
Monoculture 209
Mountain ash tree 12, 20
Mourning cypress tree
  110, 113
Mulberry tree 14, 31, 69,
  71, 119
Mustard tree 192
Mysore gum tree 156

Nagay tree 199
Nallur tamarind grove
  tree 162
Narra tree 236
National Botanical
  Research Institute,
  Lucknow 86
National Salt Satyagraha
  Memorial 231, 233
Naya Qila, Hyderabad 34,
  136–37
Neem tree 58, 77, 222, 238
Nilgiri Biosphere
  Reserve 194
Nimbaka tree 222
Nithu tree 125
Nizhal 37
Nyagrodha (banyan) tree
  12, 14, 21, 32, 34–35,
  38–39, 66, 82–83, 109,
  119, 120–21, 131, 138,
  147–48, 158–59, 226–28,
  230–32

Oak tree 17, 25, 27, 79,
  208, 228
Oldest bristlecone pine
  tree 18

Paan (betel leaves) 153
Pagoda tree 248
Pakenham, Thomas 16
Palmer's oak tree 17
Palm tree 116–19
  bizarre 118
  male 117–18
Palmyra 116

wild date 119
Pando tree 17
Pārijāta tree 22, 36, 86
Parsis (Parsees) 36, 234–35
Peepal tree 62, 72, 74, 83, 100, 105, 182, 245
Pelt, Robert Van 16
Pests 14, 38, 151
  attack 24–25
Photosynthesis 27, 72
Pichumanda (neem) tree 66, 78, 222, 224
Pilkhan tree 36, 131
Pillalamarri tree 38
Pillanjaval tree 199
Pilu (peelu) trees 192
Pine tree 18, 28, 49, 79
Pipal tree 100
Plumeria/frangipani tree 248–49
Pollination 159, 219
Pollution 14, 39, 241
Precision arboriculture 43
Primavera tree 165–66
Prometheus tree 18

Qasim Shah's chinar tree 31, 34, 47
Qilian juniper tree 19
Radiocarbon dating 19, 22–23, 27–28, 30, 122, 136
Rain tree 172, 173
Raksha Sutra Andolan 74
Raman Research Institute, Bengaluru 165–66
Rayan (khirni) tree 32, 220
Reddish wood 112
Red Silk Cotton tree 224
Redwood 12, 19–21, 23
Religions 47, 100, 113, 234
Rheede, Hendrik van 186, 249
  Hortus Malabaricus* 186, 249
  Species Plantarum* 249
Rhododendron tree 35, 79, 125

Rocky Mountain bristlecone pine tree 19
Rose plants/tree 125, 165
Rosewood trees 196, 198, 208
Royal Horticultural Society of India, Calcutta 243
Rudraksham tree 205
Rudraksh tree 207

Safeda tree 142, 155, 189
Safed Semal tree 167
Sagoon tree 208, 215
Sagun tree 208, 215
Sagwan tree 208, 215
Saka tree 208, 215
Sal tree 31
Saltbush tree 192
Sampigé (champak) tree 175, 181
Sandalwood 186
Sant Dnyaneshwar 36, 240–41
Sapthaparni trees 36
Save Rani Bagh 40
Seedling 16, 49–50, 78, 101, 142, 209, 235
Seeds 12, 17–18, 27–28, 49–50, 94, 112, 116–19, 142–43, 145, 153, 156, 159, 163, 178, 181, 191, 207, 209, 212, 219, 243–44
Semal tree 167, 224
Shajouba tree 13–35
Shalmali tree 224
Shami tree 202, 204
Shimshapa tree 196
Shirisha tree 172
Shiva Purāna* 202
Shùrapàla 66
  Vrikshàyurveda* 66
Signages 43
Sikhism 60, 68
Silk cotton tree 224
Simado tree 224
Single-stem tree 21
Sleeman, Colonel William Henry 105–06

Ramaseeana* 105
Sleeman's tree 105
Spanish Cherry tree 149
Speaking fig tree 182
Spring wood 24
Sri Mahabodhi tree 100

Tamarind tree 33, 35, 162–63
Tamboli tree 240
Teak tree 14, 24–25, 30, 33, 36, 186, 194, 196, 198, 208–09, 212, 215–16, 224
  plantation 36, 208
Teakwood 209
Temple tree 248
Tetrameles tree 212
Thimmamma marrimānu or Thimmamma's banyan tree 21, 32, 34, 138, 230
Thitpok tree 212
Toon tree 58, 77–78
Toothbrush tree 192
Trees/Plants
  aerial 110
  African 153
  age-wizened 131, 179
  American 16
  ancient 22, 39, 63, 100, 126, 204, 220, 238
  big 17, 121, 158
  conservation 15
  critically endangered 219
  curious 118
  dead-standing 25
  death of the 16, 72
  deciduous 69, 166, 240
  endangered 42, 50
  evergreen 189, 191
  exceptional 16, 22, 37
  exotic 155
  exploitation of 13
  extraordinary 236
  flowering 88, 126, 132, 165, 199
  genetically identical 17
  health 43–44
  historic 170, 219, 233, 237
  host 72

identity 18, 42
landmark 37
largest 33, 52, 139, 155, 199, 228
leguminous 174
living 18–20, 22–23, 25, 28, 31, 122
location of 18–19
loneliest 49–50
majestic 55, 71
male and female 69
medicinal and aromatic 49
morbidity and mortality 16, 38–39, 43
mother 94, 118, 130, 156, 162
non-clonal 18
notable 17
noteworthy 17
old and large trees 15–16, 20, 22, 28, 38, 40–41, 80, 136, 162, 204
oldest 18, 20, 22, 44, 80, 134, 248
oldest known living tree 18, 122
oldest living 19, 28
ornamental evergreen 149
outstanding 13, 43
preservation of 39
registers 17, 42
religious 83
remarkable 37–38, 41
royal 47
sacred 22, 62, 66, 68–69, 72, 74, 80, 83–84, 100–101, 105, 131, 134, 220, 245
sampled 80
scientifically dated 22
sexually reproducing 18
single-stemmed 18–19, 49
special 14–15, 37, 40

species 15–16, 19, 116, 125, 209, 230
specific 15, 36
stoutest 153
tallest and the largest living 20, 23
terrestrial 14
top walks 16
transplantation 43
unusual 38
widest 121
wish-granting abilities of 36–37
worship 15, 36
Tree age 18, 22–23, 25, 28, 30, 40–41, 80, 167, 202, 220, 230
determination of 23, 71, 136
exact 31, 78, 93, 222
methods for calculating 18, 23, 88
Tree of enlightenment 100
True Mahogany tree 238

Ugamaram tree 192–93
United Nations Educational, Scientific and Cultural Organization (UNESCO) 21, 100, 112, 114
'Development without Destruction' 41
World Network of Biosphere Reserves 112
United Nations Environment Programme (UNEP) 41
Upside-down tree 144

Vallabh Vad 12, 226–27
Vannimaram tree 202, 204
Vat Vriksh tree 138, 147, 158, 226, 228, 231

Vegetation 84, 179, 186, 188, 193, 219
Vilayti Siris tree 172
Vilva (wood apple) tree 66

Wadia Institute of Himalayan Geology, Dehradun 28
Walking tree 121
Warty marble tree 205
Weeping fig tree 170
Weeping (or funeral) cypresses tree 110–14
Western juniper or Bennett Juniper tree 19
West Indian or Cuban Mahogany tree 238
White fig tree 131
White kapok tree 34, 167
White Silk Cotton tree 167
White wood 198
Wight, Robert 218–19
Icones Plantarum Indiae Orientalis 218
Wild cinnamon tree 207
Wild pear tree 134–35
Wishing tree 122
Wood 24–25, 27–28, 43, 58, 64, 66, 78, 88, 112, 127, 130, 136, 142, 162, 191, 198, 201, 212, 220
Woodrow, George Marshall 243–44
Mango: Its Culture and Varieties 244
World Heritage Site 100, 114

Yellow Teak tree 194

Zonal Agricultural Research Station of the Mahatma Phule Krishi Vidyapeeth 244

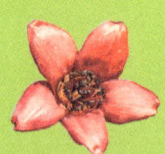

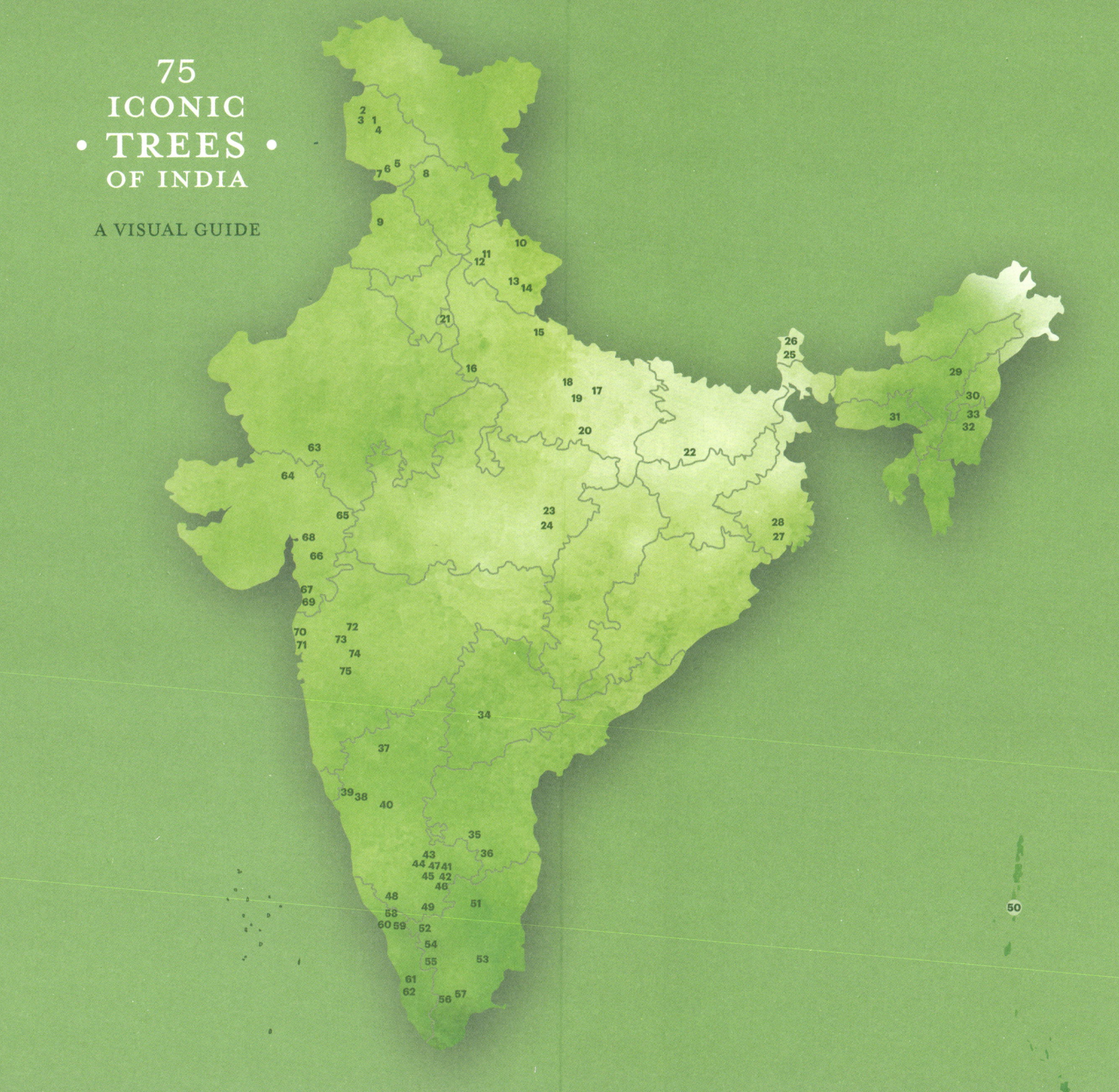

75
ICONIC
• TREES •
OF INDIA

A VISUAL GUIDE

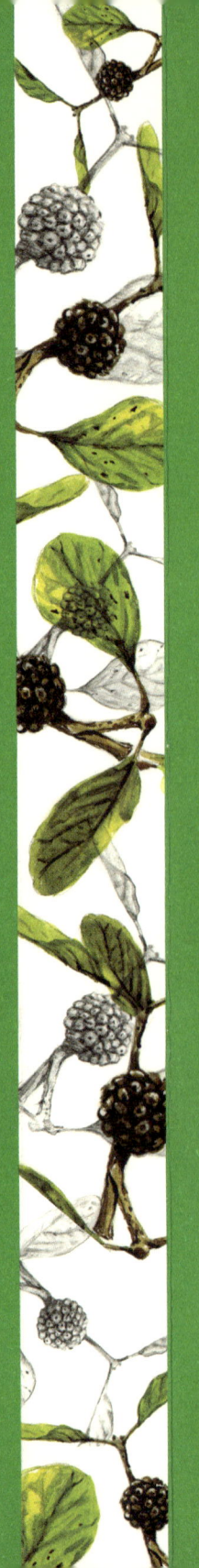

'A book for everyone – a glorious testament to iconic Indian trees. Read this book, and then go out and look for as many of these trees as you can find!'

HARINI NAGENDRA

Director, Centre for Climate Change and Sustainability, Azim Premji University

'Rich in cultural detail and full of biological interest, this engaging book is a gift to all those in awe of the great trees of the world and to everyone fascinated by the charismatic megaflora with which we share our planet.'

SIR PETER CRANE FRS

Former Director of the Royal Botanic Gardens, Kew

'A must read, *Iconic Trees of India* is a pilgrimage that we must all make to understand the beauty that surrounds us and that gives us life. S. Natesh's effort to document them brings to us the idea of what we must conserve; the trees that have witnessed all the changes around them; withstood and outlived adversity. A tribute to the living!'

SUNITA NARAIN

DG, Centre for Science and Environment

'A monumental effort by Dr. Natesh and stunning colour portraits of 75 iconic trees by artist Sagar Bhowmick that greatly enrich the narrative that deftly weaves precise science with folklore and history around these ecological monuments. I believe this compelling story has the power to draw new readers out of their technocratic metaverses to appreciate, value and conserve the remnants of India's spectacular but threatened wild nature around them.'

K. ULLAS KARANTH PH.D, F.A.SC.

Emeritus Director, Centre for Wildlife Studies

COVER: The Khirni Tree of Chirag Dilli.

FORE-EDGE: Detail of the Kadamb tree.